Springer Series in
OPTICAL SCIENCES
87

founded by H.K.V. Lotsch

Springer
Berlin
Heidelberg
New York
Hong Kong
London
Milan
Paris
Tokyo

Physics and Astronomy

ONLINE LIBRARY

http://www.springer.de/phys/

Springer Series in
OPTICAL SCIENCES

The Springer Series in Optical Sciences, under the leadership of Editor-in-Chief *William T. Rhodes*, Georgia Institute of Technology, USA, and Georgia Tech Lorraine, France, provides an expanding selection of research monographs in all major areas of optics: lasers and quantum optics, ultrafast phenomena, optical spectroscopy techniques, optoelectronics, quantum information, information optics, applied laser technology, industrial applications, and other topics of contemporary interest.
With this broad coverage of topics, the series is of use to all research scientists and engineers who need up-to-date reference books.

The editors encourage prospective authors to correspond with them in advance of submitting a manuscript. Submission of manuscripts should be made to the Editor-in-Chief or one of the Editors. See also
http://www.springer.de/phys/books/optical_science/

Editor-in-Chief
William T. Rhodes

Georgia Tech Lorraine
2-3, rue Marconi
57070 Metz, France
Phone: +33 (387) 20 3922
Fax: +33 (387) 20 3940
E-mail: wrhodes@georgiatech-metz.fr
URL: http://www.georgiatech-metz.fr
http://users.ece.gatech.edu/~wrhodes

Georgia Institute of Technology
School of Electrical and Computer Engineering
Atlanta, GA 30332-0250
Phone: +1 404 894 2929
Fax: +1 404 894 4641
E-mail: bill.rhodes@ece.gatech.edu
URL: http://www.ece.gatech.edu/profiles/
wrhodes/index.htm

Editorial Board
Toshimitsu Asakura

Hokkai-Gakuen University
Faculty of Engineering
1-1, Minami-26, Nishi 11, Chuo-ku
Sapporo, Hokkaido 064-0926, Japan
E-mail: asakura@eli.hokkai-s-u.ac.jp

Karl-Heinz Brenner

Chair of Optoelectronics
University of Mannheim
Institute of Computer Engineering
B6, 26
68131 Mannheim, Germany
Phone: +49 (621) 181 2700
Fax: +49 (621) 181 2695
E-mail: brenner@uni-mannheim.de
URL: http://www.ti.uni-mannheim.de/~oe

Theodor W. Hänsch

Max-Planck-Institut für Quantenoptik
Hans-Kopfermann-Strasse 1
85748 Garching, Germany
Phone: +49 (89) 2180 3211 or +49 (89) 32905 702
Fax: +49 (89) 32905 200
E-mail: t.w.haensch@physik.uni-muenchen.de
URL: http://www.mpq.mpg.de/~haensch

Ferenc Krausz

Vienna University of Technology
Photonics Institute
Gusshausstrasse 27/387
1040 Wien, Austria
Phone: +43 (1) 58801 38711
Fax: +43 (1) 58801 38799
E-mail: ferenc.krausz@tuwien.ac.at
URL: http://info.tuwien.ac.at/photonik/
home/Krausz/CV.htm

Horst Weber

Technische Universität Berlin
Optisches Institut
Strasse des 17. Juni 135
10623 Berlin, Germany
Phone: +49 (30) 314 23585
Fax: +49 (30) 314 27850
E-mail: weber@physik.tu-berlin.de
URL: http://www.physik.tu-berlin.de/institute/
OI/Weber/Webhome.htm

Harald Weinfurter

Ludwig-Maximilians-Universität München
Sektion Physik
Schellingstrasse 4/III
80799 München, Germany
Phone: +49 (89) 2180 2044
Fax: +49 (89) 2180 5032
E-mail: harald.weinfurter@physik.uni-muenchen.de
URL: http://xqp.physik.uni-muenchen.de

Peter Török   Fu-Jen Kao   (Eds.)

Optical Imaging and Microscopy

Techniques and Advanced Systems

With 260 Figures
Including 25 in Color

Springer

Sep/4e
PHYS

Peter Török
Imperial College London
Department of Physics
Blackett Laboratory
Prince Consort Road
London SW7 2BW
United Kingdom
E-mail: peter.torok@imperial.ac.uk

Fu-Jen Kao
Department of Physics
National Sun Yat-sen University
70 Lien Hai Road
Kaohsiung 80424
Taiwan
E-mail: fjk@mail.nsysu.edu.tw

ISSN 0342-4111

ISBN 3-540-43493-3 Springer-Verlag Berlin Heidelberg New York

Library of Congress Cataloging-in-Publication Data: Optical imaging and microscopy: techniques and advanced systems/ Peter Török, Fu-Jen Kao (eds.). p. cm. – (Springer series in optical sciences; 87) Includes bibliographical references and index. ISBN 3-540-43493-3 (alk. paper) 1. Optoelectronic devices. 2. Microscopy. 3. Imaging systems. I. Török, Peter, 1965– . II. Kao, Fu-Jen, 1961–. III. Series. TA1750.O54 2003 621.36'7–dc21 2003041225

Springer-Verlag Berlin Heidelberg New York
a member of BertelsmannSpringer Science+Business Media GmbH

http://www.springer.de

© Springer-Verlag Berlin Heidelberg 2003
Printed in Germany

Data conversion by LE-TeX, Leipzig
Cover concept by eStudio Calamar Steinen using a background picture from The Optics Project. Courtesy of John T. Foley, Professor, Department of Physics and Astronomy, Mississippi State University, USA.
Cover production: *design & production* GmbH, Heidelberg

Printed on acid-free paper   SPIN 10852035   57/3141/tr   5 4 3 2 1 0

Preface

The motivation to collect contributions from a wide variety of subjects in contemporary optics, centered around optical imaging, originates from two ideas. First, it is easy to recognise that certain fields of contemporary optics have been developing in quite a parallel manner. Sometimes workers of the different fields discover each other's contributions, but mostly they do not. One of our major goals is to show how closely these contributions are inter-related. Such an example is the development of scanning/confocal optical microscopy and optical data storage. In the former, imaging in the classical sense, occurs by scanning a tightly focused laser beam over the sample. In optical data storage imaging does not occur as the aim is to *detect* pits, rather than to image them. Nevertheless, the optical systems of these two arrangements have striking resemblance and hence their governing equations are practically the same. The second motivation of this book is to collect contributions from imaging related subjects that were not previously published in this form or they are difficult to access. Such examples are a chapter on white light interferometry, surface plasmon microscopy or the characterisation of high numerical aperture microscope objective lenses.

We are extremely pleased that we have contributions in this book from the international leaders of individual fields. It has been our privilege to work with these authors and we would like to take this opportunity to thank them all.

It now remains to acknowledge those who contribution made the publication of this book possible. First and foremost we wish to thank Miss Janey Lin and Mr Eoin Phillips for their work on the manuscript. They spent endless hours correcting typographical errors, and putting the manuscripts into a format that was suitable for publication. We are also grateful to Mr Sebastian Rahtz who answered all our questions regarding LaTeX; without his help this book would have looked fairly different. Dr Angela Lahee of Springer Verlag was kind enough to offer her help throughout this project that we all but overused. We would like to acknowledge the National Science Council of Taiwan for generous support towards the publication of this book.

The Editors hope that the Reader will derive as much joy from reading the contributions in this book as we did while working on it.

London and Kaohsiung,
January 2003

Peter Török
Fu-Jen Kao

Contents

List of Contributors

M.R. Arnison
University of Sydney
Department of Physical Optics
School of Physics A28
NSW 2006
Australia
mra@physics.usyd.edu.au

J.J.M. Braat
Optics Research Group
Faculty of Applied Sciences
Delft University of Technology
Lorentzweg 1
NL 2627 CA Delft
The Netherlands
j.j.m.braat@tnw.tudelft.nll

G.J. Brakenhoff
Biophysics and Microscopy Group
Swammerdam Institute for Life
Sciences
University of Amsterdam
P.O. Box 94062, 1090 GB Amsterdam
The Netherlands
brakenhoff@science.uva.nl

S.A. Boppart
Department of Electrical and Computer
Engineering
University of Illinois at Urbana-
Champaign
405 North Mathews Avenue
Urbana, IL 61801
USA
boppart@uiuc.edu

W.T. Cathey
Electrical and Computer Engineering
Department
University of Colorado at Boulder,
Boulder, CO 80309-0425
USA
cathey@schof.colorado.edu

D. Chana
King's College London
Department of Physics
Strand
London
WC2R 2LS
UK
deeph.chana@kcl.ac.uk

C.J. Cogswell
Electrical and Computer Engineering
Department
University of Colorado at Boulder,
Boulder, CO 80309-0425
USA
carol.cogswell@schof.colorado.edu

J.C. Dainty
Imperial College
Blackett Laboratory
London
SW7 2BZ
UK
jcd@ic.ac.uk

P. Dirksen
Philips Research Laboratories
Building WA
Professor Holstlaan 4
NL - 5656 AA Eindhoven
The Netherlands
peter.dirksen@philips.com

J. Huisken
MBL-Heidelberg
Light Microscopy Group
Cell Biophysics Programme
Meyerhofstrasse 1
D-69117 Heidelberg
Germany
jan.huisken@EMBL-Heidelberg.de

S. Inoué
Marine Biological Laboratory
7 MBL Street
Woods Hole, MA 02543-1015
USA
jmacneil@mbl.edu

A.J.E.M. Janssen
Philips Research Laboratories
Building WA
Professor Holstlaan 4
NL - 5656 AA Eindhoven
The Netherlands
peter.dirksen@philips.com

S.-h. Jiang
King's College London
Department of Physics
Strand
London
WC2R 2LS
UK

R. Juškaitis
University of Oxford
Department of Engineering Science
Parks Rd
Oxford OX1 3PJ
UK
rimas.juskaitis@eng.ox.ac.uk

F. Lagugné Labarthet
University of California, Berkeley
Department of Physics
Berkeley, CA 94720-7300
USA
lagugne@socrates.berkeley.edu

M. Müller
Biophysics and Microscopy Group
Swammerdam Institute for Life
Sciences
University of Amsterdam
P.O. Box 94062, 1090 GB Amsterdam
The Netherlands
muller@bio.uva.nl

P. Neocleous
King's College London
Department of Physics
Strand
London
WC2R 2LS
UK

M. Ohtsu
Tokyo Institute of Technology
Interdisciplinary Graduate School of
Science and Engineering,
4259 Nagatsuta, Midori-ku,
Yokohama 226-8502
Japan
ohtsu@ae.titech.ac.jp

E.R. Pike
King's College London
Department of Physics
Strand
London
WC2R 2LS
UK
erp@maxwell.ph.kcl.ac.uk

A. Rohrbach
MBL-Heidelberg
Light Microscopy Group
Cell Biophysics Programme
Meyerhofstrasse 1
D-69117 Heidelberg
Germany
rohrbach@EMBL-Heidelberg.de

M. Roy
University of Sydney
Department of Physical Optics
School of Physics A28
NSW 2006
Australia
m.roy@physics.usyd.edu.au

Y.R. Shen
University of California, Berkeley
Department of Physics
Berkeley, CA 94720-7300
USA
shenyr@physics.berkeley.edu

C.J.R. Sheppard
University of Sydney
Department of Physical Optics
School of Physics A28
NSW 2006
Australia
colin@physics.usyd.edu.au

S.A. Sherif
Blackett Laboratory
Imperial College
Prince Consort Rd.
London SW7 2BW
sherif.sherif@imperial.ac.uk

M.G. Somekh
Optical Engineering Group
University of Nottingham
School of Electrical and Electronic
Engineering
University Park
Nottigham NG2 7RD
UK
mike.somekh@nottingham.ac.uk

E.H.K. Stelzer
MBL-Heidelberg
Light Microscopy Group
Cell Biophysics Programme
Meyerhofstrasse 1
D-69117 Heidelberg
Germany
Ernst.Stelzer@EMBL-Heidelberg.de

C.-K. Sun
National Taiwan University
Institute of Electro-Optical Engi-
neering and Department of Electrical
Engineering
Taipei 10617
Taiwan
sun@cc.ee.ntu.edu.tw

P. Török
Blackett Laboratory
Imperial College
Prince Consort Rd.
London SW7 2BW
peter.torok@imperial.ac.uk

Part I

High Aperture Optical Systems and Super-Resolution

1 Exploring Living Cells and Molecular Dynamics with Polarized Light Microscopy

S. Inoué

1.1 Introduction

Here, I would like to discuss the reasons, and methods, for using polarized light microscopy for studying the organization, and function, of living cells. In illustrating these points, I will be relying on some biological examples that I am particularly familiar with, namely, those related to the events of mitosis and cell division, which represent some of the more fundamental and critical events in the life of all living organisms.

Now, why should we be interested in polarized light microscopy for the study of living cells? What is special about living cells and why is polarization microscopy particularly useful for their studies?

As biologists have learned over the last half century, the cell is highly organized not only at the microscopically resolvable micrometer level, but also at diverse ultra-structural levels, that is, down at nanometer levels, many times smaller than the classical limit of resolution of the light microscope. Furthermore, the structures in the living cells are dynamic and their molecular organizations and locations are often changing rapidly with time.

In order to follow these *dynamic* events, the light microscope is the instrument of choice so long as we can penetrate and gain the desired information at the submicroscopic levels. Fortunately, many structures inside living cells are organized into local para-crystalline arrays, and show a weak but definite optical anisotropy [1,2]. The optical anisotropy, such as birefringence can be detected and measured with polarized light microscopy, if examined with sufficiently high sensitivity, and spatial and temporal resolution.

What I and others have been able to do over the last half century is to improve the sensitivity for detecting the optical anisotropy, especially the weak birefringence retardation exhibited by structures inside functioning living cells, and to make such detection and measurement possible at reasonably high spatial and temporal resolutions. The birefringence that is detected non-invasively, in turn, signals the events taking place at the submicroscopic, molecular, and even submolecular levels.

1.2 Equipment Requirement

As the pioneer W. J. Schmidt pointed out in the 1930's, the combined use of high extinction polars (polarizers), strain-free optics, a Brace-Koehler compensator, and a bright light source is indispensable for using a polarizing microscope to detect the very low birefringence retardation (BR) exhibited by many organelles in living cells (e.g., [2]). In fact, the BR of many components of interest in the living cell typically ranges from a few nm to a small fraction of an nm, making them detectable only with very high extinction polarization optics [3,4].

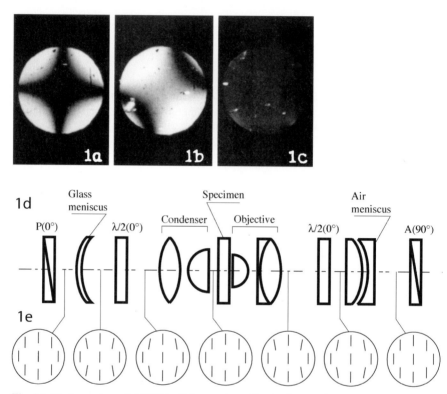

Fig. 1.1. Back aperture of 1.25 NA oil-immersion objective lens and matched NA condenser. (**a**) between crossed polars. (**b**) with analyzer rotated clockwise 3 degrees. (**c**) with crossed polars after rectification. (**d**) schematic of microscope with meniscus rectifiers. (**e**) orientation of polarized light at indicated planes. From [5]

With the small dimensions of many of the structures of interest, it is also necessary to improve the extinction factor ($I_{\parallel}/I_{\perp}$) for microscope condenser and objective lenses that provide high numerical aperture (NA), so that one can combine the high sensitivity for detecting and measuring small BR with high spatial resolution. Even using carefully selected strain-free lenses, however, the extinction factor between crossed polars generally drops exponentially as the NA is increased, primarily due to the differential transmittance of the parallel and perpendicular vectorial components of polarized light as the rays pass through oblique air-glass interfaces. This results in the "Maltese cross" seen at the back aperture of the objective lens, whereby up to several percent of the light passing between the arms of the cross escapes through the analyzer and drastically reduces the extinction factor at high NAs (Figs. 1.1 (a), 1.1 (b)). We were able to overcome this problem and gain the required high extinction factor, as well as a uniformly extinguished back aperture of high NA lenses, by introducing the polarization rectifier (Figs. 1.1 (c), 1.1 (d), 1.1 (e)) [5]. Rectification also eliminates the diffraction anomaly that, in non-rectified lenses, distorts the unit diffraction pattern (of weakly birefringent point sources) from an Airy disc to a four-leaf clover pattern (Figs. 1.2 (a), 1.2 (b)) [6,7]. An example of the sensitivity and

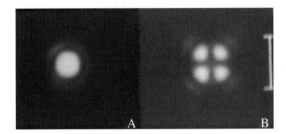

Fig. 1.2. Diffraction image of pinhole observed with 1.25 NA objective lens and condenser. (**a**) after rectification. (**b**) in the absence of rectifiers. Scale bar = 1.0 μm. From Inoué and Kubota, 1958 [6]

resolution gained by the use of rectified high NA lenses is shown in Fig. 1.3. These panels display the arrangement and distribution of DNA bundles in the living sperm head of an insect, cave cricket. The images taken under three compensator settings show, at very high optical resolution and veritable contrast, the helical arrangement of the negatively birefringent DNA bundles, as well as the tandem packing arrangement of the chromosomes. These features of living sperm structure had never before been seen in sperm of any species by any mode of microscopy.

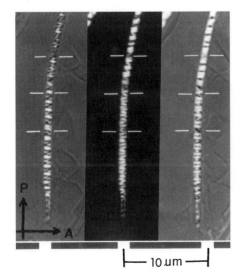

— 10 μm —

Fig. 1.3. Images of live sperm head of cave cricket. The sperm was immersed in dimethyl sulfoxide and observed between crossed polars at three compensator settings. From [8]

More recently, as an extension of the LC Pol-Scope (that I will touch on a little later), we developed another type of rectifier (Fig. 1.4). These new rectifiers reduce the ellipticity introduced by the anti-reflection coatings on the lens surfaces in addition to the differential attenuation (di-attenuation) of the s- and p- components that we primarily compensated for with the earlier rectifiers.

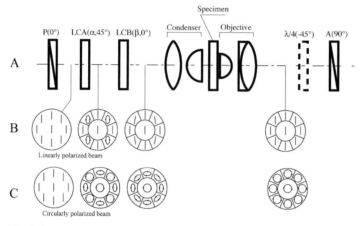

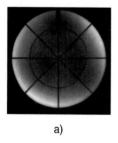

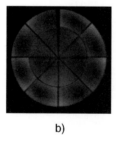

Fig. 1.4. Rectifier using sectored liquid crystal. (**a**) schematic of optical system. (**b**), (**c**) states of polarization at planes indicated. From [9]

When a microscope is equipped with oppositely rotating circular polars instead of crossed linear polars, the back aperture of non-rectified objective and condenser lenses shows a dark center surrounded by a brighter ring (Fig. 1.5). The light from the outer zone of the aperture arises from a combination of di-attenuation of the s- and p- components and elliptical polarization. The elliptical polarization for modern microscope lenses with multi-layer anti-reflection coatings tend to be considerably larger than in earlier lenses with single MgF coatings.

Fig. 1.5. Back aperture of 40×/0.85 NA lens. (**a**) observed between left and right circular polars without rectification. (**b**) in presence of sectored LC-rectifier. From [9]

Both the ellipticity and di-attenuation at high NA lens surfaces can be reduced quite effectively by using the new universal rectifier made of computer-driven sectored liquid crystal elements and $\lambda/2$ liquid-crystal element (Fig. 1.5). Figures 1.6

Fig. 1.6. Optical bench universal microscope. From [10]

and 1.7 show an optical-bench-inverted polarizing microscope that we developed for our biological and optical studies (see, e.g., pp. 158–162 in [10]). In addition to meeting the optical requirements discussed above, we prefer to use monochromatic green light (546 nm) from a 100 Watt mercury arc lamp. That satisfies the need to gain high field luminance (required since the analyzer removes much of the light) and to do so in the green spectral region, where least damage is introduced to living cells in general, despite the high intensity of illumination that the specimen receives. The uneven, very small image of the concentrated mercury arc is passed through a fiber-optic light scrambler in order to gain uniform field, and uniform full aperture, illumination [11]. (Light scramblers optimized for various microscopes are available from Technical Video, Ltd., PO Box 693, Woods Hole, MA 02543, USA; rknudson@tiac.net.)

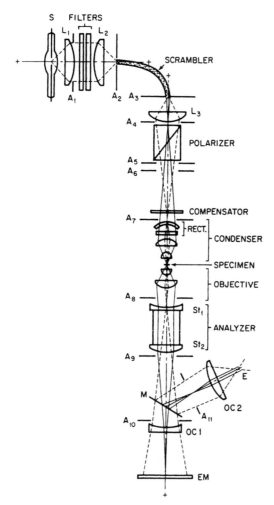

Fig. 1.7. Schematic optical path for microscope in Fig. 1.6. From [10]

1.3   Biological Examples

Before continuing with more instrumentation issues, let me introduce some examples of dynamic biological events. The first scene shows the segregation of chromosomes in a living plant cell of *Haemanthus*, the African blood lily (Fig. 1.8 and video). These scenes were taken in phase contrast by A. S. Bajer and J. Mole-Bajer in the 1950's (for on-line video see [12]), soon after Zernike's phase contrast microscope became available commercially.

In these video copies of the time-lapsed sequences, originally captured on 16 mm movie film, phase contrast clearly brings out the structure and behavior of the nucleolus and chromosomes, whose refractive indexes are somewhat greater than that of the surrounding nucleoplasm and cytoplasm. We clearly see the dissolution of the nucleoli, the condensation, splitting, and anaphase movement of the chromosomes,

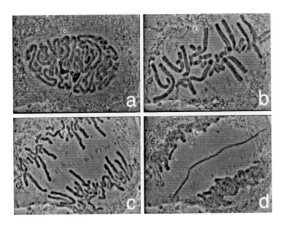

Fig. 1.8. Mitosis and cell plate formation in plant cell observed with phase contrast microscopy. After [13,14]. See also on-line version at www.molbiolcell.org. From [12]

as well as the appearance of the cell plate that separates the cell between the two daughter nuclei. The machinery responsible for the alignment and segregation of the chromosomes, and the deposition of the cell plate, were, however, undetectable with bright field or phase contrast microscopy in the living cells. The filaments that were reputed to be attached to the chromosomes, and those that deposit the cell plate, were visible only after the cells were killed with acidic fixatives and not in healthy living cells. Such observations had led to many theories over half a century as to how chromosomes move, including their autonomous motility. Since it has long been known that the proper segregation of chromosomes is essential for proper bipartitioning of hereditary factors, and errors in proper bipartitioning lead to developmental anomalies as well as cancer and other critical diseases, the mechanism of mitosis has received much attention over the years.

In 1951, using a sensitive hand-built polarizing microscope (Fig. 1.9), I was finally able to show the fibrous elements of the mitotic spindle that bring the chromosomes to the cell equator and then move them apart to the two spindle poles. In the sensitive polarizing microscope, the dynamic appearance, disappearance, and changes of the fibers (made up of weakly birefringent parallel fibrils, now known as microtubules) were clearly visible as they aligned the chromosomes to the metaphase plate and then brought them to the spindle poles (Fig. 1.10 and video). These observations on the low levels of birefringence , made directly on healthy living cells from several species of plants and animals, finally settled a 50 year controversy as to the reality of those fibers and fibrils and opened up a new era of study on the mechanism of mitosis, chromosome and organellar transport, and the establishment of cell polarity. In addition, the fluctuating birefringence that demonstrated the dynamic nature of the protein filaments opened up a new concept that those filaments (microtubules) and their subunit molecules (tubulin) were in a dynamic equilibrium (Fig. 1.11 and video). We showed not only that microtubule-based structures could be assembled in the living cell reversibly by polymerization of tubulin and then be taken apart again after finishing a particular physiological task, but also that the microtubules could be depolymerized reversibly by cold and hydrostatic pressure,

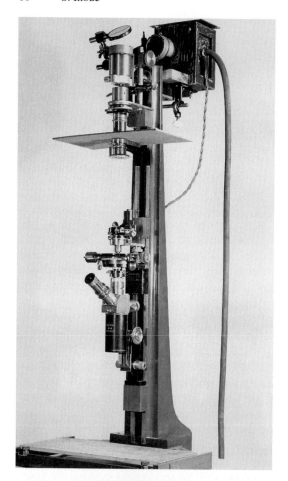

Fig. 1.9. Universal, inverted polarizing microscope designed by author in 1949. From [15]

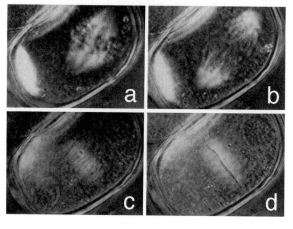

Fig. 1.10. Mitosis and cell plate formation in pollen mother cell of Easter Lily observed with sensitive polarizing microscope shown in Fig. 1.9. From [19]

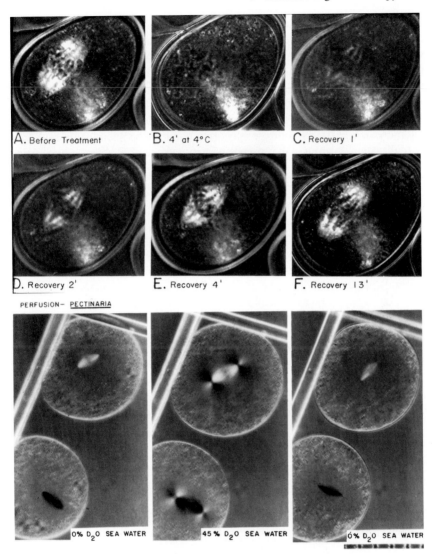

Fig. 1.11. Top. Reversible disassembly of spindle fiber microtubules induced by cold. (**a**) before chilling. (**b**) after 4 min at $-4°C$. (**c**)–(**f**): recovery at room temperature. From [19]. Bottom. Reversible assembly of spindle fiber microtubules by D_2O. Scale 10 μm. From [17]

in other words, thermodynamically review in [16]. In the meanwhile, we posited that the self-assembling microtubules could push and position cell organelles such as chromosomes, and by the very act of depolymerizing, pull objects to their destinations (review in [17,18]). Such heretic notions, that we derived by observing the birefringence changes in living cells, took many years to be fully accepted. Eventually, however, they were proven to be sound when isolated microtubules in media free of Ca^{2+} ions were disassembled reversibly by exposure to cold in the test tube [20,21]. In fact, chilling cell extracts to depolymerize microtubules, centrifuging to remove formed particles, and then warming the clear supernatant to polymerize microtubules has become the standard method for separating tubulin from other cell ingredients and for obtaining purified microtubules. By the late 1980's, a number of investigators have further been able to show that single microtubules isolated or assembled outside of the living cell could remain attached to and pull loads, such as chromosomes, to the microtubule anchoring point, as the microtubule itself was made to disassemble, even in the absence of common chemical energy sources such as ATP review in [18].

1.4   Video-Enhanced Microscopy

Returning to advances in light microscopy, one of the developments over the last couple of decades that made it possible to shed considerably more light on molecular behavior in living cells and functional cell extracts is video-enhanced microscopy. Surprisingly, there turned out to be much useful information in the microscope image that had been hidden and unexplored in the past but could now be brought forth by video enhancement. For example, the next video-enhanced DIC (Nomarski) sequence shows, in real time, the energetic growth of the acrosomal process from the tip of a sperm head triggered by its close approach to the egg (Fig. 1.12 and

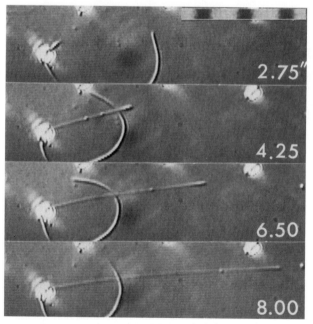

Fig. 1.12. Growth of 65 nm diameter acrosomal process from sperm of a sea cucumber. Video-enhanced DIC, elapsed time in seconds, scale bar = 10 μm. After [22]

video). The growth of this thin process, which in the electron microscope measures only 65 nm in diameter, is propelled by an explosive polymerization of actin molecules within the process, and is essential for successful fertilization in almost all animal species. In order to bring out the features previously hidden, all that was needed was to use a suitable video camera and circuits, to subtract away the stationary background optical noise (inhomogeneity), and to raise the image contrast by boosting the amplifier gain [23,24]. Thus, the faint diffraction image of single microtubules, only 25 nm wide, could now be clearly visualized with the light microscope by video-enhanced DIC. Furthermore, the gliding behavior of individual microtubules relative to the motor proteins attached to the coverslip, as well as the polarized growth, shortening, and distortion of the microtubules themselves, could be followed and quantitated at video rate. Such observations have led to the discovery and characterization of many new motor proteins and other molecules that associate with microtubules and transport organelles, as well as the microtubules themselves, in the living cell.

It should be noted in passing that, in contrast to fluorescence or dark-field microscopy, video-enhanced DIC and polarization microscopy yield rather shallow depths of field and, therefore, opportunities for serial optical sectioning and three-dimensional image construction even in the absence of confocal optics. Examples are shown in the following two video sequences. The first sequence shows real-time through-focal optical sections of live sea urchin embryo in video-enhanced DIC [25]. The second sequence shows real-time through-focal optical sections in video-enhanced polarization microscopy of a developing sea cucumber embryo (S. Inoué, unpublished data).

In addition to DIC microscopy, electronic imaging and processing in many forms are now playing important roles in confocal, fluorescence, and polarization microscopy as well. Among some of the more recent developments in polarization microscopy, I would like to touch on some of our own further contributions that include: a new type of polarized light microscope called the LC Pol-Scope, the centrifuge polarizing microscope, and observations of fluorescence polarization.

1.5   The LC Pol-Scope

The LC Pol-Scope was devised by Rudolf Oldenbourg in our Program at the Marine Biological Laboratory in Woods Hole, Massachusetts, to make possible the simultaneous imaging of weakly birefringent objects regardless of their axis orientation, and without the need for any mechanical adjustments of the specimen orientation or of the polarization optical components. With conventional polarization microscopy, the contrast of a birefringent object varies as the $\sin^2$(2× orientation angle), so that the specimen contrast drops to zero when the optical axes of the specimen come to lie parallel to the transmission axes of the crossed polars (polarizer and analyzer). With the LC Pol-Scope, not only is this angle dependence eliminated, but the brightness of the image is now directly proportional to the BR, rather than to $\sin^2$(BR), and thus directly reflects, e.g., the number of unit filaments that are packed into a fiber

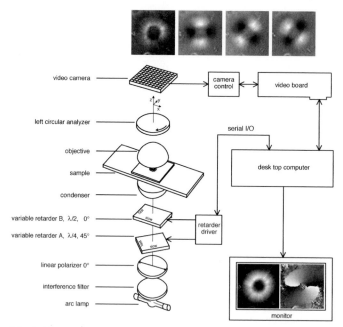

Fig. 1.13. Schematic of Oldenbourg's LC Pol-Scope. Top row: four images sent to computer. Bottom left: retardance image. Bottom right: gray scale image of azimuth angle. From [26]

whose diameter is less than the microscope's unit diffraction width [26]. As shown in the schematic of the system (Fig. 1.13), the LC Pol-Scope uses two liquid crystal (LC) retarders in place of a regular (mechanically adjusted) compensator, and a circular analyzer in place of a linear polarizing analyzer. A computer calculates and quickly displays the retardance distribution image from the four images captured by a digital camera at preset LC voltages. The next videos (Fig. 1.14 and videos) show examples of the dynamic retardance images. Figure 1.15 illustrates the sensitivity of the LC Pol-Scope for measuring the retardance of submicroscopic filaments.

1.6   The Centrifuge Polarizing Microscope

When living cells suspended in a density gradient are exposed to high acceleration in a centrifuge, the cell contents gradually stratify within the cell according to their density differences. For example, in an invertebrate egg cell, the heavy pigment and yolk granules sink to the bottom, while the lighter lipid granules and the nucleus accumulate at the top. By developing a centrifuge microscope, which allows the use of polarized light for observations of the weak cellular BR, we have been able to observe the alignment of fine structures in living cells associated with stratification of their organelles. As the organelles stratify, some membrane components become regularly oriented and exhibit birefringence (Fig. 1.16 (a), 1.16 (b), and video). Surprisingly, the birefringence disappears in a few seconds when the egg is activated,

Fig. 1.14. Newt lung epithelial cell undergoing mitosis, observed with the LC Pol-Scope. See also on-line version at www.molbiolcell.org. From [26]

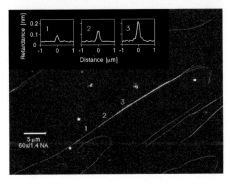

Fig. 1.15. Birefringence distribution of microtubules recorded with the LC Pol-Scope. Insets show retardance versus distance plots measured across single 1, 2, and 3 microtubules. From [27]

e.g., by the introduction of sperm or influx of Ca^{2+} ions into the cell followed by assembly of the meiotic spindle (Fig. 1.16 (c), 1.16 (d), and video). These changes that signal an early response of the egg cell to activation, as well as the emergence of polymerizing microtubules that generate the meiotic spindle in the centrifuged egg fragment, are illustrated in the video taken with the centrifuge polarizing microscope. This centrifuge polarizing microscope, which we developed in collaboration with Hamamatsu Photonics and Olympus Optical, required several technical innovations (Fig. 1.17). In brief, the specimen is carried in a special leak-proof chamber with low stress-birefringence windows; the chamber, supported in the rotor mounted on an air-spindle motor, transects the optical axes of the microscope between the objective and condenser lenses; and a frequency-doubled Nd:YAG laser fires a 6 nsec pulse exactly as the specimen comes in alignment with the microscope axis so that the image is frozen and stationary to better than 1.0 μm (Fig. 1.18 and video). Speckles that arise from the highly monochromatic, brief laser flashes are eliminated by using multiple-length optical fibers in the illumination path, and the development of interference patterns at or near the camera face plate is prevented by introducing a fiber bundle plate in front of the CCD detector. Different regions of the spinning specimen chamber are viewed by remotely driving the microscope, mounted on an x, y, z set of sleeve bearings. The polarization optics and interference fringe-free video camera provide a 1 nm retardance sensitivity for measuring birefringence, as

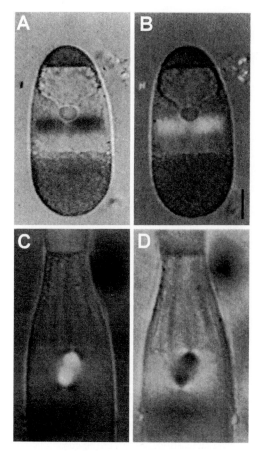

Fig. 1.16. Parchment worm egg observed in the centrifuge polarizing microscope. (**a**), (**b**) stratified cell with oil cap, very large nucleus with protruding nucleolus layered above negatively birefringent membrane layers that are oriented vertically. (**c**), (**d**) ca. 10 minutes after activation. The nuclear envelope and membrane birefringence are gone and replaced by the positively birefringent meiotic spindle. (**a**), (**c**) with compensator slow axis oriented vertically. (**b**), (**d**) horizontally. From [28]

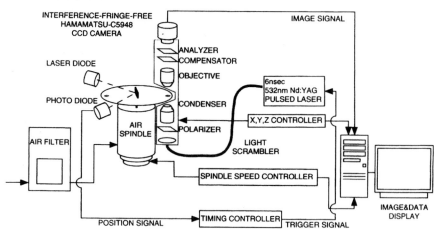

Fig. 1.17. Schematic of centrifuge polarizing microscope. From [29]

well as capability for fluorescence (532 nm excitation) and DIC observations. The system and chamber design are detailed in a recent article [29], together with examples of several applications in the study of living cells and related objects [30].

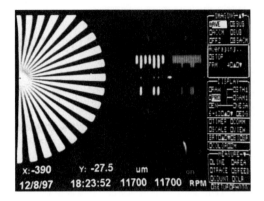

Fig. 1.18. Image of MBL-NNF test target recorded with centrifuge polarizing microscope rotating at 17000 rpm (10500× Earth's gravitational acceleration). 1 and 2 μm rulings show to the right of the Siemens test star. From [29]

1.7   Polarized Fluorescence of Green Fluorescent Protein

Finally, I would like to describe some fluorescence polarization properties of green fluorescent protein (GFP) crystals that we have been studying since last summer. GFP is a protein extracted from the light-producing organ of a jellyfish, *Aequorea* sp., and is widely used today by biologists as a sensitive fluorescent reporter [31]. GFP is a remarkably stable fluorescent protein whose gene can be expressed by

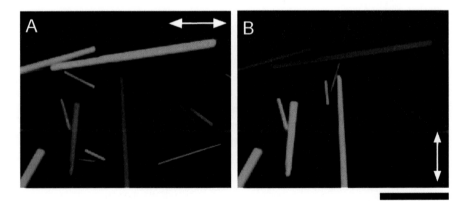

Fig. 1.19. Fluorescence anisotropy by crystals of native GFP. (**a**) polarizer transmission axis oriented horizontally. (**b**) vertically. Scale bar = 20 μm. From [32]

living cells together with other selected gene products without interfering with either gene function. It is thus used for signaling the expression, location, and interaction of a variety of proteins of interest in living cells and tissues.

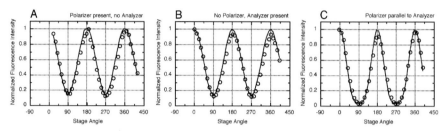

Fig. 1.20. Normalized fluorescence intensity versus GFP crystal orientation. (**a**) with polarizer but no analyzer. (**b**) with analyzer but no polarizer. (**c**) with parallel polarizer and analyzer. O: measured points; solid-line curves calculated from equations. From [32]

Our own interest in the fluorescence of this material was awakened when we examined the fluorescence emitted by crystals of purified native GFP under the polarizing microscope (Fig. 1.19). When illuminated, or examined, through a linear polar, the crystals exhibited a maximum-to-minimum ratio of fluorescence as high as 6:1 (Fig. 1.20 (a), 1.20 (b)). Furthermore, between a pair of polars *oriented parallel to each other*, the maximum-to-minimum fluorescence ratio rose to an incredibly high ratio of 30:1 (Fig. 1.20 (c)). Unexpectedly, the latter is nearly the product of the ratios measured with a polarizer or an analyzer alone. In fact, as the specimen is rotated on the microscope's revolving stage, we find that the fluorescence intensity follows a function that includes a cosine 4th power term (square of cosine squared terms) [32]. We believe these observations re-awaken the utility of polarized light microscopy for studying the arrangement of chromophores in a crystal or semi-crystalline material. Furthermore, the use of parallel polars (in other words, using both a polarizer and analyzer on a polarizing microscope, but with their axes oriented parallel rather than perpendicular to each other) should substantially improve the sensitivity for dynamically following the changes in orientation even of single fluorophores. For example, Kinosita [33] has used the orientation-dependent changes in polarized fluorescence of attached rhodamine to demonstrate the gyration of a single actin filament (ca. 12 nm in diameter) gliding on a lawn of myosin molecules. The increased sensitivity achieved by use of parallel polars instead of a single polarizing element could also prove useful for following conformational changes in portions of a single molecule tagged with appropriate fluorophores.

1.8   Concluding Remarks

In summary, while polarized light microscopy has still not been as widely used in biology as its potential indicates, it has played some critical roles in deciphering the molecular events underlying the enigmatic, yet vitally important, processes in living cells such as mitosis and the dynamic organizational roles played by microtubules. With expanded interest in biophysics and improved collaboration between optical physicists, electrical engineers, material scientists, and biologists, we look forward to further non-invasive explorations of living cells by polarized light microscopy, and to the unveiling of hidden dynamic molecular and submolecular events that underlie the physical basis of life.

References

1. H. Ambronn, A Frey: *Das Polarisationsmikroskop, seine Anwedung in der Kolloid-forschung in der Farbarei* (Akademische Verlag, Leipzig 1926)
2. W.J. Schmidt: 'Die Doppelbrechung von Zytoplasma, Karyoplasma und Metaplasma', In: *Protoplasma Monographien*, Vol. 11 (Borntraeger, Berlin 1937)
3. M.M. Swann, M. Mitchison: J. Exp. Biol. **27**, 226 (1950)
4. S. Inoué, K. Dan: J. Morphol. **89**, 423 (1951)
5. S. Inoué, W.L. Hyde: J. Biophys. & Biochem. Cytol. **3**, 831 (1957)
6. S. Inoué, H. Kubota: Nature **182**, 1725 (1958)
7. H. Kubota, S. Inoué: J. Opt. Soc. Am. **49**, 191 (1959)
8. S. Inoué, H. Sato: 'Deoxyribonucleic Acid Arrangement in Living Sperm', In: *Molecular Architecture in Cell Physiology*, ed. by T. Hayashi, A. Szent-Györgyi (Prentice Hall, New Jersey 1966) pp. 209–248
9. M. Shribak, S. Inoué, R. Oldenbourg: Opt. Eng. **43**, 943 (2002)
10. S. Inoué, K.R. Spring: *Video Microscopy – The Fundamentals* (Plenum Press, New York 1997)
11. G.W. Ellis: J. Cell Biol. **101**, 83a (1985)
12. S. Inoué, R. Oldenbourg: Mol. Biol. Cell. **9**, 1603 (1998)
13. A. Bajer, J. Molè–Bajer: Chromosoma **7**, 558 (1956)
14. A. Bajer, J. Molè–Bajer: J. Cell Biol. **102**, 263 (1986)
15. S. Inoué: *Studies of the Structure of the Mitotic Spindle in Living Cells with an Improved Polarization Microscope PhD Thesis* (Princeton University, New Jersey 1951)
16. S. Inoué: J. Cell Biol. **91**, 131s (1981b)
17. S. Inoué, H. Sato: J. Gen. Physiol. **50**, 259 (1967)
18. S. Inoué, E.D. Salmon: Mol. Biol. Cell. **6**, 1619 (1995)
19. S. Inoué: 'Organization and Function of the Mitotic Spindle', In: *Primitive Motile Systems in Cell Biology*, ed. by R.D. Allen, N. Kamiya (Academic Press, New York 1964) pp. 549–598
20. R. Weisenberg: Science **177**, 1196 (1972)
21. J.B. Olmsted, G.G. Borisy: Biochemistry **14**, 2996 (1975)
22. L.G. Tilney, S. Inoué: J. Cell Biol. **93**, 820 (1982)
23. R.D. Allen, J.L. Travis, N.S. Allen, H. Yilmas: Cell Motil. **1**, 275 (1981)
24. S. Inoué: J. Cell Biol. **89**, 346 (1981)
25. S. Inoué, T.D. Inoué: Biol. Bull. **187**, 232 (1994)
26. R. Oldenbourg: Nature **381**, 811 (1996)
27. R. Oldenbourg, E.D. Salmon, P.T. Tran: Biophys. J. **74**, 645 (1998)
28. S. Inoué: FASEB J. **13**, S185 (1999)
29. S. Inoué, R.A. Knudson, M. Goda, K. Suzuki, C. Nagano, N. Okada, H. Takahashi, K. Ichie, M. Iida,K. Yamanaka: J. Microsc. – Oxford **201**, 341 (2001)
30. S. Inoué, M. Goda, R.A. Knudson: J. Microsc. – Oxford **201**, 357 (2001)

31. S. Inoué, M. Goda: Biol. Bull. **201**, 231 (2001)
32. S. Inoué, O. Shimomura, M. Goda, M. Shribak, P.T. Tran: P. Natl. Acad. Sci. USA **99**, 4272 (2002)
33. K. Kinosita Jr.: FASEB J. **13**, S201 (1999)

2 Characterizing High Numerical Aperture Microscope Objective Lenses

Rimas Juškaitis

2.1 Introduction

Testing and characterization of high quality lenses have been perfected into fine art with the advent of lasers, phase-shifting interferometers, CCD cameras and computers. A bewildering array of techniques is described in Malacara's classical reference book on the subject [1]. Several of these techniques, in particular the Twyman–Green interferometer and the star test, are applicable to testing of microscope objective lens.

Characterizing high-numerical aperture (NA) objective lenses presents unique challenges. Many of the standard approaches, including Twyman–Green interferometry, are in fact comparative techniques. They require a reference object – an objective or a concave reflective surface – of the same or larger numerical aperture and of perfect (comparatively speaking) optical quality. This is problematic. Even if two lenses of the same type are available a problem still remains of appropriating the measured aberrations to the individual lenses. The star test, which is absolute, hits a similar problem in that the Airy disc produced by the lens being tested is impossible to observe directly, and hence it has to be magnified by a lens with a similar or better resolution, i.e. higher NA. Immersion lenses create further complications. All tests described in this Chapter are free from these problems. They are absolute and use a small point scatterer or a flat mirror to create a reference wavefront against which the lens aberrations are checked. Together with advanced interferometric techniques and processing algorithms this results in a range of techniques suitable for routine characterization of all available microscope objective lenses.

2.1.1 Disclaimer

A few words have to be said regarding identity of the lenses used throughout this chapter. Just as in a movie, which despite being a work of fiction uses real life people as actors, I had to use flesh-and-blood microscope objective lenses in order to verify the techniques and gather some "typical" data. The choice of lenses was typically dictated by what was available in our lab at a particular moment. Inevitably these were not necessarily the top-of-the-range and most up-to-date specimens. Being acutely aware of how sensitive the lens manufacturers can be about publishing such data – positive or otherwise – I feel it prudent to precede this Chapter with the following disclaimers:

Török/Kao (Eds.): Optical Imaging and Microscopy, Springer Series in Optical Sciences
Vol. 87 – © Springer-Verlag, Berlin Heidelberg 2003

All characters in this chapter are entirely fictitious. Any resemblance to a living lens is purely coincidental. Any attempts to match data published here with real lenses and to infer any generalizations about their respective manufacturers are to be undertaken entirely at readers' risk.

and also:

No lenses were harmed during these experiments.

2.1.2   Objective Lens Basics

Before describing specific lens testing techniques it might be useful to repeat here a few basic facts about microscope objective lenses in general. Modern objective lenses are invariably designed for infinite conjugate ratio, i.e. the object of observation is placed in the front focal plane and its image is formed at infinity. In order to obtain a real intermediate image a separate lens, called the tube lens, is used. The focal length of this lens F (which ranges from 165 mm for Zeiss and 180 mm for Olympus to 200 mm for Leica and Nikon) together with the magnification of the objective M gives the focal length of the objective $f = F/M$.

One of the basic postulates of aberration-free lens design in that it has to obey the Abbe's sine condition. For a microscope objective treated as a thick lens this can be interpreted as the fact that its front principal surface is actually a sphere of radius f centered at the front focus. Any ray leaving the focus at an angle α to the optical axis is intercepted by this surface at the height $d = f \sin \alpha$ and emerges from the back focal plane parallel to the axis at the same height, as shown in Fig. 2.1. For immersion lenses this has to be multiplied by the refractive index of the immersion fluid n.

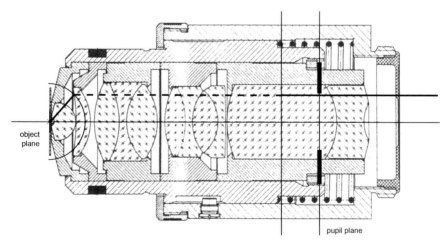

Fig. 2.1. Schematic diagram of a typical high NA planapochromat objective lens. Principal surfaces, aperture stop and marginal ray are indicated

In most high NA objective lenses the back focal plane, also called the pupil plane, is located inside the lens and is not therefore physically accessible. Fortunately lens designers tend to put an aperture stop as close to this plane as possible, which greatly simplifies the task of identifying the pupil plane when reimaging it using an auxiliary lens. Any extra elements, such as phase rings in phase contrast lenses or variable aperture iris diaphragms, will also be located in the back focal plane.

The physical aperture of an objective lens D is related to its numerical aperture $n \sin \alpha$ via

$$D = \frac{2Fn \sin \alpha}{M}. \tag{2.1}$$

Ultimately it is limited by the size of the objective thread. For a modern low magnification high NA immersion lens with, say, $n \sin \alpha = 1$ and $M = 20$, D can be as large as 20 mm. This is one of the reasons why some lens manufacturers (notably Leica and Nikon) have now abandoned the former golden standard of RMS thread and moved to larger thread sizes.

2.2   Point Spread Function

A perfect lens transforms a plane wave front into a converging spherical wave. Any deviations from this ideal behaviour, if they are not too dramatic, can be described by introducing a complex Pupil function $P(\rho, \theta)$, where ρ is the normalized radial coordinate in the pupil plane and θ is the azimuthal angle in the same plane. Both amplitude and phase aberrations can be present in a lens, but it is the latter that usually play the dominant role. The amplitude aberrations are typically limited to some apodization towards the edge of the pupil; these are discussed in more detail in Sect. 2.5.2.

The optical field distribution produced by this (possibly aberrated) converging wave is termed the Point Spread Function, or PSF, of the lens. This distribution can be obtained in its most elegant form if dimensionless optical coordinates in lateral

$$v = \frac{2\pi}{\lambda} n \sin \alpha \sqrt{x^2 + y^2} \tag{2.2}$$

and axial

$$u = \frac{8\pi}{\lambda} n \sin^2 \frac{\alpha}{2} z \tag{2.3}$$

directions are used. In these coordinates the intensity distribution in the PSF is independent of the NA of the lens, and the surface $u = v$ corresponds to the edge of the geometric shadow. The actual focal field distribution in these newly defined cylindrical coordinates is given by [2]:

$$h(u, v, \psi) = A \exp\left[\frac{iu}{4\sin^2 \frac{\alpha}{2}}\right] \int_0^1 \int_0^{2\pi} P(\rho, \theta)$$

$$\times \exp\left\{-i\left[v\rho \cos(\theta - \psi) + \frac{u\rho^2}{2}\right]\right\} \rho \, d\rho \, d\theta. \tag{2.4}$$

The exponential term in front of the integral is nothing else than a standard phase factor of a plane wave $2\pi nz/\lambda$.

For the aberration-free case $P = 1$ and the integral over θ can be calculated analytically to give $2\pi J_0(v\rho)$. Equation (2.4) now simplifies to

$$h(u, v) = 2\pi A \exp\left[\frac{iu}{4\sin^2\frac{\alpha}{2}}\right]\int_0^1 \exp\left(-\frac{iu\rho^2}{2}\right)J_0(v\rho)\rho\,d\rho. \tag{2.5}$$

This equation is readily calculated either numerically, or using Lommel functions. The resulting intensity distributions are well know and can be found, for instance, in [2]. Not only that, but also PSFs subjected to various aberrations have been calculated countless times in the past and are instantly recognizable to most microscopists. It is precisely for this reason that a relatively straightforward measurement of the PSF can frequently provide an instant indication of what is wrong with a particular objective lens.

Equations (2.4) and (2.5) are, of course, scalar approximations. This approximation works remarkably well up to angular apertures of about 60°. Even above these angles the scalar approximation can be safely used as a qualitative tool. For those few lenses that seem to be beyond the scalar approach (and for the rigorous purists) there is always an option to use a well developed vectorial theory [3].

2.2.1   Fibre-Optic Interferometer

Requirement to measure both amplitude and phase of the PSF calls for an interferometer-based setup. The Fibre-optic interferometer, Fig. 2.2, that was eventually chosen for the task has several important advantages. It is an almost common-path system, which dramatically improves long-term stability. It is also a self-aligning system: light coupled from the fibre to the lens and scattered in the focal region is coupled back into the fibre with the same efficiency. For convenience the whole setup is built around a single-mode fibre-optic beam splitter, the second output of which is index-matched in order to remove the unwanted reflection. A He-Ne laser operating at 633 nm is used as a light source. The whole setup bears cunning resemblance to a confocal microscope. In fact, *it is* a confocal microscope, or at least can be used as such [4]! Provided that light emerging from the fibre overfills the pupil of the objective lens the former acts as an effective pinhole ensuring spatial filtering of the backscattered light. Thus if the object can be regarded as a point scatterer, then the amplitude of the optical signal captured by the fibre

$$R \sim h^2 = |h|^2 \exp\left[2i\arg(h)\right], \tag{2.6}$$

i.e. its magnitude is equal to the intensity of PSF whereas the phase is twice the phase of the PSF. In order to measure both these parameters light reflected back along the fibre from the tip is used as a reference beam r. Wavefronts of both signal and reference beams are perfectly matched in a single mode fibre – this is the

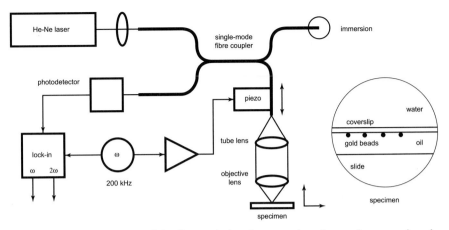

Fig. 2.2. Schematic diagram of the fibre-optic interferometer based setup for measuring objective PSFs

beauty of fibre-optic interferometer. Furthermore, the fibre tip is dithered to introduce a time-varying phase shift between the interfering beams $\phi(t) = \phi_0 \cos \omega t$. The interference signal reaching the photodetector is now given by

$$I = |r + R|^2 = r^2 + |R|^2 + 2r\left[\text{Re}(R)\cos(\phi_0 \cos \omega t) - \text{Im}(R)\cos(\phi_0 \cos \omega t)\right],$$
(2.7)

were r was assumed to be real for simplicity. It is now a simple matter to extract both $\text{Re}(R)$ and $\text{Im}(R)$ from this signal by using lock-in detection technique. The signal of (2.7) is multiplied by $\cos \omega t$ and the result is low-pass filtered to give

$$I_1 = r\text{J}_1(\phi_0)\text{Im}(R),$$
(2.8)

whereas synchronous demodulation with $\cos 2\omega t$ yields

$$I_2 = r\text{J}_0(\phi_0)\text{Re}(R).$$
(2.9)

By appropriately adjusting the modulation amplitude ϕ_0 it is easy to achieve $\text{J}_1(\phi_0) = \text{J}_2(\phi_0)$ and, by substituting (2.6), to calculate

$$h \sim \sqrt{I_1^2 + I_2^2} \exp\left(\frac{i}{2}\arctan\frac{I_1}{I_2}\right).$$
(2.10)

Thus the goal of obtaining both the amplitude and phase of the PSF of the objective lens has been achieved. Of course, in order to obtain full 2- (2-D) or 3-dimensional (3-D) PSF corresponding scanning of the object, the point scatterer, is still required.

2.2.2   PSF Measurements

In order to demonstrate the effectiveness of this method in detecting small amounts of aberrations it was tested on an special kind of objective lens. This 60× 1.2 NA water immersion plan-apochromat was developed for deconvolution applications and

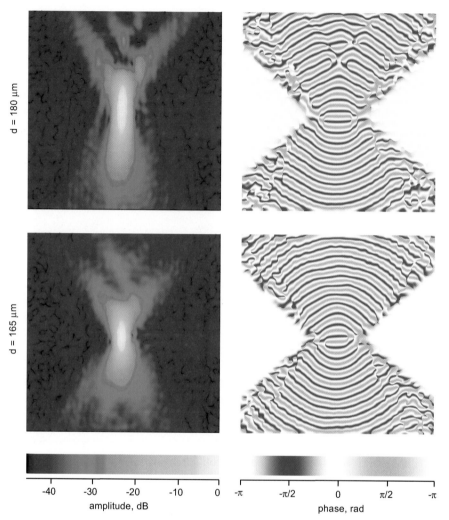

Fig. 2.3. The amplitude and phase of the effective PSF for 60× 1.2 NA water immersion lens with correction collar. Results for two different collar settings are shown. Image size in both x (horizontal) and z (vertical) are 5 μm

hence was specifically designed to have a well-corrected PSF. It was also equipped with a correction collar to compensate for cover glass thicknesses in the range 0.14–0.21 mm. 100 nm colloidal gold beads mounted beneath a Nr 1.5 coverslip of nominal thickness 0.17 mm acted as point scatterers in this case. The coverslip was in turn mounted on a microscope slide and a gap between them was filled with immersion oil so as to eliminate reflection from the back surface of the coverslip. The size of the bead was carefully chosen experimentally in order to maximize the signal level without compromising the point-like behaviour. Indeed a control experiment using 40 nm beads yielded similar results to those presented below but with a vastly inferior signal-to-noise ratio.

In principle this apparatus is capable of producing full 3-D complex PSF data sets. It was found however that in most cases x–z cross-sections provided sufficient insight into the aberration properties of the lens without requiring too long acquisition times. Such results are shown in Fig. 2.3 for two settings of the correction collar. In order to emphasize the side lobe structure the magnitude of the PSF in displayed in dB with the peak value taken to be 0 dB. It can be seen that a collar setting of 0.165 mm gives a near-perfect form to the PSF. The axial side lobes are symmetric with respect to the focal plane and the phase fronts away from this plane quickly assume the expected spherical shape. On the other hand a small 10% deviation from the correct setting already has quite pronounced effect on the PSF in the bottom row of Fig. 2.3. The symmetry is broken, axial extent of the PSF has increased by about 30% and distinct phase singularities appeared on the phase fronts. Everything points towards a certain amount of uncompensated spherical aberration being present in the system. It is interesting to note that the phase map of the PSF seems to be more sensitive to the aberrations than the magnitude. This can be used as an early warning indicator of the trouble. It also shows the importance of measuring both the magnitude and phase of the PSF.

Although so far the measured PSF has been described in purely qualitative terms some useful quantitative information about the objective lens can also be extracted from this data. One parameter that can be readily verified is the objective's NA. Axial extent of the PSF is more sensitive to the NA than its lateral shape. Using the axial section of the PSF is therefore the preferred method to determine the NA. Besides, the interference fringes present in the z-scan provide a natural calibration scale for the distance in z. The actual measurement was obtained by finding the best fit to the curve in Fig. 2.4. A somewhat surprising result of this exercise was that the

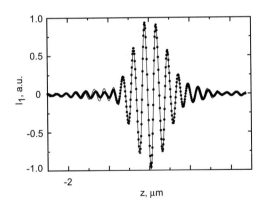

Fig. 2.4. Measured (*dots*) and calculated (*line*) amplitude axial responses for the same lens

best fit corresponded to NA of 1.15, rather than the nominal value of 1.2. This is not a coincidence: such discrepancies were found with other high NA objective lenses as well. The reason for this kind of behaviour will become clear in Sect. 2.4.4.

2.3   Chromatic Aberrations

Chromatic aberrations constitute another class of aberrations that can adversely affect the performance of any microscopy system. These aberrations are notoriously difficult to overcome in high NA objective lens design. The reason, at least in part, is relative uniformity of dispersion properties of common glasses used in objective lenses. Ingenious solutions have been found at an expense of dramatic increase of the number of lens elements – typically to more than a dozen in apochromats. Even then part of the correction may need to be carried out in the elements external to the objective.

Lateral and axial colour, as they are called by lens designers, are usually treated as separate chromatic aberrations. The former, which manifests itself as the wavelength-dependant magnification, is easy to spot in conventional microscopes as colouring of the edges of high contrast objects. Lateral chromatic aberration is also the more difficult to correct of the two. Traditionally this has been done by using the tube lens, or even the ocular, to offset the residual lateral colour of the lens. Some latest designs claim to have achieved full compensation within the objective lens itself, – claims to be treated with caution. In any case the correct testing procedure for the lateral colour should include at least a matched tube lens. The simplest test would probably be to repeat the experiments described in Sect. 2.2.2 for several wavelengths at the edge of the field of view and record the shift of the lateral position of the point image.

In confocal microscopy, where the signal is determined by the overlap of the effective excitation and detection PSFs, the loss of register between them should lead to reduction of signal towards the edge of the field of view. It has to be said, though, that in most confocal microscopes almost always only a small area around the optical axis is used for imaging, hence this apodization is hardly ever appreciable. Axial colour on the other hand is rarely an issue in conventional microscopy, but it can be of serious consequence for confocal microscopy, especially when large wavelength shifts are involved, such as in multiphoton or second and third harmonic microscopy. Mismatch in axial positions of excitation and detection PSFs can easily lead to degradation or even complete loss of signal even in the centre of the field of view. Below we describe a test setup which uses this sensitivity of confocal system to characterise axial chromatic aberration of high NA objective lenses.

2.3.1   Apparatus

Ideally one could conceive an apparatus similar to that in Fig. 2.2, whereby the laser is substituted with a broadband light source. One problem is immediately obvious: it is very difficult to couple any significant amount of power into a single-mode fibre from a broadband light source, such as an arc lamp. Using multiple lasers provides only partial (and expensive) solution. Instead it was decided to substitute the point scatterer with a plane reflector. Scanning the reflector axially produces the confocal signal [6]:

$$I = \left[\frac{\sin u/2}{u/2} \right]^2 .$$

(2.11)

The maximum signal is detected when the plane reflector lies in the focal plane. This will change with the wavelength if chromatic aberration is present. Using mirror instead of a bead has another advantage: the resulting signal is one-dimensional, function of u only, and hence a dispersive element can be used to directly obtain 2-D spectral axial responses without the necessity of acquiring multiple datasets at different wavelengths.

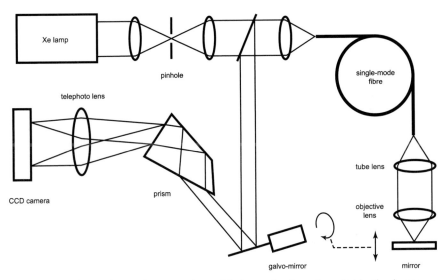

Fig. 2.5. Experimental setup for measuring axial chromatic aberration of objective lenses

The resulting apparatus, depicted in Fig. 2.5 and described in more detail in [7], is again based around a fibre-optic confocal microscope setup, but the interferometer part is now discarded. Instead, a monochromator prism made of SF4 glass is introduced to provide the spectral spread in the horizontal direction (i.e. in the image plane). Scanning in the vertical direction was introduced by a galvo-mirror moving in synchronism with the mirror in the focal region of the objective lens. The resulting 2-D information is captured by a cooled 16-bit slow-scan CCD (charge coupled device) camera. A small-arc Xe lamp is used as a light source providing approximately 0.2 μW of broadband visible radiation in a single-mode fibre. This is sufficient to produce a spectral snapshot of a lens in about 10 s.

2.3.2   Axial Shift

Typical results obtained by the chromatic aberration measurement apparatus are shown in Fig. 2.6. Since the raw images are not necessarily linear either in z or λ a form of calibration procedure in both coordinates is required. To achieve this the arc lamp light source was temporarily replaced by a He-Ne and a multi-line Ar$^+$ lasers. This gave enough laser lines to perform linearisation in λ. As a bonus coherent laser radiation also gave rise to interference fringes in the axial response with the reflection from the fibre tip acting as a reference beam, just as in the setup of Sect. 2.2.1. When a usual high NA objective lens was substituted with a low NA version these fringes covered the whole range of the z scan and could be readily used to calibrate the axial coordinate. The traces shown in Fig. 2.6 have been normalized to unity

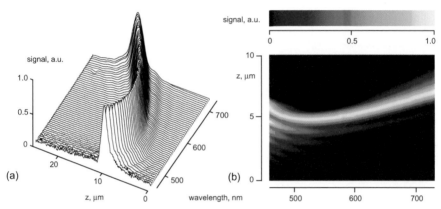

Fig. 2.6. Experimental results from (**a**) 32× 0.5 NA plan-achromat objective, displayed as 3-D plot, and (**b**) 50× 0.8 NA plan-achromat in pseudo-colour representation. Note the change in z scale

at each individual wavelength. The presence of longitudinal chromatic aberration is clearly seen in both plots. Their shapes are characteristic for achromats in which the longitudinal colour is corrected for two wavelengths only. It is interesting also to note the change of the shape of the axial response with the wavelength. This is noticeable for 32× 0.5 NA plan-achromat but it becomes particularly dramatic for 50× 0.8 NA plan-achromat, Fig. 2.6b. Clearly the latter lens suffers from severe spherical aberration at wavelengths below 550 nm, which results in multiple secondary maxima at one side of the main peak of the axial response.

An interesting problem is posed by the tube lens in Fig. 2.5. This lens may well contribute to the longitudinal chromatic aberration of the microscope as a whole and therefore it is desirable to use here a proper microscope tube lens matched to the objective. In fact it transpired that in some cases the tube lens exhibited significantly larger axial colour than the objective itself. This is hardly surprising: the tube lens is typically a simple triplet and three elements are hardly enough to achieve any sophisticated colour correction.

In this experiment, however, the main task was to evaluate the properties of the objective lens itself. A different approach was therefore adopted. The same achromatic doublet (Melles Griot 01 LAO 079) collimating lens was used with all objective lenses. Because the chromatic aberrations of this lens are well documented in the company literature these effects could be easily removed from the final results presented in Fig. 2.7. Figure 2.7b presents the same data but in a form more suited to

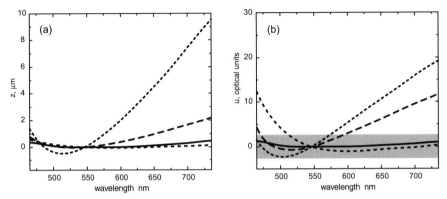

Fig. 2.7. Wavelength dependence of the axial response maxima represented in physical (**a**) and optical (**b**) axial coordinates. The four traces correspond to: 32× 0.5 NA plan-achromat (*dotted line*), 50× 0.8 NA plan-achromat (*long dashes*), 100× 1.4 NA plan-apochromat (*short dashes*), and the same lens stopped down to 0.7 NA (*solid line*)

confocal microscopy whereby the chromatic shift is now expressed in optical units (2.3). The half width of the axial response to a plane mirror is given by 2.78 optical units at all wavelengths and for all NAs. This region is also shown in the figure. The zero in the axial shift is arbitrarily set to correspond to $\lambda = 546$ nm for all the objectives tested.

As could be expected these results show improvement in performance of apochromats over achromats. They also show that none of the tested objectives (and this includes many more not shown in Fig. 2.7 for fear of congestion) could meet the requirement of having a spectrally flat – to within the depth of field – axial behaviour over entire visible range. This was only possible to achieve by stopping down a 1.4 NA apochromat using a built-in aperture stop – the trick to be repeated several times again before this Chapter expires.

2.4   Pupil Function

Pupil function is the distribution of the phase and amplitude across the pupil plane when the lens is illuminated by a perfect spherical wave from the object side. It is related in scalar approximation to the PSF via a Fourier-like relationship (2.4). It would appear therefore that they both carry the same information and therefore the choice between them should be a simple matter of convenience. Reality is a bit more complicated than that. Calculating the Pupil function from the PSF is an ill-posed problem and therefore very sensitive to noise. Measurements of the Pupil function provide direct and quantitative information about the aberrations of the lens – information that can only be inferred from the PSF measurements.

The trouble with mapping the Pupil function is that a source of a perfect spherical wave is required. Such thing does not exist but, fortunately, the dipole radiation approximates such wave rather well, at least as far as phase is concerned. The approach described in this Section is based on using small side-illuminated scatterers as sources of spherical waves. The actual Pupil function measurement is then performed in a phase-shifting Mach–Zender interferometer in a rather traditional fashion.

2.4.1 Phase-Shifting Interferometry

The experimental setup depicted in Fig. 2.8 comprised a frequency doubled Nd$^+$:YAG laser which illuminated a collection of 20 nm diameter gold beads deposited on the surface of high refractive index glass prism acting as dipole scatterers. Because the laser light suffered total internal reflection at the surface of the prism

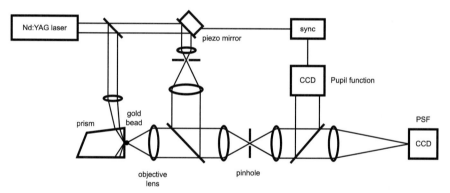

Fig. 2.8. Phase shifting Mach–Zehnder interferometer used for the Pupil function measurements

no direct illumination could enter the objective lens. The gold scatterers converted the evanescent field into the radiating spherical waves which were collected by the lens and converted into plane waves. These waves were then superimposed on a collimated reference wave. A 4-f lens system was then used to image the pupil plane of the lens onto a CCD camera. A pinhole in the middle of this projection system served to select a signal from a single scatterer. The size of this pinhole had to be carefully controlled as not to introduce artifacts and degrade resolution in the image of the Pupil function. A second CCD camera was employed to measure the PSF at the same time.

One of the mirrors in the reference arm of the interferometer was mounted on a piezo-electric drive and moved in synchronism with the CCD frame rate to produce successive interferograms of the pupil plane shifted by $2\pi/3$ rad

$$I_l \sim \left| r + P(\rho,\theta) \exp\left(i\frac{2\pi l}{3}\right) \right|^2, \qquad l = 0, 1, 2. \tag{2.12}$$

Using these three measurements the phase component of the Pupil function was then calculated as

$$\arg[P(\rho,\theta)] = \arctan \frac{\sqrt{3}(I_1 - I_2)}{I_1 + I_2 - 2I_0}. \tag{2.13}$$

The lens together with the prism were mounted on a pivoting stage which could rotate the whole assembly around the axis aligned to an approximate location of the pupil plane. Thus the off-axis as well as on-axis measurements of the Pupil function could be obtained. A set of such measurements is presented in Fig. 2.9 which clearly demonstrates how the performance of the lens degrades towards the edge of the field of view. Not only appreciable astigmatism and coma are introduced, but also vignetting becomes apparent. The presence of vignetting would be very difficult to deduce from the measurement of the PSF alone as it could be easily mistaken for astigmatism. Not that such vignetting is necessarily an indication that something is wrong with the lens: it can well be deliberate technique employed by the lens designer to block off the most aberrated part of the pupil.

2.4.2 Zernike Polynomial Fit

Traditionally the phase aberrations of the Pupil functions are described quantitatively by expanding them using a Zernike circle polynomial set:

$$\arg[P(\rho,\theta)] = \sum_{i=1}^{\infty} a_i Z_i(\rho,\theta), \tag{2.14}$$

where a_i are aberration coefficients for corresponding Zernike polynomials $Z_i(\rho,\theta)$. Significant variations between different modifications of Zernike polynomials exist. In this work a set from [9] was used. The first 22 members of this set are listed in Table 2.1 together with their common names. This list can be further extended. In practice, however, expansion the beyond second order spherical aberration is not very reliable due to experimental errors and noise in the measured image of $P(\rho,\theta)$.

The determination of the expansion coefficients a_i should, in principle, be a simple procedure, given the Zernike polynomials are orthonormal. Multiplying the measured Pupil function by a selected polynomial and integrating over the whole pupil area should directly yield the corresponding aberration coefficient. The real life is a bit more complicated, especially when processing the off-axis data, such as shown in Fig. 2.9. One obstacle is vignetting: the standard Zernike set is no longer

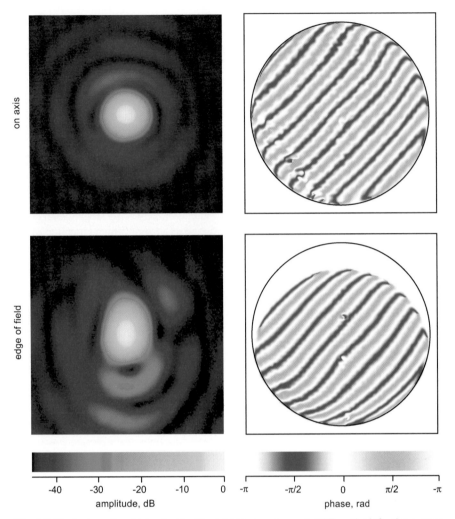

on axis

edge of field

-40　　-30　　-20　　-10　　0	-π　　-π/2　　0　　π/2　　-π
amplitude, dB	phase, rad

Fig. 2.9. Typical results obtained with 40× 1.2 NA Apochromat, PSF on the left, phase maps of corresponding pupil functions on the right. Performance both on axis, top, and at the edge of the field of view, bottom, is shown

orthonormal over a non-circular pupil. Even without vignetting the 2π phase ambiguity poses a problem. Before the expansion procedure can be applied the phase of the Pupil function has to be unwrapped – not necessarily a trivial procedure.

An entirely different expansion technique was developed to overcome these difficulties. This technique is based on simulated iterative wavefront correction routine, originally conceived to be used in adaptive optics applications together with a diffractive optics wavefront sensor [10]. The essence of the method is that small simulated amounts of individual Zernike aberrations are applied in turns to the

Table 2.1. Orthonormal Zernike circle polynomials

i	n	m	$Z_i(r,\theta)$	*Aberration term*
1	1	0	1	piston
2	1	1	$2r\cos\theta$	tilt
3	1	−1	$2r\sin\theta$	tilt
4	2	0	$\sqrt{3}(2r^2-1)$	defocus
5	2	2	$2\sqrt{3}r^2\cos 2\theta$	astigmatism
6	2	−2	$2\sqrt{3}r^2\sin 2\theta$	astigmatism
7	3	1	$2\sqrt{2}(3r^3-2r)\cos\theta$	coma
8	3	−1	$2\sqrt{2}(3r^3-2r)\sin\theta$	coma
9	3	3	$2\sqrt{2}r^3\cos 3\theta$	trefoil
10	3	−3	$2\sqrt{2}r^3\sin 3\theta$	trefoil
11	4	0	$\sqrt{5}(6r^4-6r^2+1)$	primary spherical
12	4	2	$\sqrt{10}(4r^4-3r^2)\cos 2\theta$	
13	4	−2	$\sqrt{10}(4r^4-3r^2)\sin 2\theta$	
14	4	4	$\sqrt{10}r^4\cos 4\theta$	
15	4	−4	$\sqrt{10}r^4\sin 4\theta$	
16	5	1	$2\sqrt{3}(10r^5-12r^3+3r)\cos\theta$	
17	5	−1	$2\sqrt{3}(10r^5-12r^3+3r)\sin\theta$	
18	5	3	$2\sqrt{3}(5r^5-4r^3)\cos 3\theta$	
19	5	−3	$2\sqrt{3}(5r^5-4r^3)\sin 3\theta$	
20	5	5	$6\sqrt{3}r^5\cos 5\theta$	
21	5	−5	$6\sqrt{3}r^5\sin 5\theta$	
22	6	0	$\sqrt{7}(20r^6-30r^4+12r^2-1)$	secondary spherical

measured Pupil function. After each variation the in-focus PSF is calculated and the whole process iteratively repeated until the Strehl ratio is maximized. The final magnitudes of the Zernike terms are then taken to be (with opposite signs) the values of the Zernike expansion coefficients of the experimentally measured aberrated Pupil function. This procedure is reasonably fast and sufficiently robust, provided that the initial circular aperture can still be restored from the vignetted pupil.

The power of this technique is demonstrated in Fig. 2.10 where a 40× 1.2 NA water immersion lens was investigated at three different settings of the correction collar. As expected, adjusting the collar mainly changes the primary and secondary spherical aberration terms. Variations in other terms are negligible. The optimum compensation is achieved close to $d = 0.15$ mm setting, were small amounts of both

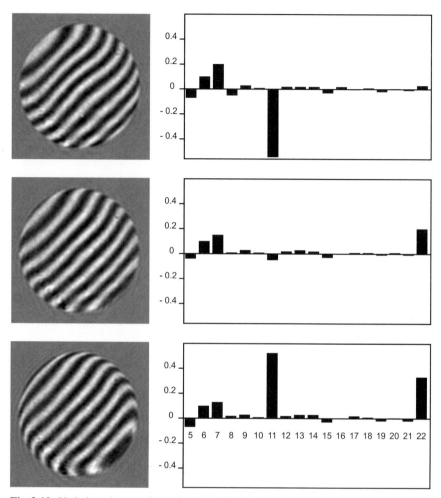

Fig. 2.10. Variations in wavefront aberration function expressed via Zernike modes when correction collar of a water immersion lens is adjusted

aberrations with opposite signs cancel each other. The usefulness of the Zernike expansion is further demonstrated by the fact that the main residual term in this case was the defocus, which, although not an aberration itself, could be easily mistaken upon visual inspection of the interference pattern for a spherical aberration.

2.4.3   Restoration of a 3-D Point Spread Function

Nowhere the power of the Pupil function approach to the objective lens characterization is more apparent than in cases when full three-dimensional shape of the PSF needs to be determined. Such need may arise, for instance, when using deconvolution techniques to process images obtained with a confocal microscope.

As is clear from (2.4) not only an in-focus PSF can be calculated from a measured Pupil function, but the same can be done for any amount of defocus, by choosing an appropriate value for the axial coordinate u. Repeating the process at regular steps in u yields a set of through-focus slices of the PSF. These can then be used to construct a 3-D image of the PSF much in the same manner as 3-D images are obtained in a confocal microscope. Compared to the direct measurement using a point scatterer, vis Sect. 2.2.2, advantages of this approach are clear. A single measurement of the Pupil function is sufficient and no scanning of the bead in three dimensions is required. Consequently, exposures per image pixel can be much longer. As a result this method provides much improved signal-to-noise ratio in the final rendering of the PSF allowing even faintest sidelobes to be examined.

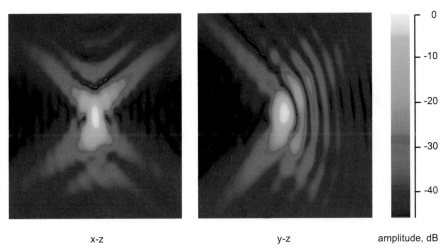

x-z   y-z   amplitude, dB

Fig. 2.11. Three-dimensional PSF restored from pupil function data, shown here via two meridional sections

Obviously, presenting a complete 3-D image on a flat page is always going to be a problem but, as Fig. 2.11 shows, even just two meridional cross sections of a 3-D PSF provide infinitely more information than a plain 2-D in-focus section of the same PSF at the bottom of Fig. 2.9. Thus, for instance, the $y - z$ section clearly shows that the dominant aberration for this particular off-axis position is coma. Comparing the two sections is also possible to note different convergence angles for the wavefronts in two directions – a direct consequence of vignetting.

2.4.4   Empty Aperture

Testing objective lenses with highest NAs (1.4 for oil, 1.2 for water immersion) one peculiar aberration pattern is frequently encountered. As shown in Fig. 2.12, the lens is well corrected over up to 90-95% of the aperture radius, but after that we see a runaway phase variation right to the edge. Speaking in Zernike terms residual spherical aberration components of very high order are observed. Because of this high-order it appears unlikely that the aberrations are caused by improper immersion fluid or some other trivial reasons. These would manifest themselves via low-order spherical as well. More realistically, this is a design flaw of the lens.

The portion of the lens affected by this feature varies from few to about 10 percent. For the lens in Fig. 2.12 the line delimiting the effective aperture was somewhat arbitrarily drawn at NA=1.3. What is undeniable is that the effect is not negligible. In all likelihood this form of aberration is the reason for the somewhat mysterious phenomenon when a high NA lens exhibits a PSF, which is perfect in all respects except for a somewhat reduced NA. This was the case in Sect. 2.2.2 and also described by other researchers [11].

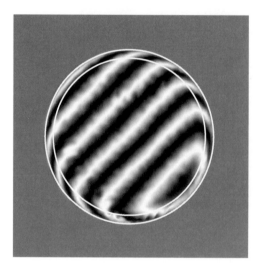

Fig. 2.12. Pupil function of 63×1.4 NA oil immersion lens with both nominal (outer ring) and effective working (inner ring) apertures indicated

It is quite clear that competitive pressures push the lens designers towards the boundary (and sometimes beyond the boundary) of the technical possibilities of the day. Few years ago no microscope lens manufacturer could be seen without a 1.4 NA oil immersion lens when the Joneses next door were making one. The plot is likely to be repeated with the newly emerging 1.45 NA lenses. It is also true that a hefty premium is charged by the manufacturers for the last few tenths in the NA. It is quite possible that in many cases this is a very expensive *empty* aperture, which, although physically present, does not contribute to the resolving power of the lens.

This discussion may seem to be slightly misguided. Indeed, many people buy high NA lenses not because of their ultimate resolution, but because of their light gathering efficiency in fluorescence microscopy. This property is approximately proportional to NA^2 and therefore high NA lenses produce much brighter, higher contrast images. At the first glance it may seem that the aberrated edge of the pupil will not affect this efficiency and hence buying a high NA lens, however aberrated, still makes sense. Unfortunately, this is not true. Because the phase variation at the edge is so rapid, the photons passing through it reach the image plane very far from the optical axis. They do not contribute to the main peak of the diffraction spot, instead they form distant sidelobes. In terms of real life images it means that the brightness of the background, and not the image itself, is increased. Paradoxically, blocking the outermost portion of the pupil would in this case *improve* the image contrast!

The last statement may well be generalized in the following way: in the end the only sure way of obtaining a near-perfect high NA objective lens is to acquire one with larger-than-required nominal NA and then stop it down. Incidentally, in certain situations this may be happening even without our deliberate intervention. Consider using an oil immersion lens on a water-based sample: no light at NA > 1.33 can penetrate the sample anyway and hence the outer aperture of the lens is effectively blocked.

2.5 *Esoterica*

In this Section a few more results obtained with the Pupil function evaluation apparatus are presented. These need not necessarily be of prime concern to most microscope users but might be of interest to connoisseurs and, indeed, could provide further insight into how modern objective lenses work.

2.5.1 Temperature Variations

Many of microscopists would recall seeing 23 °C on a bottle of immersion oil as a temperature at which the refractive index is specified, typically $n = 1.518$ at $\lambda = 546$ nm. But how important this "standard lab temperature" is to the performance of high NA lenses? In order to answer this question the Pupil function of a 100× 1.4 NA oil immersion lens was measured at a range of temperatures and the results were processed to obtain the variation of the primary aberration coefficients with the temperature.

The major factor in the degradation of the imaging qualities of the immersion lenses with temperature is the variation of the refractive index of the immersion oil, usually at the level of $dn/dT = 3$–4×10^{-4}. The effect of this change is similar to that of introducing a layer of a refractive-index-mismatched material between the lens and the sample. The resulting aberrations are well understood, their exhaustive analysis can be found in [12]. In short, spherical aberrations of various orders will be generated; the relative weight of higher order terms rises dramatically with the NA of the lens. This is corroborated by the experimental data in Fig. 2.13 which show steady, almost linear variation in both primary and secondary spherical aberration terms with temperature. Less predictably the same plot also registers significant variations in the other first order aberrations: coma, astigmatism and trefoil. Because of their asymmetry these aberrations can not be explained by the oil refractive index changes. They are probably caused by small irregular movements of individual elements within the objective lens itself.

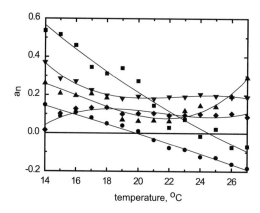

Fig. 2.13. Temperature variation of primary (■) and secondary (●) spherical aberrations as well as magnitudes of astigmatism (▲), coma (▼) and trefoil (◆). The latter three were defined as $a = \sqrt{a_s^2 + a_c^2}$, where a_s^2 and a_c^2 are the sin and cos components of the corresponding aberrations

Strictly speaking aberrations caused by the refractive index changes in the immersion fluid should not be regarded as lens aberrations. In practice, however, the lenses are designed for a particular set of layers of well defined thicknesses and refractive indexes between the front of the lens and the specimen. Any change in these parameters upsets the fine optical phase balance in the system and results in aberrated PSF. This might be an argument why it makes sense to treat the immersion medium as being a part of the objective lens. The temperature effect depends dramatically of the type of the immersion fluid used. Water with $dn/dT \approx 8 \times 10^{-5}$ is far less sensitive; dry lenses, of course, are not affected at all. Large working distance lenses will be at a disadvantage too due to longer optical paths in the immersion fluid.

Closer analysis of Fig. 2.13 reveals that the aberrations are indeed minimized around 23 °C. where the combined primary and secondary spherical aberrations are close to their minimum. A small but noticeable hysteresis effect was also noted when after a few temperature cycles the aberration coefficients failed to return to their low initial values. It is tempting to draft this effect to the explanation of the fact that imaging properties of even the best of lenses always deteriorate with age – although accidental straining during experiments is still likely to remain the prevailing factor.

2.5.2 Apodization

So far the emphasis of this investigation was on measuring the phase aberrations. This is justified by the fact that the phase deviations from an ideal spherical wavefront have considerably more impact on the shape of the PSF than similar imperfections in amplitude. Nevertheless for completeness sake it might be interesting now to have a closer look at the apodization effects occurring in high NA lenses. Using dipole radiation as a probe offers unique advantages in this task. This is because the angular intensity distribution of the dipole radiation is well defined. Therefore any deviations from perfect lens behaviour should be easy to spot.

Let's assume that the polarization vector of a dipole situated in the focus is aligned in x direction. Angular intensity distributions in the x-z and y-z (i.e. meridional and equatorial) planes will be given by, respectively, $I_x \sim \cos^2 \alpha$ and $I_y = $ const. Due to purely geometric reasons these distributions will change when light propagates to the pupil plane even if the lens is perfect. With reference to Fig. 2.1 and to the sine condition $d = nf \sin \alpha$ it is not too difficult to show that an extra factor of $\sec \alpha$ is has to be introduced when going from the object to the pupil side of the objective in order to satisfy the energy conservation law. This factor was well known since the early days of high NA lens theory [3]. Intensity distributions in the pupil plane therefore should therefore look like $I_x \sim \cos \alpha$ and $I_y \sim \sec \alpha$ or, with the help of sine condition:

$$I_x \sim \frac{nf}{\sqrt{(nf)^2 - d^2}}, \qquad I_y \sim \frac{\sqrt{(nf)^2 - d^2}}{nf}. \qquad (2.15)$$

An experiment to measure these distributions was carried out on the setup of Fig. 2.8 by simply blocking the reference beam and capturing the pupil intensity image alone. To produce the results shown in Fig. 2.14 images of 8 individual scatterers were acquired and averaged in the computer. Intensity distributions in the two principal planes were then extracted. They follow the theoretical predictions rather well up to about half of the pupil radius. After that apodization is apparent which increases gradually reaching about 30–50% towards the edge of the pupil.

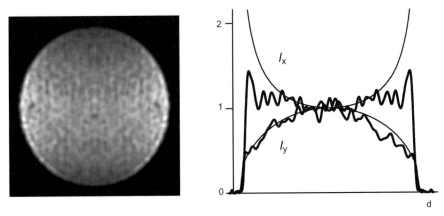

Fig. 2.14. Apodization effects on an image of a dipole viewed at the pupil plane of a 100× 1.4 NA lens. Theoretical predictions (*thin line*) according to (2.15)

The cause of this apodization in all likelihood are the Fresnel losses in the elements of the objective lens. Broadband antireflection coatings applied to these elements are less effective at higher incidence angles that the high aperture rays are certain to encounter. Because of the nature of these losses they are likely to be very individual for each particular type of objective lens. It is also worth noting slight polarization dependence of the losses, which contributes to polarization effects described in the following section.

2.5.3 Polarization Effects

Polarization effects encountered when imaging a dipole with a high NA lens have been covered elsewhere [13] (For a general discussion also see Chapter 1). For the purposes of this investigation they are interesting inasmuch as the imperfections of the lens contribute to these effects. When an image of a small scatter is viewed between the crossed polarizers characteristic "clover leaf" pattern emerges. An image of the pupil plane, Fig. 2.15, is particularly telling. It shows, that the only rays that travel very close to the edge of the pupil pass through the polarizer. This happens

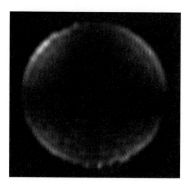

Fig. 2.15. Polarization effects on an image of a dipole viewed at the pupil plane

because the polarization of light in these four segments is slightly rotated from its original x direction. The reason for this rotation is twofold. First of all the radiation of a dipole is linearly polarized in the meridional plane, which can only be approximated by a uniformly polarized plane wave for small angles α. When this spherical wavefront is straightened by the lens and projected into the pupil plane only light propagating in x-z and y-z possess perfect x polarization, the remainder is rotated to some degree. The degree of rotation increases dramatically with higher aperture angles. This effect is fundamentally geometrical; its close cousin is a well know problem of how to comb a hairy ball.

The second reason for the polarization rotation is more prosaic: Fresnel losses. These tend to be higher for s than for p polarized light. For a beam travelling in a plane bisecting the x-z and y-z planes overall tendency would be to rotate the polarization towards the radial direction. Hence this effect seems to work in the opposite direction to that caused by the geometric factor, which favours the azimuthal direction.

2.6 Conclusion

The range of various experimental setups and techniques dedicated to the characterization of the high NA objective lenses could be continued. For instance the lateral chromatic aberration has not been considered so far. One has to be practical, however, and draw a line at some point.

From the tests described in this Chapter the measurement of the Pupil function provides the most detailed and insightful information about the capabilities of a particular lens. In many cases two or three Pupil functions measured across the field of view and, perhaps, tabulated in Zernike coefficient form would be more than sufficient to predict the lens performance in most practical situations. It is disheartening to think of how much wasted time, frustration and misunderstandings could be avoided if such information were to be supplied with the objective lenses by their manufacturers.

My overall conclusion is that the vast majority of currently designed objective lenses perform really well. Any imperfections visible in a microscope image are far more likely to be a result of a sloppy practice (for example tilted coverslip, incorrect immersion fluid etc.) than a fault of the lens itself. That said, cutting-edge designs are always going to be a problem and the very highest NA lenses should be approached with caution. It is also worth pointing out that elements of the microscope other than the objective lens may also be a factor in imaging quality. The tube lens is one such element of particular concern. Having evolved little over the last years this lens is typically just a simple triplet with too few elements to achieve the aberration correction on par with that of an all-singing all-dancing objective lens. This situation is further exacerbated by the advent of a new breed of low-magnification high NA objectives with their enormous back apertures.

Acknowledgment

I would like to thank my colleague M.A.A. Neil for numerous contributions to this work, in particular the implementation of the Zernike polynomial fit routine and processing of the relevant data.

References

1. D. Malacara: *Optical shop testing* (John Wiley & Sons, New York 1992)
2. M. Born, E. Wolf: *Principles of optics* (Pergamon Press, Oxford 1998)
3. B. Richards, E. Wolf: P. Roy. Soc. Lond. A **253**, 358 (1959)
4. T. Wilson, R. Juškaitis, N.P. Rea, D.K. Hamilton: Opt. Commun. **110**, 1 (1994)
5. R. Juškaitis, T. Wilson: J. Microsc. – Oxford **189**, 8 (1997)
6. T. Wilson, C.J.R. Sheppard: *Theory and practice of scanning optical microscopy* (Academic Press, London 1984)
7. R. Juškaitis, T. Wilson: J. Microsc. – Oxford **195**, 17 (1999)
8. R. Juškaitis, M.A.A. Neil, T. Wilson: 'Characterizing high quality microscope objectives: a new approach', In: *Proceedings of SPIE, San Jose Jan. 28–Feb. 2, 1999*, ed. by D. Cabib, C.J. Cogswell, J.-A. Conchello, J.M. Lerner, T. Wilson (SPIE, Washington 1999) pp. 140–145
9. V.N. Mahajan: Appl. Optics **33**, 8121 (1994)
10. M.A.A. Neil, M.J. Booth, T. Wilson: J. Opt. Soc. Am. A **17**, 1098 (2000)
11. S.W. Hell, P.E. Hanninen, A. Kuusisto et al.: Opt. Commun. **117**, 20 (1995)
12. P. Török, P. Varga, G. Németh: J. Opt. Soc. Am. A **12**, 2660 (1995)
13. T. Wilson, R. Juškaitis, P.D. Higdon: Opt. Commun. **141**, 298 (1997)

3 Diffractive Read-Out of Optical Discs

Joseph Braat, Peter Dirksen, Augustus J.E.M. Janssen

3.1 Introduction

An extensive literature is available on optical data storage. Because of the multi-disciplinary character of optical storage, the subjects range from optics, mechanics, control theory, electronics, signal coding and cryptography to chemistry and solid-state physics. Several books [2]–[9] have appeared that are devoted to one or several of the subjects mentioned above. This chapter will be limited to purely optical aspects of the optical storage systems. In Section 3.2 of this chapter we first present a short historic overview of the research on optical disc systems that has led to the former video disc system. In Section 3.3 we briefly review the general principles of optical storage systems that have remained more or less unchanged since the start some thirty years ago. An interesting feature of the DVD-system is its standardised radial tracking method that we will treat in some detail in Section 3.4. The appearance of new generations with higher spatial density and storage capacity has triggered solutions for the backward compatibility of the new systems with the existing ones. In Section 3.5 we pay attention to these backward compatibility problems that are mainly caused by the smaller cover layer thickness, the shorter wavelength of the laser source and the higher numerical aperture of the objective in the new generations. Finally, in Section 3.6 we indicate how an efficient modelling of the optical read-out system can be done. Important features like the radial cross-talk and inter-symbol interference can be studied with an advanced model of the optical read-out and the resulting signal jitter, an important quality factor for a digital signal, is efficiently obtained. A new approach is presented that can speed up the calculations so that the modelling can really serve as an interactive tool when optimizing the optical read-out.

3.2 Historic Overview of Video and Audio Recording on Optical Media

Optical data storage uses the principle of the point-by-point scanning of an object to extract the recorded information on the disc. In another context, such an approach was used in a somewhat primitive way in the 'flying-spot' film scanner. This apparatus was developed in the early fifties of the last century when a need was felt to transform the pictorial information on film into electrical information to be recorded on magnetic tape. The flying-spot in the film scanner was obtained with the aid of

Török/Kao (Eds.): Optical Imaging and Microscopy, Springer Series in Optical Sciences
Vol. 87 – © Springer-Verlag, Berlin Heidelberg 2003

a cathode ray tube and its resolving power was adequate for the resolution encountered on the standard cinema film format. In Fig. 3.1 the focusing lens L_o is shown that produces the scanning spot on the film. The light is partially transmitted and diffracted by the information on the film surface and the detector collects the transmitted light via the collecting lens L_c. The required resolution in the film plane was of the order of 20 to 30 µm and this was not an extreme demand neither for the electron optics nor for the 'objective' lens L_o. The flying-spot scanner was not analysed in depth for some time, it just functioned correctly for the purpose it had been developed for and produced coloured images with the required TV-resolution. In

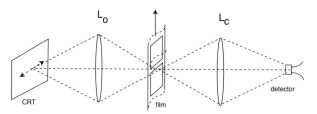

Fig. 3.1. Schematic layout of a flying-spot film scanning device

1961 a patent was attributed to M. Minsky [10] where he describes the principle of a scanning microscope with a resolving power comparable to the standard microscope. Minsky even proposed to use several scanning spots in parallel, thus realising a large-field scanning . The practical use of his (confocal) scanning microscope was limited because of the inefficient use of the power of the (thermal) light source.

A further step towards the optical data storage principle as we know it nowadays was taken by a research group at Stanford University that focused on the recording of video information on a record with the size of a standard audio Long Play disc [11]. The group developed a disc-shaped carrier with data tracks that carried the directly modulated video signal. After a few years the research was abandoned because of the cross-talk problem between the information in neighbouring tracks and because of the bad signal to noise ratio of the detected signal. The belief in a commercial success as a consumer product dwindled away and the sponsor, the 3M-company, stopped funding.

In parallel to this early research on direct signal recording, several groups worked on the principle of storing pictures on a disc, either directly or via holographic encoding. Of course, the latter method was believed to be much more robust because of the far-field recording of the pictorial information. However, one has to conclude that the various prototypes of 'pictorial' video-players that were developed at the end of the sixties and the early 1970's were not satisfactory because of the inefficient use of the available recording surface, the difficult mechanics in the case of the 'direct picture' approach and the colour and speckle problem for the holographic devices.

3.2.1   The Early Optical Video System

At the end of the 1960's, various disciplines and techniques were present at Philips Research Laboratories in the Netherlands that enabled a breakthrough in the field of optical data storage. Important factors enabling this breakthrough were

- availability of expertise in precision optics and control mechanics,
- research on a consumer-type of HeNe-laser where the delicate problems of alignment and stability of such a coherent source were solved in such a way that relatively cheap mass-production could be envisaged,
- work on efficient modulation techniques for video recording on magnetic tape,
- a strong belief in the market for educational applications (teaching, instruction, continuous learning).

In an early stage, some decisions [12] were taken (reflective read-out, information protection via the disc substrate) that have proven to be of vital importance for the success of 'robust' optical data storage. In 1975, several companies (Philips, Thomson [13], Music Corporation of America [14]), later joined by Pioneer, have united to establish a standard for the video disc system (VLP = Video Long Play) so that worldwide disc exchange would be possible. The most important standardised features were

- HeNe laser source, $\lambda = 633$ nm, numerical aperture (NA) of the read-out objective equal to 0.40, track pitch of information spiral is 1.6 µm, disc diameter is 30 cm, rotational velocity is 25 or 30 Hz depending on the television standard.
- frequency modulation of the analog video signal resulting in optical information pits with a (small) pit length modulation; the average pit length typically is 0.6 µm with comparable 'land' regions in between.
- recording of two video frames per revolution (constant angular velocity, CAV) or recording with constant linear velocity (CLV). The first choice allowed for the still-picture option without using a frame memory (not yet available at that time), the second choice yielded discs with the longest playing time (up to two hours).

Further research on the video disc system (later called the Laser Disc system) has centred on an increase of the recording density, e.g. by

- halving the track pitch to 0.8 µm and keeping the cross-talk at an acceptable level by introducing a two-level pit depth that alternates from track to track [15]
- using (high-frequency) track undulation for storing extra information, e.g. colour and/or sound channels.

Other research efforts have resulted in the introduction of the semiconductor laser ($\lambda = 820$ nm) and the simplification of the optics, among others by introducing aspherical optics for the scanning objective [16].

Further potential changes in disc definition resulting from this research on higher density have not led to an improved video disc standard; but the semiconductor laser source and the cheaper optics have been readily introduced at the beginning of the 1980's.

In parallel to the video application, storage of purely digital data has been pursued in the early stages of optical storage. Various systems using large-size optical discs have been put on the professional market while the recording of data was made possible by the development of high-power semiconductor lasers. In the first instance, the data were irreversibly 'burnt' in the recording layer. Later research has revealed the existence of suitable phase-change materials [17,18] that show an optical contrast between the amorphous and crystalline phase and permit rapid thermally induced switching between both phases. Other optical recording materials rely on the laser-induced thermal switching of magnetic domains [7]; the optical read-out of the domains with opposite vertical magnetisation is done using the magneto-optical Kerr effect.

3.2.2   The Origin of the CD-System

The Laser Disc video system knew a limited success due to its restricted playing time and its lack of the recording option which made direct competition with the well-established video cassette recorder rather difficult. Other opportunities for the optical storage technique were already examined for several years and the replacement of the audio Long Play disc by an optical medium was actively researched. An enormous extra capacity becomes available when switching from the mechanical LP needle to the micron-sized diffraction-limited optical 'stylus'. The approximately hundred times larger storage capacity on an equally sized optical disc has been used in two ways, first by reducing the size of the new medium and, secondly, by turning towards digital encoding of the audio signal which, at that time, asked for a very substantial increase in signal bandwidth. After a certain number of iterations where disc size, signal modulation method, digital error correction scheme and playing time were optimised, the actual Compact Disc standard was established in 1980 by the partners Philips and Sony. The disc diameter of 12 cm was established to accommodate for the integral recording on a single disc of the longest version of the ninth symphony of Beethoven (74 minutes playing time, Bayreuth 1951 recording by the Berliner Philharmoniker, conductor Wilhelm Fuertwangler).

The worldwide market introduction of the CD-system took place at the end of 1982 and the beginning of 1983. Apart from the infrared semiconductor laser source and the corresponding miniaturisation of the optics, no really fundamental changes with respect to the preceding laser disc video system were present regarding the optical read-out principles. The CD system turned out to be simpler and more robust than its video signal predecessor. This was mainly due to the digitally encoded signal on the CD-disc that is much less sensitive to inter-track cross-talk than the former analog video signal.

3.2.3 The Road Towards the DVD-System

For a long period, the application of the CD-system has been restricted to the digital audio domain. In an early stage, a read-only digital data storage system with identical capacity (650 MByte) has been defined (CD-ROM) but the market penetration of this system has been relatively late, some 8 to 10 years after the introduction of the audio CD. Once the personal computer application took off, the demand for recordable and rewritable CD-systems (CD-R and CD-RW) immediately followed. At the same time, the computer environment asked for an ever increasing data capacity to handle combination of pictorial and data content (e.g. for games). Spurred by the quickly developing market of video (using the lower quality digital MPEG1 video standard), a serious need was felt for an improved optical data storage standard. The DVD-standard (see Fig. 3.2) has been defined as of 1994 by a consortium of 10 companies and offers a seven times higher capacity than the CD-format. Because of its multi-purpose application, the acronym DVD stands for Digital Versatile Disc and both MPEG2 video information, super-audio signals and purely digital data can be recorded on this medium. Again, the optical principles used in the DVD-system have not fundamentally changed; a relatively new radial tracking method was introduced and the backward compatibility with the existing CD-family asked for various provisions. In recent years, the DVD-medium has also developed into a

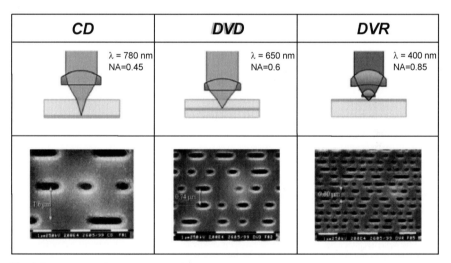

Fig. 3.2. A sketch of the light paths for the CD, DVD and DVR or Blu-ray Disc system and electron microscope photographs with identical magnification of the information pits of the three systems

family with recordable and rewritable versions. The main innovation in the DVR or Blu-ray Disc system is the use of the deep blue semiconductor laser; combined with a drastically increased value of the numerical aperture of the objective, a forty times higher density with respect to the CD-system is achieved.

Even if the basics of optical disc systems have not changed, an enormous effort in research and development has been needed to adapt the media and the playing units to the drastically increased demands with respect to spatial density, scanning speed and data rate. Another less visible development is the digitisation of the opto-mechanical and electronic control systems in the player. Without the easily programmable digitised servo and control systems, the multi-purpose disc player/recorder that easily accepts a broad range of media would be impossible.

In this chapter we will restrict ourselves to the purely optical aspects of an optical disc player/recorder. More general overviews can be found in the various textbooks mentioned in the introductory section.

3.3   Overview of the Optical Principles of the CD- and the DVD-System

In this section we briefly recall the read-out method to extract the stored information from the optical disc. The information is optically coded in terms of a relief pattern on a read-only disc while recordable and rewritable discs employ recording layers that influence both the phase and the amplitude of the incident light. The arrangement of the information is along tracks that have to be followed, both in the recorded and the 'blank' or pre-formatted situation. To this purpose, a bipolar signal is optically derived that can be connected to a servo mechanism to keep the objective head on track during reading or recording. Simultaneously, an error signal is needed to correct for any deviation of the information layer from the optimum focus position.

3.3.1   Optical Read-Out of the High-Frequency Information Signal

In Fig. 3.3, the typical data formats to be read in an optical disc player are shown. The typical track spacing on CD is $1.6\,\mu m$ and the pit length varies between 0.9 and $3.3\,\mu m$ with typical increments of $0.3\,\mu m$. The data disc in Fig. 3.3 (b) has a different coding scheme that initially looked more appropriate for the hole-burning process by avoiding the length modulation. In Fig. 3.3 (c) the CD-ROM format is present on a phase change disc with the low-contrast regions of micro-crystalline and amorphous structure. The disc in Fig. 3.3 (d) is not compatible with the standard (or DVD) reading principle because of the polarisation-sensitive read-out.

The reading of the information is done with the aid of the scanning microscope principle that is sketched in Fig. 3.4. As it is well-known from the literature, the scanning microscope and the classical microscope are very closely related regarding their resolution and optical bandwidth [19,20]. In terms of maximum transmitted spatial frequency, the classical microscope attains a value of $2\,NA/\lambda$ in the case of incoherent illumination via the condensor system (the numerical aperture of the condensor C then needs to be equal to or larger than the numerical aperture NA of the objective O_I). The scanning microscope achieves the same maximum frequency transfer when the detector D at least captures the same aperture NA as offered by the

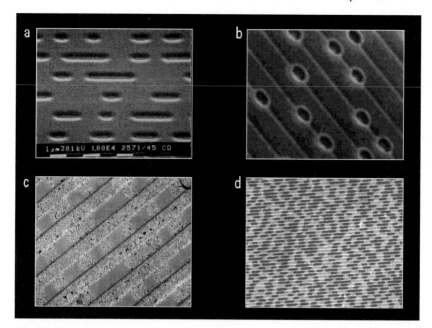

Fig. 3.3. Typical data formats present on optical storage media (**a**) CD-information recorded as a relief pattern in the surface of a disc obtained via injection molding (**b**) 'burned' data on a digital data disc (**c**) amorphous islands (gray) arranged along a pre-formatted track in a micro-crystalline 'sea' (phase-change recording) (**d**) polarisation microscope photograph of a magneto-optical layer with recorded CD-like information

scanning objective O_S (see heavy dashed lines in Fig. 3.4 (a). In the case of a reflective scanning system, the equal aperture of O_S and D is obtained in a straightforward manner because of the double role of O_S as scanning objective and collecting aperture for the detector.

The optical transfer function of a scanning microscope has been visualised in Figs. 3.5–3.6. The diffraction of a plane wave with propagation angle α by a (one-dimensional) periodic structure (frequency $v = 1/p$) leads to a set of distinct plane waves with propagation direction angles α_m given by

$$\sin \alpha_m = \sin \alpha + m\lambda v ,\tag{3.1}$$

where m is the order number of the diffracted plane wave.

In the case of an impinging focused wave (aperture angle u), the diffracted field is a superposition of diffracted spherical waves. The incident spherical wave can be considered to be the superposition of a set of plane waves with an angular distribution ranging form $-u$ to $+u$. Note that in the case of a spherical wave, the diffracted orders can not easily be separated in the far field at a large distance from the periodic structure. This is in particular the case for low spatial frequencies and this is in line with the intuitive feeling that a finely focused wave is not capable of sharply discriminating a large period. As soon as the period becomes comparable with the

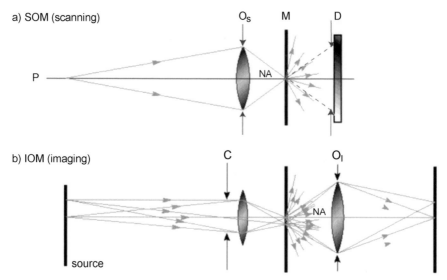

Fig. 3.4. The scanning optical microscope *SOM* (**a**) and the 'classical' imaging optical microscope *IOM* with a full field illumination (**b**). In the scanning microscope, the detector has been schematically drawn and is optically characterised by its angular extent (dashed cone) with respect to the object. In the figure, the angular extent or 'aperture' of the detector has been drawn equal to the numerical aperture *NA* of the objective O_S, a situation that corresponds to standard reflective read-out of an optical disc

lateral extent of a focused wave (typically of the order of $\lambda/2NA$), the diffracted spherical waves are fully separated. However, we do not need to fully separate the diffracted spherical orders to draw conclusions about the presence and the position of the periodic structure. In the regions where e.g. the spherical waves with order number $m = 0$ and $m = 1$ overlap, the two waves are brought to interference and the phase difference between them will depend on the lateral position of the periodic structure with respect to the focal point of the incident spherical wave (note that the phase of a first diffracted order shows a phase shift ϕ that equals $2\pi\nu\Delta$ if the periodic structure suffers a lateral shift of Δ).

The strength of the interfering signal depends on the size of the overlapping region and on the usefully captured signal by the detector. In the case of a centred detector, the signal depends on its lateral extent and it is normally expressed in terms of the angular beam extent ($NA = \sin u$) of the incident focused beam.

In Fig. 3.6, the normalised frequency transfer function has been depicted as a function of the commonly called coherence factor $\gamma = NA_D/NA$ (NA_D is the numerical aperture of the detector as defined by geometrical conditions or by a collecting optical system in front of the detector, see [19]). The maximum detectable frequency is $2NA/\lambda$. In this case ($\gamma \geq 1$), the optical system is approaching the fully incoherent limit which means that from the complex amplitude transmission function of an object only the modulus part is detectable. At low values of γ, leading

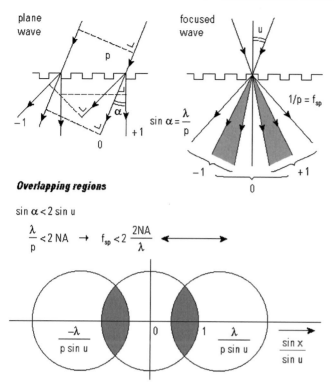

Fig. 3.5. The diffraction of a plane wave and a focused wave by a one-dimensional periodic structure. The useful region in the far-field for signal detection is given by the regions of overlap of zeroth and first orders

to the limit of 'coherent' read-out, the phase part of an object, even if it is a weakly modulated function, can also be detected in principle. Note that on an optical disc the relief pattern mainly affects the phase part of the optical transmission or reflection function. Because of the strong height modulation with respect to wavelength, the phase modulation can also be detected in the (almost) incoherent mode at $\gamma = 1$. If the phase modulation switches between 0 and π, the phase modulation even becomes a pure amplitude modulation with extreme values of ± 1. The choice of the value of γ has been fixed at unity from the very beginning of optical storage. In contrast with e.g. optical lithography where a certain minimum transfer value is required and where it can be useful to trade off the maximum transmitted frequency against a better modulation via a choice $\gamma < 1$, the transfer of an optical disc system can be improved electronically afterwards. This electronic equalisation has proved to be powerful and thus permits to virtually use the entire optical pass-band up to the optical cut-off frequency of $2\,NA/\lambda$; in practice, frequencies turn out to be detectable up to 90% or 95% of the cut-off frequency. The typical optical frequencies in a

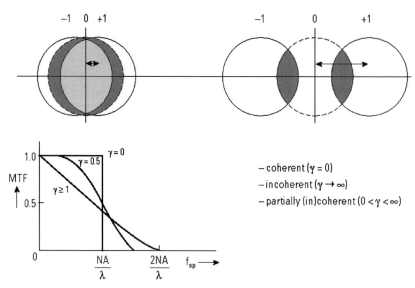

Fig. 3.6. The normalised optical transfer function as a function of the coherence factor γ, determined by the numerical aperture ratio of detector and scanning objective

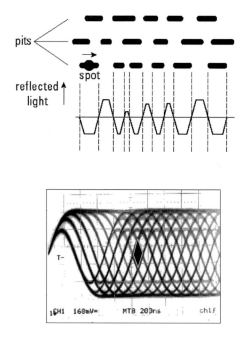

Fig. 3.7. The pit structure on an optical disc and a schematic drawing of the detected signal that shows a typical maximum slope defined by the spot size λ/NA. An oscilloscope image of a large sequence of signal sections is shown below when the scope is triggered on an arbitrary positive transient in the signal; this signal is commonly called the digital 'eye-pattern' and should preferably not contain transients that are close to the centre of a digital eye (see the drawn lozenge in the figure)

CD-system ($\lambda = 785$ nm, $NA = 0.45$) are 55% of the cut-off frequency for the radial period and approximately 50% for the highest frequency in the direction of the tracks (this is the 3T–3T alternation, the shortest land and pit distances (0.9 µm) on a CD-disc given the clock length of 0.3 µm). Of course, the binary relief pattern in the track of a CD may contain higher frequencies than this highest fundamental frequency. In the DVD-system ($\lambda = 650$ nm, $NA = 0.60$) the optical pass band is used in a more aggressive way and the typical frequencies are shifted to 75% of the cut-off frequency $2 NA/\lambda$.

The typical signal that is obtained from the high-frequency detector in an optical disc player is shown in Fig. 3.7. A well-opened digital 'eye' is essential for a correct read-out of the stored information.

3.3.2   Optical Error Signals for Focusing and Radial Tracking of the Information

Focus Error Signal Various methods have been proposed for deriving a focus error signal that can be fed to e.g. the coils of a miniature linear electromotor so that an active positioning of the objective with respect to the spinning disc is achieved. In

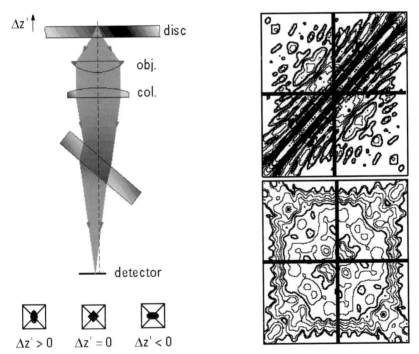

Fig. 3.8. An optical light path using the astigmatic focusing method by exploiting the astigmatism introduced by a plane parallel plate in the returning converging beam (*left*). The calculated diffraction patterns corresponding to a focal line and to the best focus position are depicted in the right-hand part of the figure, together with the separation lines of the focus quadrant detector (*horizontal* and *vertical* in this picture)

Fig. 3.8 the commonly used astigmatic method [21] is shown. The light reflected from the disc passes through a relatively thick plane parallel plate (the beam splitter on the way forth from the source to the disc). Depending on the convergence angle of the returning beam and on the thickness of the plate, a certain amount of astigmatism is introduced that generates two focal lines along the optical axis with a 'best' focal region in between them. Projected back onto the information surface and taking into account the axial magnification of the combination of collimator and objective, one aims at a distance between the astigmatic lines of some 10 to 15 µm. The best focus position corresponds to a balanced intensity distribution on the four detectors and should be made to coincide with optimum high-frequency read-out of the optical disc. In the right part of Fig. 3.8 a contour picture is shown of the diffraction image in the best focus position and in the case when one of the line images is focused on the detector.

Another candidate for the detection of a focus defect is the Foucault knife-edge test. Although its sensitivity can be higher than that of the astigmatic method, the position tolerances of the knife edge are somewhat critical. A circularly symmetric version of the knife-edge method is the 'ring-toric' lens method [22]. Another option is the 'spot-size' detection method that measures the variation of beam cross-section as a function of defocus. All these methods are capable of delivering a signal with changing polarity as a function of the sign of the focus error. Another important property of an error signal is the maximum distance through which the signal is well above the noise, the so-called effective acquisition range. For both methods, a typical value of 30 to 50 µm can be expected. More refined methods use the detected high-frequency information to monitor the absolute focus-setting and to correct for any drift in the detector balance or the mechanical settings of the light path.

Radial Error Signal The methods indicated in Fig. 3.9 have been devised for the detection of the radial position of the reading or recording light spot with respect to the centre of a track. The first method uses two auxiliary light spots generated by means of a grating (twin-spot method [23]). The light spots have a quarter track spacing off-set with respect to the central light spot. The reflected light belonging to each auxiliary spot is collected in the focal region on separate detectors and the difference in average signal from each detector provides the desired bipolar tracking signal that is zero on-track. An AC-variant of this method with the auxiliary spots uses a small lateral oscillation of the main spot itself. The detection of the phase (either 0 or π) of the oscillation in the detected AC-signal yields the required tracking error signal. The oscillation can be imposed on the scanning spot in the optical player but, preferably, a so-called track wobble is given to the track with a known phase so that no active wobble is needed in the player. Of course, the track wobble should not generate spurious information in the high-frequency signal band nor should it spoil other information (track number, address information) that is stored in an 'undulation' or lateral displacement of the information track. Both methods described above retrieve the tracking information from a refocused light distribution on one or more detectors that are optically conjugate with the information layers and

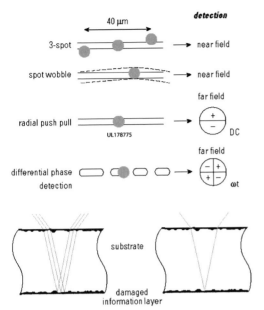

Fig. 3.9. The various options for deriving a bipolar optical tracking signal. The label 'near field' here means that the detector arrangement is positioned in the image plane of the information layer. In the case of 'far field' detection, the detector is effectively located in the limiting aperture of the objective (or in an image of this diaphragm)

for this reason they are called near-field methods. A well-known far-field method is the so-called Push-Pull method [21] that uses a duo-detector in subtractive mode for radial tracking (see Fig. 3.10). The method is based on the fact that in the presence of a tracking error (the centre of the scanning spot hits e.g. the edges of the information pits) the average propagation direction is slightly bent away from the perpendicular direction. For relatively small values of the phase shift of the light reflected by the pits, the difference signal of the duo-detector shows a change of sign when the scanning spot crosses the information track. If we define the phase depth $\Delta\phi$ of the information pits as the phase shift suffered by the light on reflection from the information pits with respect to the light reflected from non-modulated land regions, we observe that the difference signal of the duo-detector reduces to zero if $\Delta\phi$ equals π. The obvious reason is that a phase shift of π between pits and land regions has reduced the information structure to effectively an amplitude structure with no directional phase information available. For this reason, the Push-Pull method is critical with respect to the phase shift induced by the information pits and it also behaves in a rather complicated way once the optical 'effects' have a combined amplitude and phase character (phase-change recording material). A far-field method that, in a first instance, does not suffer from the restrictions on the phase depth $\Delta\phi$ is the differential phase (or time) detection where the phase relations between signal components from a far-field quadrant detector at the instantaneous signal frequency ω are used to derive a tracking signal [24]. A tracking method based on this principle has been standardised for the DVD-system and will be discussed in more detail in a following section.

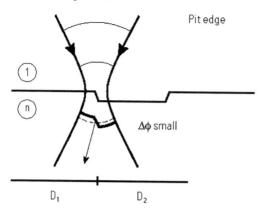

Fig. 3.10. The phase change introduced by a pit edge to the focused wave (here shown in transmission). The average propagation direction of the transmitted beam is slightly tilted causing unbalance between the detectors D_1 and D_2

The possible perturbations of the near-field or far-field methods is correlated with the conjugate position of the respective detectors. In Fig. 3.9, lower part, the near-field methods should be sensitive to damage of the information layer while the far-field based methods should be more sensitive to damage or scratching of the upper surface (non-info) of the disc. Because of the limited bandwidth of the tracking error signals, their (average) information is collected from a relatively large disc area (a stripe on the info-layer, an elongated circle on the non-info surface). Because of the spatial averaging effect, no substantial difference of preference is found for the near- or far-field methods.

3.3.3 Examples of Light Paths

The various methods for deriving the high-frequency signal and the optical error signals have to be incorporated in a light path that has wide tolerances both during manufacturing and product life (optical damage, thermal and mechanical drifts). In Fig. 3.11, left-hand side, a classical light path for CD is shown with the light from the laser diode source coupled in via a semi-transparent plate towards the collimator-objective combination. The reflected light is transmitted through the beam splitter towards a double-wedge prism that separates the far field region in order to accommodate the detection of a Push-Pull radial tracking signal. The double wedge also serves as 'knife-edge' for each beam half thus producing two complementary knife-edge images for focus detection. This light path is made from simple, easily available components. The tightest tolerance regarding temperature and mechanical drift is found in the relative position of source and detector with respect to the beam splitter.

In the right-hand figure, the beam splitting action is obtained with the aid of a holographic element. The relatively thick substrate carries a grating on the lower surface for generating the three spots needed for radial tracking. The upper surface contains a more complicated structure, the holographic or diffractive element. At the way forth to the disc, the three beams generated by the grating continue as the zeroth order of the holographic structure and they are captured by the collimator and

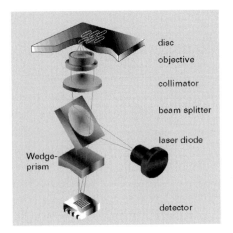

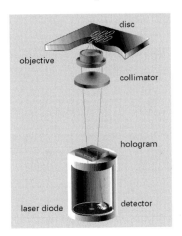

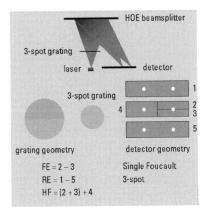

Fig. 3.11. An optical light path using separate optical components for achieving the beam separation and the generation of the focus and radial error signals (left). In the right-hand picture, the splitting function and the generation of the error signals is obtained by means of holographic or diffractive optical element. In the lower picture, the various beams are shown that propagate towards the detector after traversal of the holographic beam splitter *HOE*

objective. Consequently, three light spots are focused on the disc. On the way back, the light diffracted by the hologram propagates towards the detector region and a far-field splitting is obtained by sending one half circle of the beam cross-section to the detectors 2 and 3 and the other half to detector 4. The auxiliary spots produced by the grating are directed towards the detectors 1 and 5. A focus error signal is obtained by taking the difference between detectors 2 and 3 while the Push-Pull radial tracking error signal can be obtained by the signal combination $(2 + 3) - 4$. Moreover, an extra twin-spot radial error signal is obtained from the detector difference signal 1–5 (the radial offset of each spot is one quarter of a pitch). The high frequency signal is derived from the sum of the detectors 2, 3 and 4. The attractiveness of the hologram solution is its mechanical and environmental stability. The directions of the diffracted orders are rather insensitive to hologram tilt and displacement and the source and detector are mounted on one single substrate. A wavelength change will shift the diffracted spots along the dividing line between detectors 2 and 3 but does not cause an offset.

3.4   Radial Tracking for DVD

Some disadvantages of the radial tracking methods had become manifest while using them for the CD-system. The twin-spot method requires an extra grating with a rather delicate angular orientation and, in the recording mode, precious optical power (some 20%) is spread out to the auxiliary twin spots. The Push-Pull method is sensitive to the phase depth $\Delta\phi$ of the information and does not work in the presence of pure amplitude information patterns. The spot or track wobble method, due to its very nature, consumes precious space in the radial direction and is likely to increase the cross-talk between tracks when compared to the other methods at equal density. For this reason, the Differential Phase Detection (*DPD*) method [24], originally devised for the analog video disc system, has been revisited to adapt it to the digital signals on the high-density DVD disc. It is now commonly called the Differential Time Detection (*DTD*) method.

3.4.1   A Diffraction Model for the DPD and DTD Tracking Signal

An easy picture for the understanding of the differential phase (or time) detection method is obtained when the information in the tangential direction is replaced by a single frequency, so that, together with the radial periodicity, a purely two-dimensional grating is present on the disc. In Fig. 3.12 the various quasi-DC and AC signals are depicted that are present in the far-field on a quadrant detector (quadrants A, B, C and D). Each modulation term depends on the diffraction direction. Radially diffracted orders show a (varying) intensity level in an overlapping region with the zeroth order that is proportional to $\cos(\psi_{10} \pm \phi_r)$. The reference phase ψ_{10} between the first and zeroth orders is determined by the disc structure; it varies from $\pi/2$ for very shallow structures to π if the phase depth $\Delta\phi$ of the optical effects attains the value of π. The quantity $\phi_r = 2\pi r/q$ is proportional to the tracking error r (q is the track pitch). The sign of ϕ_r depending on the order number ± 1 of the diffracted order. In the same way, the overlapping regions in the tangential direction show an intensity variation according to $\cos(\psi_{10} \pm \omega t)$, caused by the scanning with uniform speed of the tangentially stored periodic pattern. Obliquely diffracted orders show a mixed phase contribution and are carrying the interesting information for deriving the tracking error.

Collecting the interference terms on the various quadrants, we obtain

$$S_A(t, \phi_r) = \cos(\omega t + \psi) + \alpha \cos(\omega t - \phi_r + \psi)$$
$$S_B(t, \phi_r) = \cos(\omega t - \psi) + \alpha \cos(\omega t + \phi_r - \psi)$$
$$S_C(t, \phi_r) = \cos(\omega t - \psi) + \alpha \cos(\omega t - \phi_r - \psi)$$
$$S_D(t, \phi_r) = \cos(\omega t + \psi) + \alpha \cos(\omega t + \phi_r + \psi) , \qquad (3.2)$$

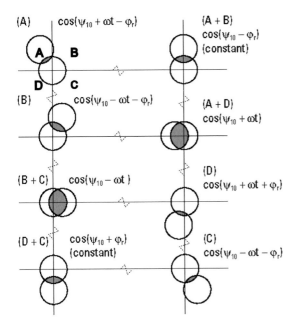

Fig. 3.12. Overlapping regions in the far-field quadrants A, B, C and D with typical modulation terms in the DC-domain and in the high-frequency domain (frequency ω). In each of the eight examples shown, the letters are printed (between brackets) of the quadrants in the far field where a certain harmonic modulation term of the signal is detected. The dependence on the disc translation is given by the term ωt and a radial displacement by the phase ϕ_r. If no translation term ωt is present, the signal only shows variations of average intensity (denoted by 'constant') due to radial position. The tangential direction is chosen horizontally in the figure

where all possible reference phase angles between zeroth and first diffracted orders, for reasons of simplicity, have been replaced by a single phase angle ψ and where α is a factor less than unity that accounts for the relatively small contribution of the diagonal orders with respect to the tangential orders.

With this approximation, the various high-frequency signals are now given by:

$$S_{CA}(t, \phi_r) \propto +(1 + \alpha \cos \phi_r) \cos \psi \cos(\omega t)$$
$$S_{tPP}(t, \phi_r) \propto +(1 + \alpha \cos \phi_r) \sin \psi \sin(\omega t)$$
$$S_{rPP}(t, \phi_r) \propto +\alpha \sin \phi_r \sin \psi \cos(\omega t)$$
$$S_{dPP}(t, \phi_r) \propto -\alpha \sin \phi_r \cos \psi \sin(\omega t) \ . \tag{3.3}$$

The subscripts here denote the signals according to the detector schemes Central Aperture ($CA = A + B + C + D$), tangential Push-Pull ($tPP = A - B - C + D$), radial Push-Pull ($rPP = A + B - C - D$) and diagonal Push-Pull ($dPP = A - B + C - D$).

For the derivation of a tracking signal, a reference signal with a well-defined time-dependent phase is needed for comparison with a signal whose phase depends on the tracking error. As a reference signal we can choose the Central Aperture (CA) signal or the tangential Push-Pull signal. If we take the CA-signal and multiply it with the dPP-signal after a phase shift of the latter over $\pi/2$ we obtain after low-pass filtering:

$$S_1(\phi_r) \propto -\alpha \cos^2 \psi \left(\sin \phi_r + \frac{1}{2} \alpha \sin 2\phi_r \right). \tag{3.4}$$

The multiplication of the tangential PP-signal and the diagonal PP signal directly yields after low-pass filtering:

$$S_2(\phi_r) \propto -\alpha \sin \psi \cos \psi \left(\sin \phi_r + \frac{1}{2} \alpha \sin 2\phi_r \right). \tag{3.5}$$

Inspection of the signals $S_1(\phi_r)$ and $S_2(\phi_r)$ shows that they provide us with the required bipolar signal of the tracking error ϕ_r. The factor in front, containing ψ, depends on the phase depth $\Delta\phi$ and makes $S_2(\phi_r)$ less apt for discs with relatively deep optical structures while $S_1(\phi_r)$ performs optimum in that case.

Other possible combinations are obtained when combining the high-frequency radial Push-Pull signal with either the CA- or the tPP-signal. One easily deduces that the following signals result after multiplication, phase-shifting and low-pass filtering of, respectively, rPP and CA $S_3(\phi_r)$ and rPP and tPP Radial $S_4(\phi_r)$:

$$S_3(\phi_r) \propto \alpha \sin \psi \cos \psi \left(\sin \phi_r + \frac{1}{2} \alpha \sin 2\phi_r \right). \tag{3.6}$$

$$S_4(\phi_r) \propto \alpha \sin^2 \psi \left(\sin \phi_r + \frac{1}{2} \alpha \sin 2\phi_r \right). \tag{3.7}$$

The ψ-dependence of especially $S_4(\phi_r)$ makes it unsuitable for standard optical discs but especially favourable for extra shallow disc structures.

3.4.2   The Influence of Detector Misalignment on the Tracking Signal

The single-carrier model can be equally used for studying the effects of misalignment of the quadrant detector with respect to the projected far-field pattern, the so-called beam-landing effect. This effect is produced during quick access of a remote information track. The sledge (coarse movement) and the scanning objective (precise positioning) are actuated together and this action produces a lateral displacement of the objective with respect to the fixed collimator and detector that can amount to some 10% of the beam foot-print on the detector. As a consequence, the whole diffraction pattern is shifted on the detector in the radial direction (see the arrow 'spot offset' in Fig. 3.13). Taking into account such a mechanically induced offset with a size of ϵ (normalised with respect to the half-diameter of the beam

Single carrier model

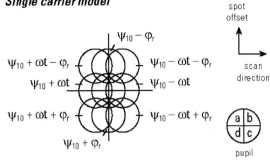

spot offset

scan direction

pupil

Radial Error extraction

Multiplication methods:

original:	$RE = (a - b + c - d) \cdot (a + b + c + d)_{90°}$
Philips:	$RE = (a - d) \cdot (b + c)_{90°} + (c - b) \cdot (a + d)_{90°}$
JVC:	$RE = a \cdot b - c \cdot d$
MEI:	$RE = (a - b + c - d) \cdot (a + d - b - c)$

Time difference methods:

DTD2:	$RE = \phi (a + c, b + d)$
DTD2a:	$RE = \phi (a, d) = \phi (c, b)$
DTD4:	$RE = \phi (a, d) + \phi (c, b)$
DTD4a:	$RE = \phi (a, b) + \phi (c, d)$

Fig. 3.13. Various methods for the extraction of a Differential Time radial error signal using the signals present in the far-field quadrants

foot-print), we write the signals corresponding to the interference terms in the overlapping regions of the far field as

$$S_A(t, \phi_r) = (1 + \epsilon) \cos(\omega t + \psi) + \alpha \cos(\omega t - \phi_r + \psi)$$
$$S_B(t, \phi_r) = (1 + \epsilon) \cos(\omega t - \psi) + \alpha \cos(\omega t + \phi_r - \psi)$$
$$S_C(t, \phi_r) = (1 - \epsilon) \cos(\omega t - \psi) + \alpha \cos(\omega t - \phi_r - \psi)$$
$$S_D(t, \phi_r) = (1 - \epsilon) \cos(\omega t + \psi) + \alpha \cos(\omega t + \phi_r + \psi). \tag{3.8}$$

The beam-landing effect has only been applied to the purely tangential diffraction orders; the diagonal orders are rather far away from the horizontal dividing line of the quadrants.

Two of the four basic signals derived from the quadrants A, B, C and D depend on the radial misalignment and they become:

$$S_{rPP}(t, \phi_r) \propto \alpha \sin \phi_r \sin \psi \cos(\omega t) - \epsilon \cos \psi \cos(\omega t)$$
$$S_{dPP}(t, \phi_r) \propto -\alpha \sin \phi_r \cos \psi \sin(\omega t) - \epsilon \sin \psi \sin(\omega t) \ . \tag{3.9}$$

The four possible DPD-signals now become

$$S_1(\phi_r) \propto -\alpha \cos^2 \psi \left(\sin \phi_r + \frac{1}{2}\alpha \sin 2\phi_r \right) - \epsilon \sin \psi \cos \psi (1 + \alpha \cos \phi_r)$$

$$S_2(\phi_r) \propto -\alpha \sin \psi \cos \psi \left(\sin \phi_r + \frac{1}{2}\alpha \sin 2\phi_r \right) - \epsilon \sin^2 \psi (1 + \alpha \cos \phi_r)$$

$$S_3(\phi_r) \propto \alpha \sin \psi \cos \psi \left(\sin \phi_r + \frac{1}{2}\alpha \sin 2\phi_r \right) - \epsilon \cos^2 \psi (1 + \alpha \cos \phi_r)$$

$$S_4(\phi_r) \propto \alpha \sin^2 \psi \left(\sin \phi_r + \frac{1}{2}\alpha \sin 2\phi_r \right) - \epsilon \sin \psi \cos \psi (1 + \alpha \cos \phi_r). \tag{3.10}$$

The signals S_1 and S_4 show a beam landing sensitivity that is maximum for the $\lambda/4$ pit depth ($\psi \approx 3\pi/4$); for deep pits (optical depth is $\lambda/2$) or for amplitude structures ($\psi = \pi$) the beam landing sensitivity is zero.

An interesting combination of the signals S_1 and S_4 is given by

$$S_5(\phi_r) = S_1(\phi_r) - S_4(\phi_r) \propto \left(\sin \phi_r + \frac{1}{2}\alpha \sin 2\phi_r \right). \tag{3.11}$$

The detection scheme can be written as follows:

$$S_5(\phi_r) = [B - C][A + D]^{90°} - [A - D][B + C]^{90°}, \tag{3.12}$$

where the index 90° implies that the 90 degrees phase shifted version of the signal has to be taken. The signal S_5 is fully independent of pit depth and does not suffer at all from beam landing off-set.

Another possible linear combination is given by

$$S_6(\phi_r) = S_2(\phi_r) - (1 + s)S_3(\phi_r) \propto$$
$$\propto -(1 + s/2) \sin \psi \cos \psi \left(\sin \phi_r + \frac{1}{2}\alpha \sin 2\phi_r \right)$$
$$+ \epsilon/2 \left[\sin^2 \psi - (1 + s) \cos^2 \psi \right] (1 + \alpha \cos \phi_r). \tag{3.13}$$

The beam landing influence on this signal is reduced to zero by the condition

$$s = \tan^2 \psi - 1 . \tag{3.14}$$

The detection scheme for S_6 is

$$S_6(\phi_r) = [CD - AB] + (s/4) \left[(C + D)^2 - (A + B)^2 \right]. \tag{3.15}$$

In the particular case that $\psi = 3\pi/4$, the signal is free from beam landing effects when using its simplest form $[CD - AB]$; the corresponding tracking signal is obtained after low-pass filtering of this high-frequency signal product.

3.4.3   The DTD Tracking Signal for the DVD-System

Shifting from analog to digital signals with their associated clock reference signal, it is evident that the measured phase shifts will now be replaced by time differences between the clock signals associated with specific digital signals derived from the quadrant detector.

The principle of differential time detection can be explained with the aid of Fig. 3.14. The focused light spot (centre in M) is scanning an information pit slightly off-track (distance v_0) in the direction of the arrow. The leading and the trailing edge

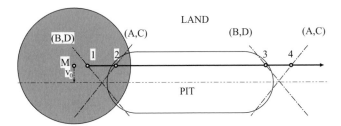

SCANNING SPOT

Fig. 3.14. A schematically drawn focused spot (gray area) is scanning an optical pit with a tracking error v_0. The positions 1 and 3 mark the front and end of the pit as they are detected by the detector pair (B, D); the positions 2 and 4 correspond with the beginning and end of the pit when detected by the diagonal detector pair (A, C)

of the pit induce diffraction of the light perpendicularly to the hatched lines (A, C) and (B, D). When the centre M of the scanning spot is in the position 1, the detector pair (B, D) perceives the leading edge of the pit and the intensity on this detector pair will go down. A short time later, the detector pair (A, C) will measure the passage of the leading edge because the intensity goes down at the position 2. Once the scanning spot has advanced towards the pit end, the trailing edge will be detected at the positions 3 and 4 by a rise in intensity on respectively the detector pairs (B, D) and (A, C). By measuring the time difference between the information signal on the diagonal detector pairs, a bipolar function of the tracking error v_0 can be obtained. Further away from the edges, in the longer pit sections and land regions, the summed intensities on the two quadrant pairs are virtually equal and there is no contribution to the DTD-signal.

The standard DTD-signal is derived by first summing the intensities on a diagonal pair of detector and then detecting the time shift by electronic means. It is also possible to compare the phase difference between individual quadrants and we conclude that the following set of detection schemes is possible:

- $\tau_{(A+C)} - \tau_{(B+D)}$
  The time difference between the sum signal from quadrants $A + C$ and the sum signal from quadrants $B + D$ is taken. This is the standard DTD-method based on a diagonal difference signal and the denomination is *DTD2*-signal.

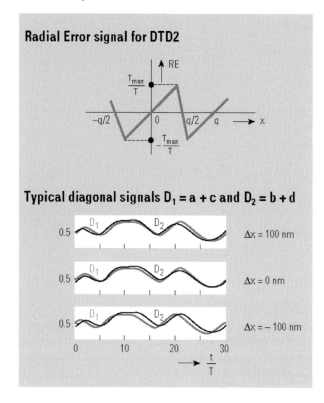

Fig. 3.15. The mutual shift in time of the diagonal quadrant signals D_1 and D_2 as a function of the tracking error and the resulting radial error signal, schematically drawn in the upper part of the figure as a function of the radial off-track position with periodicity q

- $(\tau_B - \tau_C) + (\tau_D - \tau_A)$

  This difference signal requires four independent high-frequency detectors; on track, each difference signal $(D - A)$ and $(B - C)$ is identical zero. The common denomination is *DTD4*-signal

- $(\tau_B - \tau_A) + (\tau_D - \tau_C)$

  This signal is comparable to the preceding one; the time differences $B - A$ and $D - C$ are not zero on track but depend, among others, on the optical pit depth.

3.4.4 The DTD2 and the DTD4 Signal in the Presence of Defocus

In the case of the standard analog *DPD*-signal, we have seen that the phase reference value ψ between the zeroth order and the first order diffracted light played an important role via a $\cos^2 \psi$-dependence. It is a well-known fact that the effective phase difference between zeroth and first order light is affected by a defocusing. Zernike's discovery of phase contrast was triggered by his observation that a phase grating image showed beautiful contrast in a defocused setting. This focus-dependence of ψ means that the the amplitude of the standard *DPD*-signal but also the *DTD2*-signal will vary when defocus is present. If $\psi = \pi$, the signal amplitude is symmetric around focus and the maximum is found in the optimum focus setting. If $\psi < \pi$, the tracking signal will show a maximum value in a defocused situation and can even become zero in focus. Special *DPD*-signals ($S_5(\phi_r)$ and $S_6(\phi_r)$) remained unaffected by the value of ψ and they also show a behaviour that, in a first instance, is independent of defocus. It turns out that for the *DTD4*-signal, where the time difference is evaluated independently for each quadrant detector, a comparable robustness holds with respect to defocus. In Fig. 3.16 the *DTD2* and *DTD4* signals have been de-

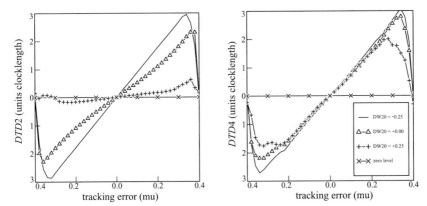

Fig. 3.16. The radial tracking signals *DTD2* and *DTD4* with different amounts of defocus. The defocus parameter *DW20* is expressed in wavefront aberration at the pupil rim in units of the wavelength λ of the light. A value of 0.25 corresponds to a defocusing Δz_s of one focal depth (typically 0.8 μm for a DVD-system). The tracking error signal has been numerically computed using a long sequence of pits obeying the standard EFM (*E*ight to *F*ourteen *M*odulation) scheme [25]. The disk layout and the reading conditions are those encountered in the DVD-system but the depth of the information pits has been reduced to 60 nm ($\psi \approx 3\pi/4$) instead of the common 100 nm. *drawn line*: Δz_s=-0.8 µm, *triangles*: Δz_s=0, *crosses*: Δz_s=+0.8 µm

picted for a rather shallow disk pattern with DVD-density. It is clear from the figure that there is a pronounced asymmetry around focus for the *DTD2*-signal while the *DTD4*-signal virtually remains unaffected by defocus; in particular the slope at the central zero set-point of the servo system remains unaltered on defocusing.

We conclude by reminding that the introduction of the DTD-tracking method in the CD-system was hampered by a signal instability that appeared when neighbouring track portions showed comparable information with a fixed mutual position over a distance that was within the bandwidth of the radial servo system. An example of such a problematic situation was found in the fixed digital pattern used to represent absolute silence on a CD audio disc. In the DVD-system, this fixed pattern problem has been avoided by an appropriate extra coding step that suppresses the prolonged appearance of fixed patterns.

3.5   Compatibility Issues for the DVD-and the CD-System

The compatibility issues between the two standards *CD* and *DVD* are caused by the change in substrate thickness, the change in wavelength and, to a lesser degree, the change in numerical aperture. The combination of smaller wavelength and higher *NA* would lead to a density increase by a factor of 2.5. Further gain in density has been obtained by the higher frequency choice in the DVD-system with respect to the available optical pass-band; this has led to a total factor of five. The final factor of 7.2 has been achieved by more efficient modulation schemes and error correction techniques.

In Table 3.1 the specifications of both systems have been tabulated and we observe that the substrate thickness has been halved for *DVD*. The reason for this reduction in thickness was the specification on disc tilt during playback. The higher *NA* and shorter wavelength of the DVD-system causes a two times larger comatic aberration of the read-out spot at equal tilt angle. The amount of coma being linearly dependent on the substrate thickness, a reduction of thickness was required. The 600 μm compromise between tolerances and mechanical stability was solved by bonding two substrates together, leading to a doubled capacity. As it appears

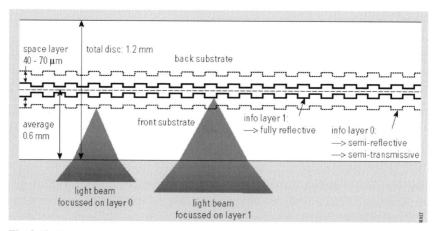

Fig. 3.17. Cross-section of a double-sided DVD-disc with an extra semi-transparent information layer (dotted) on each side

from Fig. 3.17, the doubled capacity can be further increased by introducing in both substrates a buried information layer that is approximately semi-transparent. The distance between the buried layer and the fully reflecting information layer is specified in a margin of 40 to 70 μm to avoid a too large value of spherical aberration when reading the layers at an *NA* of 0.60 in red light (λ=650 nm).

Table 3.1. Table with the most relevant specifications of the *CD*- and the *DVD*-system

CD and DVD Specifications

	CD	DVD
Disc diameter	120 mm	120 mm
Disc thickness	1.2 mm	1.2 mm
Disc structure	Single substrate	Two bonded 0.6 mm substrates
Laser wavelength	780 nm (infrared)	650 and 635 nm (red)
Numerical aperture	0.45	0.60
Track pitch	1.6 µm	0.74 µm
Shortest pit/land length	0.83 µm	0.4 µm
Reference speed	1.2 m/sec CLV	4.0 m/sec CLV
Data layers	1	1 or 2
Data capacity	Approx. 680 Mbytes	Single layer: 4.7 Gbytes Dual layer: 8.5 Gbytes
Reference user rate	Mode 1: 153.6 kbytes/sec Mode 2: 176.4 kbytes/sec	1.108 kbytes/sec 1.1 Mbytes/sec

Video format

	Video CD	DVD-video
Video data rate	1.44 Mbits/sec (video, audio)	1 to 10 Mbits/sec variable (video, audio, subtitles)
Video compression	MPEG1	MPEG2
Sound tracks	2 Channel-MPEG	Mandatory (NTSC) 2-channel linear PCM; 2-channel/5.1-channel AC-3 Optional: up to 8 streams of data available
Subtitles	Open caption only	Up to 32 languages

3.5.1   The Substrate-Induced Spherical Aberration

In a DVD player, the most critical task is the reading of the high-density DVD disc with an *NA* value of 0.60 and with light of a wavelength of typically 650 nm. Although a CD or CD-ROM disc could also be read with the same wavelength, this does not necessarily apply to a CD-R or a CD-RW disc. The correct optical contrast between recorded optical effects and the land regions is only guaranteed at the original CD-wavelength of 780 nm. The reading of a CD-type disc thus first requires an extra source and, secondly, a solution needs to be found for the reading through a substrate that is 600 µm thicker than prescribed in the specification for the

DVD-objective. The thicker substrate causes spherical aberration in the CD-reading spot due to the fact that the marginal rays are more refracted than the paraxial rays, thus enlarging the focal distribution in the axial direction away from the objective. Simultaneously, a serious lateral smearing out of the light intensity distribution is observed and the correct reading of the CD-information is prohibited by this aberration phenomenon. In Fig. 3.18, images of intensity distributions at various focus settings are shown for a focused wave that suffers from spherical aberration. The picture at the height of the chain-dotted line is found at the axial position corresponding to the paraxial focus; approximately two pictures up, the marginal focus position is found where the aberrated marginal rays cut the optical axis. The amount of wavefront spherical aberration W_S, written as a power series of the numerical aperture ($NA = \sin \alpha$), and normalised with respect to the wavelength of the light is given by

$$W_S = \frac{d}{\lambda} \left\{ \left(\frac{n^2 - 1}{8n^3} \right) \sin^4 \alpha + \left(\frac{n^4 - 1}{16n^5} \right) \sin^6 \alpha + \left(\frac{5(n^6 - 1)}{128n^7} \right) \sin^8 \alpha + \cdots \right\},$$

(3.16)

where n is the refractive index of the disc substrate material and d the thickness deviation.

If we use the numbers $\lambda = 650$ nm, $d = 600$ μm, $n = 1.586$ (refractive index of polycarbonate) and $\sin \alpha = 0.60$, the numerical values of the three factors in the power series of $\sin \alpha$ in (3.16) amount to, respectively, 5.68, 1.43 and 0.36 in units of wavelengths. At a first sight, knowing that optical disc read-out has to be done in diffraction-limited conditions, these values are far too large. In Fig. 3.19 we have depicted the behaviour of the central intensity along the axis and in Fig. 3.20 we present calculated intensity distributions at axial positions between the paraxial and the 'best' focus (halfway the paraxial and the marginal focus). From Fig. 3.20, it is obvious that the reading of a CD-disc is impossible close to the 'best' focus setting (e.g. using the light distribution from the second row, last picture) . The reading spot is very much degraded and almost all energy has drifted towards strong diffraction rings; at read-out, the diffraction rings will yield an unacceptable intersymbol interference and the digital eye pattern fully disappears. However, the asymmetry with respect to the 'best' focus position shows a peculiar effect: towards the paraxial region, the strong diffraction rings disappear and a central lobe is observed superimposed on a reduced intensity background. The central lobe is much broader than in the best focus but could have an acceptable size for reading a low-density CD-disc when the background intensity is somehow made inoffensive by the read-out method. The distribution that most resembles a standard reading spot (appropriate half width, low level of side lobes) is found around the third position of Fig. 3.20. The focus setting corresponds to an offset from the paraxial focus of some 6 to 8 focal depths (one focal depth ≈ 0.8 μm) and yields an intensity distribution that should be adequate for CD read-out. In Fig. 3.19, this interesting axial position is approximately halfway up-hill the first maximum.

Given the low peak intensity, an important amount of low-intensity aberrated stray light can be expected around the central peak and this background should be

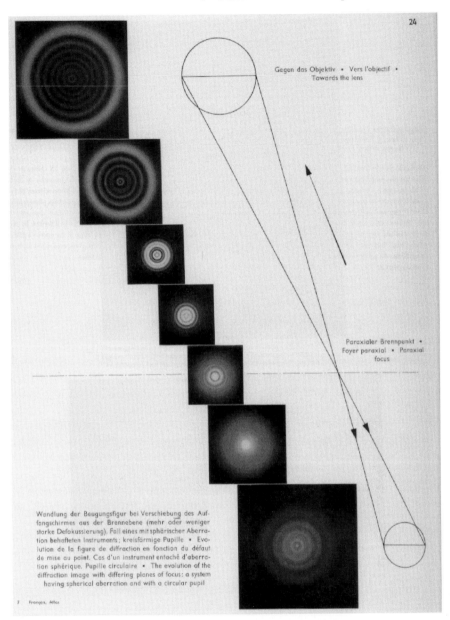

24

Gegen das Objektiv • Vers l'objectif • Towards the lens

Paraxialer Brennpunkt • Foyer paraxial • Paraxial focus

Wandlung der Beugungsfigur bei Verschiebung des Auffangschirmes aus der Brennebene (mehr oder weniger starke Defokussierung). Fall eines mit sphärischer Aberration behafteten Instruments; kreisförmige Pupille • Evolution de la figure de diffraction en fonction du défaut de mise au point. Cas d'un instrument entaché d'aberration sphérique. Pupille circulaire • The evolution of the diffraction image with differing planes of focus: a system having spherical aberration and with a circular pupil

Françon, Atlas

Fig. 3.18. Intensity profiles at different positions along the optical axis for a pencil suffering from spherical aberration (from: M. Cagnet, M. Françon, J.-C. Thrierr: *Atlas of optical phenomena* (Springer-Verlag, Berlin 1962)). In the case of a DVD-player reading a CD-disc, the extra substrate thickness introduces spherical aberration with a sign such that the lower part of the figure is closest to the objective and the light travels in the direction given by the large arrow in the upper part of the figure

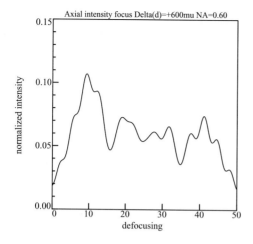

Axial intensity focus Delta(d)=+600mu NA=0.60

Fig. 3.19. A plot of the axial intensity of a pencil suffering from spherical aberration due to a substrate thickness increase of 600 μm ($NA = 0.60$, $\lambda = 650$ nm). The paraxial focus is at $z = 0$, the best focus at $z = 24$ μm and the marginal focus at $z = 48$ μm

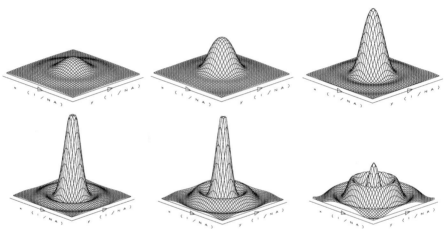

Fig. 3.20. 3D plots of the intensity distribution (from left to right) going away from the paraxial focus towards the best focus in axial steps of 3 μm (one focal depth equals 0.81 μm ($NA = 0.60$, $\lambda = 650$ nm). The maximum intensity in the figures is about 0.10 with respect to 1.0 for the diffraction-limited case. The length and width of the square plot area are $4\lambda/NA$ (4.33 μm). The upper right figure represents the optimum focus setting for reading a CD-disc

prevented from reaching the detector. This can be done by introducing an aperture in the light path towards the detector, or, alternatively, by using a detector of finite size; the strongly aberrated rays, after double passage through the mismatched substrate, will have a large transverse aberration in the image plane and miss the quadrant detector. The effect on the calculated digital eye pattern of the effective reduction of the detector aperture NA_D is illustrated in Fig. 3.21. These graphs show the resulting digital eye pattern when the detection diameter in the far field is reduced from e.g. 80% to 55% at a focus setting with $z = +6$ μm from the paraxial focus P. In the case of a digital disc system, the quality of the detector signal is generally assessed

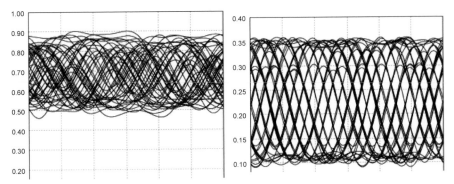

Fig. 3.21. A plot of the digital eye that is detected when focusing at a position 6 μm away from the paraxial focus (left-hand figure: far-field detection region is 80% of full aperture; right-hand figure: 55% of the aperture is effectively used for detection. Note the loss in signal strength in the second picture

by means of the so-called root mean square digital jitter. This quantity is obtained by binarising the analog signal read from a disc and comparing it with the desired binary wave form. The root mean square value of the shifts of the transients in the detected signal with respect to the corresponding positions in the originally recorded signal is called the digital jitter of the signal. The digital jitter Δ_d is expressed in the time domain or in the spatial domain and counted in units of the clock length T of the digital signal. In the right-hand figure, the residual jitter amounts to 6.7%; this value is obtained in the case of a maximum decorrelation between neighbouring tracks. When the information between the tracks is fully correlated, the jitter goes down to 3% and in practice one finds values of 5.5 to 6% in optimum CD read-out conditions.

A further reduction in jitter is possible when the residual spherical aberration present over the central area is corrected in the objective by tailoring the surface profile of the first aspheric surface of the single element objective lens. In this case, a wavefront correction is applied to the central 55% region while the outer region is left unaltered and this leads to a further reduction of the jitter in the detected signal down to 5% for a CD-disc.

3.5.2   The Effective Optical Transfer Function

The changes at read-out by a reduction of the far-field detection area to e.g. 55% of the total beam diameter can be expressed in terms of a modified MTF (modulation transfer function) of the optical read-out system. A reduction of the detection aperture is equivalent to a more coherent or partially coherent detection of the information on the disc and this increases the response at the mid-spatial frequencies although, simultaneously, the cut-off frequency of the optics is reduced. This latter effect is not very important when reading a CD-disc at $NA = 0.60$ and $\lambda = 650$ nm because the cut-off frequency is far too high (1850 mm^{-1} instead of 1150 mm^{-1} for a standard CD-player).

In Fig. 3.22 we observe that at the optimum focus setting the resulting MTF (curves with triangles) is not much different from the standard MTF encountered in a classical CD-player with $NA = 0.45$ and $\lambda = 785$ nm (the cut-off frequency would be found at an axial value of 1.15).

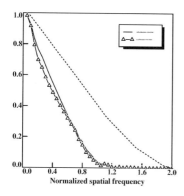

Fig. 3.22. MTF-curves of the read-out with reduced NA value at the detection side. The horizontal axis is expressed in units NA/λ. In the figure the optimum MTF obtained by selecting a relative aperture of 55% in the reflected beam (*triangles*) is further improved by introducing a correction for the residual spherical aberration over the central section of the objective (*drawn line*). For comparison, the optimum transfer function at full detection aperture has been sketched (*dashed line*)

In the figure we also show the slight improvement that is possible when the residual spherical aberration present over the central area of the reading beam is corrected in the objective (drawn line). Some 5 to 10% improvement in MTF-value is possible at the mid CD-frequencies and this becomes visible also in the reduction of the bottom jitter. Moreover, the absence of residual spherical aberration over the effective detection area leads to a better symmetry of the read-out signal around the optimum focus setting.

3.5.3   The Two-Wavelength Light Path

As we remarked before, recordable or rewritable CD-discs require a read-out at the appropriate wavelength of $\lambda = 780 - 840$ nm and the corresponding numerical aperture amounts to 0.45. The partially coherent read-out at $NA_D = 55\%$ that was possible at the red wavelength with a slight objective modification at the most, does not yield a sufficiently large optical bandwidth at $\lambda = 780$ nm. A solution is found by a further tailoring of a ring section of the surface profile of the first aspheric surface of the objective, thereby reducing the spherical aberration at CD read-out. The objective then can be considered to consist of three regions. The central paraxial region contributes to both DVD and CD read-out. A relatively narrow annular region corrects the CD read-out but is effectively lost for the DVD read-out. The outer region is only useful for the DVD read-out; this part of the read-out bundle becomes heavily aberrated at CD read-out and is then lost for detection [26]. The rather low light efficiency at CD read-out that was observed in Fig. 3.21 does not occur once a second independent source is used for CD read-out. The far-field of the second source can be adjusted to the maximum aperture ($0.45 \leq NA \leq 0.50$) needed for CD read-out.

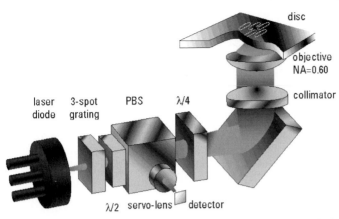

Fig. 3.23. The layout of an optical light path for DVD and CD with the possibility to inject an extra read/write beam of different colour via a second beam splitter (not shown in the figure), either before the beam splitter or in the parallel beam section after the collimator. The polarisation-sensitive light path with quarter-wave plate and polarising beam splitter (*PBS*) is used to maximise writing power

Other compatibility solutions have been proposed based on a diffractive struc-ture applied to the first aspheric surface of the objective and capable of correcting the spherical aberration in the first diffraction order [27]. More advanced solutions have recently been described in [28].

3.6 Efficient Calculation Scheme for the Detector Signal

In this section we will treat the problem of the efficient calculation of the detector signal in an optical disc player. The reliable and numerically efficient modelling of detected signals is a very welcome addition to the experimental verification of new ideas. With a reliable modelling tool, one obtains a quick estimate of tolerances and robustness of new options for recording schemes and read-out methods. In this section we present a numerical procedure that carries the mathematical analysis of the read-out process as far as possible and thus gains precious time as compared to the purely numerical approach. Especially in the case of optics with general pupil shape and aberrations, the analytical approach can bring an important reduction in calculation time.

3.6.1 Optical Configuration and the FFT-Approach

The light path of an optical disc player is well represented by a scanning microscope of Type I according to the terminology given in [20]. In the schematic drawing of a light path according to Fig. 3.24, the detector could also be positioned at the location of the aperture of O' and in many modelling approaches, this is effectively done to

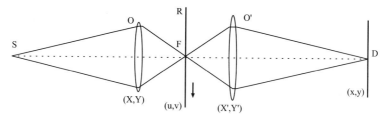

Fig. 3.24. Schematic drawing illustrating the progation of the light from the source S via the objective O towards the disc surface in the focal plane F of O. The disc presents a (spatially modulated) reflection function $R(u, v)$. After reflection the diffracted light is captured by the same objective (here denoted by O') and an image of the disc is produced at the location of the detector plane D. The pupil coordinates (X, Y) and (X', Y') are commonly normalized with respect to the half-diameter of the objective aperture. The disc coordinates (u, v) and the detector coordinates (x, y) are normalized with respect to the diffraction units λ/NA in both planes

minimize the numerical burden. But the way in which the high-frequency and the optical error signals are derived is more complicated and the diffraction step from the aperture of O' to the composite detectors in plane D is essential if a detailed knowledge of the intensity distribution on the detectors is needed. As an example we take the hologram beamsplitter of Fig. 3.11 where each beam propagating to the detector plane effectively originates from one half of the collecting aperture at O'. Another example is the light path introducing astigmatism in Fig. 3.8. The intensity distribution on the detector in best focus is some 20 times larger in linear measure as compared to the standard diffraction image. It is important to know the distribution of the high-frequency signal in the various quadrants in both the in-focus situation and the defocused case.

The standard way to calculate a detector signal is the repeated application of a propagation step from an object or image plane to an aperture and vice versa, in optical terms by propagating from the near field to the far field and back. The degree of sophistication with which the propagation is carried out is not crucial in the sense that each propagation step can be reduced to a Fourier transform. Especially when a scalar diffraction approach is sufficient, a single Fourier transform operation per step is adequate. For an efficient numerical implementation, the Fourier integral is replaced by a Fast Fourier Transform (FFT) that is faster with respect to execution time.

The assessment of the digital signal is carried out by means of the digital jitter Δ_d. As a rule of thumb, one can say that the *rms* jitter preferably should not exceed 10% of the clock length. At a value of 15%, prolonged over large signal portions, the digital signal fully breaks down and cannot be reconstructed without errors. The fact that some statistics are needed to calculate a reliable jitter value means that an extended track portion with a digitally modulated signal has to be scanned. For a reliable reconstruction of the optical signal, the sample points where the detector signal is calculated should be rather densely spaced, e.g. with a spatial increment of 0.05 to 0.1 of the diffraction unit λ/NA. For each sample point at the disc, the two FFT-steps from the disc to the aperture O' and to the detector plane D have to be executed, provided that the incident amplitude distribution at the focal point F is stored in memory. Several thousands of FFT-steps can be needed for obtaining one single value of Δ_d.

3.6.2 The Analytic Approach

The analysis given in this subsection uses the material presented in references [29] and [2]. In order to accommodate for more general signal sequences without loosing the advantage of dealing with periodic structures, we define a larger unit cell (length p) that contains several optical effects in the tangential direction and, if needed, several neighbouring tracks in the radial direction with a formal radial period of q (see Fig. 3.25).

In the presence of the double periodic disc structure, the complex amplitude in the entrance aperture of the collecting objective O' (see Fig. 3.24) is given by the sum of a certain number of diffracted waves according to

$$A'(X', Y') = \sum_{m,n} \rho_{m,\hat{n}} \, \exp\left\{ 2\pi i \left[\left(\frac{m}{p}\right) u + \left(\frac{\hat{n}}{q}\right) v \right] \right\} f\left(X' - \frac{m}{p}, Y' - \frac{\hat{n}}{q} \right), \qquad (3.17)$$

where m is the tangential order number and the quantity $\hat{n}$ equals the radial order number $(n - ms/p)$ of a disc with a track-to-track shift s of the information. The argument of the exponential function is proportional to the mutual displacement (u, v) of the scanning spot and the disc structure in, respectively, the tangential and radial direction. The factor $\rho_{m,\hat{n}}$ is equal to the complex amplitude of the diffracted wave with the corresponding order number $(m, \hat{n})$. The complex function $f(X', Y')$ stands for the lens function and carries the information about the lens transmission function (e.g. pupil shape, Gaussian filling factor of the incoming beam) and the lens aberration. Figure 3.26 shows the zeroth diffracted order and a general diffracted order (dotted circles) as they are located in the pupil of the collecting objective O'. The coordinates (X', Y') in the exit pupil are linearly related to the sines of the angles in the far-field of the disc structure if the collecting objective O' satisfies the sine condition of geometrical optics [30].

The imaging step from the exit pupil of O' to the detector plane is another Fourier transform from the coordinates (X', Y') to the detector plane coordinates (x, y). The lens function of O' is denoted by $g(X', Y')$; in the special case of a reflective system, we put $g(X', Y') = f(-X, -Y)$. The complex amplitude $A''(X', Y')$ in the exit pupil of the collecting objective becomes the product of $A'(X', Y')$ and the lens function $g(X', Y')$. If the disc is uniformly rotated, resulting in a local linear speed of s_0, the period $1/p$ is transformed into a temporal frequency $f_t = s_0/p$. We

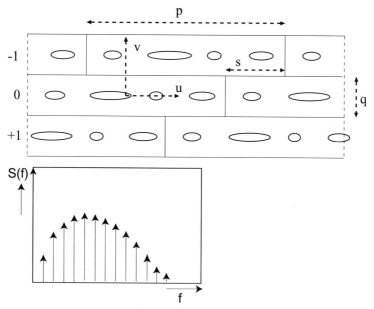

Fig. 3.25. The disc structure possesses a double periodicity. The fundamental frequency in the radial direction is given by $1/q$ with q in principle being equal to the pitch of the information track. The periodicity in the tangential direction is a 'synthetic' one. Within one period of length p several optical effects are present that obey the modulation rule of the optical structure with respect to the lengths of the optical effects and the land sections in between. The tangential information is repeated in the neighbouring tracks but a shift (distance s) has been provided to avoid directly correlated cross-talk. The tangential repetition with period p leads to a typical sampled power spectrum $S(f)$ of the digital information on the disc (lower graph); the distance between the frequency sample points along the normalised frequency axis is given by $\Delta f = \lambda/p(NA)$

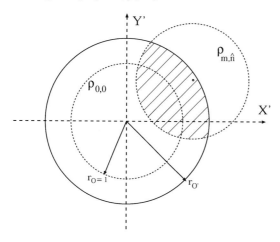

Fig. 3.26. Position of the diffracted orders of the disc structure with double periodicity in the exit pupil of the collecting objective O' with half diameter $r_{O'}$. The zeroth order with complex amplitude $\rho_{0,0}$ has been shown and a general diffraction order $(m, \hat{n})$. The hatched area indicates which part of the order $(m, \hat{n})$ is transmitted to the detector plane

finally find for the complex amplitude in the detector plane

$$B(x, y; t, v) = \iint_{O'} A''(X', Y') \exp\{2\pi i \,(X'x + Y'y)\} \, dX' dY'$$

$$= \sum_{m,n} \rho_{m,\hat{n}} \exp\left\{2\pi i \left[m\left(f_t t - \frac{sv}{pq}\right) + \frac{nv}{q}\right]\right\}$$

$$\times \iint_{O'} g(X', Y') f\left(X' - \frac{m}{p}, Y' - \frac{n - \frac{ms}{p}}{q}\right)$$

$$\times \exp\{2\pi i \,(X'x + Y'y)\} \, dX' dY' \,, \tag{3.18}$$

where the integral is formally carried out over the full area of the exit pupil of O'.

In (3.18) we separate the integral over the exit pupil area O', that only depends on the imaging properties of the lenses, from the part that is scanning-dependent and carries information about the structure of the disc via the factors $\rho_{m,\hat{n}}$. The expression for $B(x, y; t, v)$ now reads

$$B(x, y; t, v) = \sum_{m,n} \rho_{m,\hat{n}} \exp\left\{2\pi i \left[\left(m\left(f_t t - \frac{sv}{pq}\right) + \frac{nv}{q}\right)\right]\right\} F_{m,\hat{n}}(x, y), \tag{3.19}$$

where the factor $F_{m,\hat{n}}(x, y)$ equals the integral over X' and Y' in (3.18).

The detected quantity is the intensity, obtained by multiplying $B(x, y; t, v)$ with its complex conjugate and this yields

$$I(x, y; t, v) = |B(x, y; t, v)|^2 =$$

$$= \sum_{m,n} \sum_{m',n'} \rho_{m,\hat{n}} \, \rho^*_{m',\hat{n}'} \, F_{m,\hat{n}}(x, y) \, F^*_{m',\hat{n}'}(x, y)$$

$$\times \exp\left\{2\pi i \left[(m - m')\left(f_t t - \frac{sv}{pq}\right) + \frac{(n - n')v}{q}\right]\right\} \,. \tag{3.20}$$

In an optical disc system we use special detector geometries, e.g. a quadrant detector. The detector signal is obtained by integrating the intensity over the detector area, taking into account a locally varying detection sensitivity function $S(x, y)$ and this yields a kth detector signal according to

$$S_{D_k}(t, v) = \sum_{m,n} \sum_{m',n'} \rho_{m,\hat{n}} \, \rho^*_{m',\hat{n}'}$$

$$\times \exp\left\{2\pi i \left[(m - m')\left(f_t t - \frac{sv}{pq}\right) + \frac{(n - n')v}{q}\right]\right\}$$

$$\times \iint_{D_k} S(x, y) \, F_{m,\hat{n}}(x, y) \, F^*_{m',\hat{n}'}(x, y) dx dy, \tag{3.21}$$

where D_k denotes the area of the kth detector. In the expression for the detector signal we have separated the quantities related to the disc structure from the integral that only contains information on the imaging process by the objective and collecting lens. This means that one single effort in evaluating the integral in (3.21) can serve for the many times repeated calculation of detector signals corresponding to different structures on the disc that fit into the basic period of length p. In this way we can gather statistical data on the digital jitter Δ_d with a seriously reduced numerical effort; in practice, a factor of the order of hundred is observed.

3.6.3   The Harmonic Components of the Detector Signal

With the double periodicity present on the disc, the detector signal can be written as a double Fourier series where the uniform scanning speed s_0 transforms the periodic components in the tangential direction into time harmonic components with a fundamental frequency $f_t = s_0/p$.

Using (3.21), we formally obtain the temporal harmonic components $A_{\mu,k}$ and $B_{\mu,k}$ from detector D_k according to

$$
\begin{aligned}
A_{0,k}(v) \;=\;& P_{1,k}(0,0)\\
&+2\sum_{\kappa=1}^{\kappa_{max}}\left\{P_{1,k}(0,\kappa)\cos\left(2\pi\frac{\kappa}{q}v\right)+P_{2,k}(0,\kappa)\sin\left(2\pi\frac{\kappa}{q}v\right)\right\}\\[2mm]
A_{\mu,k}(v) \;=\;& 2\left\{P_{1,k}(\mu,0)\cos\left(2\pi\frac{s}{p}\frac{\kappa}{q}\mu\right)+P_{2,k}(\mu,0)\sin\left(2\pi\frac{s}{p}\frac{\kappa}{q}\mu\right)\right\}\\
&+2\sum_{\kappa=1}^{\kappa_{max}}\left\{\left[\{P_{1,k}(\mu,\kappa)+Q_{1,k}(\mu,\kappa)\}\cos\left(2\pi\frac{s}{p}\frac{v}{q}\mu\right)\right.\right.\\
&\qquad\qquad\left.+\{P_{2,k}(\mu,\kappa)+Q_{2,k}(\mu,\kappa)\}\sin\left(2\pi\frac{s}{p}\frac{v}{q}\mu\right)\right]\cos\left(2\pi\frac{\kappa}{q}v\right)\\
&\qquad\quad+\left[\{P_{2,k}(\mu,\kappa)-Q_{2,k}(\mu,\kappa)\}\cos\left(2\pi\frac{s}{p}\frac{v}{q}\mu\right)\right.\\
&\qquad\qquad\left.\left.-\{P_{1,k}(\mu,\kappa)-Q_{1,k}(\mu,\kappa)\}\sin\left(2\pi\frac{s}{p}\frac{v}{q}\mu\right)\right]\sin\left(2\pi\frac{\kappa}{q}v\right)\right\}\\[2mm]
B_{\mu,k}(v) \;=\;& 2\left\{-P_{2,k}(\mu,0)\cos\left(2\pi\frac{s}{p}\frac{\kappa}{q}\mu\right)+P_{1,k}(\mu,0)\sin\left(2\pi\frac{s}{p}\frac{\kappa}{q}\mu\right)\right\}\\
&+2\sum_{\kappa=1}^{\kappa_{max}}\left\{\left[\{-P_{2,k}(\mu,\kappa)-Q_{2,k}(\mu,\kappa)\}\cos\left(2\pi\frac{s}{p}\frac{v}{q}\mu\right)\right.\right.\\
&\qquad\qquad\left.+\{P_{1,k}(\mu,\kappa)+Q_{1,k}(\mu,\kappa)\}\sin\left(2\pi\frac{s}{p}\frac{v}{q}\mu\right)\right]\cos\left(2\pi\frac{\kappa}{q}v\right)\\
&\qquad\quad+\left[\{P_{1,k}(\mu,\kappa)-Q_{1,k}(\mu,\kappa)\}\cos\left(2\pi\frac{s}{p}\frac{v}{q}\mu\right)\right.\\
&\qquad\qquad\left.\left.+\{P_{2,k}(\mu,\kappa)-Q_{2,k}(\mu,\kappa)\}\sin\left(2\pi\frac{s}{p}\frac{v}{q}\mu\right)\right]\sin\left(2\pi\frac{\kappa}{q}v\right)\right\},
\end{aligned}\tag{3.22}
$$

where the harmonic coefficients A and B, that depend on the off-track distance v of the scanning spot, generate the detector signal with the aid of the expression

$$
S_{D_k}(t,v) = A_{0,k}(v) + \sum_{\mu=1}^{\mu_{max}}\left\{A_{\mu,k}(v)\cos(\mu f_t t) + B_{\mu,k}(v)\sin(\mu f_t t)\right\},\tag{3.23}
$$

where $\mu_{max} f_t$ is the maximum frequency transmitted by the optical read-out system.

The calculation of the detector signal $S_{D_k}(t, v)$ requires the evaluation of the coefficients $P_{j,k}(\mu, \kappa)$ and $Q_{j,k}(\mu, \kappa)$, $j = 1, 2$, that can be written as

$$P_{1,k}(\mu, \kappa) = \text{Re}\left\{\sum_m \sum_n \rho_{\mu+m,\hat{\kappa}+\hat{n}} \, \rho^*_{m,\hat{n}} \, Z_k(\mu + m, \hat{\kappa} + \hat{n}; m, \hat{n})\right\}$$

$$P_{2,k}(\mu, \kappa) = \text{Im}\left\{\sum_m \sum_n \rho_{\mu+m,\hat{\kappa}+\hat{n}} \, \rho^*_{m,\hat{n}} \, Z_k(\mu + m, \hat{\kappa} + \hat{n}; m, \hat{n})\right\}$$

$$Q_{1,k}(\mu, \kappa) = \text{Re}\left\{\sum_m \sum_n \rho_{\mu+m,\hat{n}} \, \rho^*_{m,\hat{\kappa}+\hat{n}} \, Z_k(\mu + m, \hat{n}; m, \hat{\kappa} + \hat{n})\right\}$$

$$Q_{2,k}(\mu, \kappa) = \text{Im}\left\{\sum_m \sum_n \rho_{\mu+m,\hat{n}} \, \rho^*_{m,\hat{\kappa}+\hat{n}} \, Z_k(\mu + m, \hat{n}; m, \hat{\kappa} + \hat{n})\right\} \qquad (3.24)$$

The factor $Z_k(m, \hat{n}; m', \hat{n}')$ is the integral that was present in (3.21) and is written

$$Z_k(m, \hat{n}; m', \hat{n}') = \iint_{D_k} S(x, y) \, F_{m,\hat{n}}(x, y) \, F^*_{m',\hat{n}'}(x, y) \mathrm{d}x\mathrm{d}y. \qquad (3.25)$$

The efficient calculation of the Z_k-functions is the subject of the remainder of this chapter. Especially the numerical evaluation of the basic function $F_{m,n}(x, y)$ can be time consuming and we will show how a further analysis of this function can improve the efficiency of the final evaluation.

3.6.4   The Representation of the Function $F_{m,n}(x, y)$

The function $F_{m,n}(x, y)$ is the Fourier Transform of the product of the displaced objective lens function $f(X' - m/p, Y' - n/q)$ and the collecting lens function $g(X', Y')$ where the integration area is defined by the half diameter $r_{O'}$ of the collecting objective ($r_{O'}$ is commonly normalised to unity). The standard way to calculate the Fourier transform is using a numerical two-dimensional fast Fourier Transform (FFT). In this paragraph we will show that the use of orthogonal functions in the diffracting exit pupil and in the image plane where the detectors are situated can lead to an explicit expression for the detector intensity. This expression is then readily integrated over the detector area defined by D_k.

The function to be Fourier transformed generally is a complex function that is nonzero on a typical domain within the exit pupil of O' like the hatched area in Fig. 3.26. Orthogonal functions with uniform weighting on the circular exit pupil area of O' (pupil radius normalised to unity) are the well-known Zernike polynomials [31]. These polynomials are normally used to represent the argument of the complex pupil function, in particular the aberration or the focus defect of an imaging optical system. Here we propose to extend the use of the Zernike polynomials and to represent the complete complex function $A(\rho, \theta) \exp[i\Phi(\rho, \theta)]$ in the exit pupil by a set of Zernike polynomials yielding complex expansion coefficients a_{kl} according to

$$A(\rho,\theta)\exp[i\Phi(\rho,\theta)] = \sum_{k,l} \alpha_{kl}\, R_k^l(\rho)\cos l\theta \;, \tag{3.26}$$

where (ρ,θ) are the polar coordinates in the exit pupil, $A(\rho,\theta)$ the amplitude and $\Phi(\rho,\theta)$ the phase of the pupil function. $R_k^l(\rho)$ is the radial part of the Zernike polynomial with radial degree k and azimuthal degree l ($k \geq l \geq 0$, $k-l$ even). We have limited ourselves to the cosine-polynomials, thus restricting the functions to those that are symmetric with respect to $\theta = 0$. The expansion of a general function would require an extra set of coefficients b_{kl} associated with the sine-polynomials.

The expansion in (3.26) of $A\exp(i\Phi)$ can be obtained, for instance, by a least squares fit with a finite series of polynomials, a procedure which is common practice for expanding Φ itself. It can be shown that, with the transition from cartesian coordinates (x,y) to normalised polar coordinates (r,ϕ), the complex amplitude U in the detector plane is given by

$$U(r,\phi) = 2\sum_{k,l} a_{kl}\, i^l\, V_{kl}\cos l\phi \quad, \tag{3.27}$$

where

$$V_{kl} = \int_0^1 \rho\exp(i f_z\rho^2)\, R_k^l(\rho)\, J_l(2\pi\rho r)\mathrm{d}\rho \tag{3.28}$$

for integers $k,l \geq 0$ with $k-l \geq 0$ and even (J_l denotes the Bessel function of the first kind of order l). A quadratic phase factor $\exp(i f_z\rho^2)$ has been added that is needed in the case of a defocusing of the image plane with respect to the optimum (paraxial) focus.

The Bessel series representation of V_{kl} is given in [32,33] and reads

$$V_{kl} = \exp(i f_z)\sum_{s=1}^{\infty}(-2i f_z)^{s-1}\sum_{j=0}^{p_z} v_{sj}\frac{J_{l+s+2j}(t)}{st^s} \tag{3.29}$$

with v_{sj} given by

$$v_{sj} = (-1)^{p_z}(l+s+2j)\binom{l+j+s-1}{s-1}\binom{j+s-1}{s-1}\binom{s-1}{p_z-j}\bigg/\binom{q_z+s+j}{s}, \tag{3.30}$$

for $s = 1,2,\ldots$; $j = 0,\ldots,p_z$. In (3.29) we have set

$$t = 2\pi r\;, \qquad p_z = \frac{k-l}{2}\;, \qquad q_z = \frac{k+l}{2}\;. \tag{3.31}$$

For the number S of terms to be included in the infinite series over s we have the following rule. It can be shown that, when $S=25$, the absolute truncation error is of the order 10^{-6} for all f_z,t,k,l specified by

$$|f_z| \leq 2\pi, \qquad t \leq 20, \qquad 0 \leq p_z \leq q_z \leq 6 \quad. \tag{3.32}$$

In the absence of defocusing ($f_z = 0$), the expansion of the complex pupil function in terms of Zernike polynomials leads to the relationship

$$U(r, \phi) = 2 \sum_{k,l} a_{kl}\, i^l\, (-1)^{\frac{k-l}{2}}\, \frac{J_{k+1}(2\pi r)}{2\pi r}\, \cos l\phi \,. \tag{3.33}$$

This analytic result for the in-focus amplitude can be obtained using formula (39), p. 772 in [31]

$$\int_0^1 \rho R_k^l(\rho) J_l(2\pi\rho r) d\rho = (-1)^{\frac{k-l}{2}} \frac{J_{k+1}(2\pi r)}{2\pi r} \,, \tag{3.34}$$

but it also follows as the limiting case for $f_z \to 0$ of (3.28)–(3.29).

3.6.5   Orthogonality in Pupil and Image Plane

Having seen the one-to-one relationship between the orthogonal Zernike polynomials in the pupil and the corresponding Bessel functions of the first kind in the image plane (in-focus situation), we are left with the evaluation of the inner products of the Bessel functions to check their orthogonality and establish a normalisation factor.

We define the inner product of two functions from the series in (3.33) as

$$\begin{aligned}
I_{k,l;k',l'} &= C \int_0^{2\pi} \int_0^{\infty} U_{kl} U_{k'l'}\, r\, dr\, d\phi \\
&= 4C\, i^{k+k'} \int_0^{2\pi} \int_0^{\infty} \frac{J_{k+1}(2\pi r) J_{k'+1}(2\pi r)}{4\pi^2 r^2}\, \cos l\phi \cos l'\phi\, r\, dr\, d\phi \\
&= \frac{C}{\pi}\, i^{k+k'} \int_0^{\infty} \frac{J_{k+1}(2\pi r) J_{k'+1}(2\pi r)}{r}\, dr\ \delta(l, l') \,,
\end{aligned} \tag{3.35}$$

with $\delta(l, l')$ the Kronecker symbol ($= 2$ if $l = l'=0$, $= 1$ if $l = l' \neq 0$ and equals 0 if $l \neq l'$). The common conditions for the indices of Zernike polynomials apply too (both $k - l$ and $k' - l'$ are even).

For the calculation of the definite integral over r we use the special result for the integral of the product of two Bessel functions in paragraph 11.4.6 of [34]

$$\int_0^{\infty} t^{-1} J_{v+2n+1}(t) J_{v+2m+1}(t) dt = 0 \qquad (m \neq n)$$

$$= \frac{1}{2(2n + v + 1)} \qquad (m = n), \tag{3.36}$$

with the restriction $v + n + m > -1$.

Applied to (3.35) with $v + 2n = k$, $v + 2m = k'$ and $t = 2\pi r$, we find the result

$$\begin{aligned}
I_{k,l;k',l'} &= \frac{C}{\pi} \qquad\qquad (k = k' = l = l' = 0) \\
&= \frac{(-1)^k}{2(k+1)} \frac{C}{\pi} \qquad (k = k' \neq 0;\ l = l' \neq 0)
\end{aligned} \tag{3.37}$$

For normalisation purposes, we multiply the basic functions with $\sqrt{\pi}$ so that the value of $I_{00;00}$ equals unity.

The Evaluation of the Function $Z_k(m, \hat{n}; m', \hat{n}')$ The quantity $Z_k(m, \hat{n}; m', \hat{n}')$ was the key quantity to be calculated for the evaluation of the harmonic signal components (see (3.25)).

Using the orthogonal expansion for the functions $F_{m,n}(x, y)$ in the exit pupil of the objective O' according to (3.33) we obtain

$$
Z_j(m, \hat{n}; m', \hat{n}') = \sum_{kl} \sum_{k'l'} a_{kl} a_{k'l'}'^* \, i^{l-l'} \, (-1)^{\frac{k+k'-l-l'}{2}}
$$
$$
\times \iint_{D_j} S_r(r, \phi) \frac{J_{k+1}(2\pi r) J_{k'+1}(2\pi r)}{4\pi^2 r^2} \cos l\phi \cos l'\phi \; r dr d\phi, \quad (3.38)
$$

where the asterisk denotes the complex conjugate function and the detector sensitivity function S_r has been written in polar coordinates.

In the general situation with an arbitrary detector sensitivity function, the integrals above have to be evaluated numerically. In the whole analytic process described until now, the calculation of the coefficients a_{kl} of the Zernike expansion of the complex function in the exit pupil of O' is the crucial step with respect to speed and convergence and this aspect is actively researched at this moment.

3.7   Conclusion

In this chapter, we have described some important aspects of the optical disc systems that are actually widely used, the CD- and the DVD-system. The subject of optical storage being vast, we have treated some restricted topics that are of interest for the understanding of the CD- and DVD-system themselves and the exchange of media between these two standardized platforms of optical recording. A detailed analysis has been given of the signal detection in an optical disc system. This is of particular interest when simulations are used as a predictive tool, in parallel to experimental research. The efficiency of the calculations is greatly enhanced when the data belonging to the information structure on the disc are separated from the imaging process in the optical recording unit; a procedure to achieve this separation has been described in this chapter and an important reduction in calculation time is obtained. We have also indicated that, by using Zernike polynomials to represent complex amplitude functions, the analytic calculation process can be pursued further and the speed in evaluating the detector signal is again improved.

Because of space limitations, we have not discussed recent developments in optical storage like the Blu-ray Disc system and still other research activities like near-field optical recording that aim at super high-density discs. With respect to these new developments, the analysis given in this chapter has to be adapted to the vectorial diffraction aspects of high-aperture image formation.

References

1. G. Bouwhuis, J. Braat: 'Recording and reading of information on optical disks', In: *Applied Optics and Optical Engineering*, ed. by R.R. Shannon, J.C. Wyant (Academic, New York 1983) pp. 22–45

2. G. Bouwhuis, J. Braat, A. Huijser, J. Pasman, G. van Rosmalen, K. Schouhamer Immink: *Principles of optical disc systems* (Adam Hilger, Bristol 1984)
3. J. Isailovic: *Videodisc and optical memory systems* (Prentice Hall, Englewood Cliffs 1985)
4. A. Marchant: *Optical recording; a technical overview* (Addison Wesley, Boston 1990)
5. G. Sincerbox, J. Zavislan: *Selected Papers on Optical Storage* (SPIE Optical Engineering Press, Bellingham 1992)
6. K. Schwartz: *The physics of optical recording* (Springer, Berlin 1993)
7. M. Mansuripur: *The physical principles of magneto optical data recording* (Cambridge University Press, Cambridge 1995)
8. E.W. Williams: *Textbooks in electrical and electronic engineering Vol 2, The CD-ROM and optical recording systems* (Oxford University Press, Oxford 1996)
9. S. Stan: *The CD-ROM drive, a brief system description* (Kluwer Academic Publishers, Dordrecht 1998)
10. M. Minsky: Microscopy Apparatus , U.S. Patent 3,013,467 (1961)
11. Ph. Rice et al.: J. Soc. Motion Pict. T. **79**, 997 (1970)
12. K. Compaan, P. Kramer: Philips Tech. Rev. **33**, 178 (1973)
13. G. Broussaud, E. Spitz, C. Tinet, F. LeCarvenec: IEEE T. Broadc. Telev. **BTR–20**, 332 (1974)
14. G.W. Hrbek: J. Soc. Motion Pict. T. **83**, 580 (1974)
15. J. Braat: Appl. Optics **22**, 2196 (1983)
16. J. Haisma, E. Hugues, C. Babolat: Opt. Lett. **4**, 70 (1979)
17. M. Terao, T. Nishida, Y. Miyauchi, S. Horigome, T. Kaku, N. Ohta: P. Soc. Photo–Opt. Inst. **695**, 105 (1986)
18. T. Ohta, et al.: Jap. J. Appl. Phys. **28**, 123 (1989)
19. W.T. Welford: J. Microsc. – Oxford **96**, 105 (1972)
20. T. Wilson, C.J.R. Sheppard: *Theory and practice of the scanning optical microscope* (Academic Press, London 1984)
21. C. Bricot, J. Lehureau, C. Puech, F. Le Carvennec: IEEE T. Consum. Electr. **CE–22**, 304 (1976)
22. M. Mansuripur, C. Pons: P. Soc. Photo–Opt. Inst. **899**, 56 (1988)
23. G. Bouwhuis, P. Burgstede: Philips Tech. Rev. **33**, 186 (1973)
24. J. Braat, G. Bouwhuis: Appl. Optics **17**, 2013 (1978)
25. J. Heemskerk, K. Schouhamer Immink: Philips Tech. Rev. **40**, 157 (1982)
26. N. Arai, H. Yamazaki, S. Saito: Method for recording/reproducing optical information recording medium, optical pickup apparatus, objective lens and design method of objective lens , U.S. Patent 6,243,349 (2001)
27. Y. Komma, K. Kasazumi, S. Nishino, S. Mizumo: P. Soc. Photo–Opt. Inst. **2338**, 282 (1994)
28. B.H.W. Hendriks, J.E. de Vries, H.P. Urbach: Appl. Optics **40**, 6548 (2001)
29. H.H. Hopkins: J. Opt. Soc. Am. **69**, 4 (1979)
30. W.T. Welford: *The aberrations of optical systems* (Adam Hilger, Bristol 1986)
31. M. Born, E. Wolf: *Principles of Optics* (Pergamon Press, Oxford 1970)
32. A.J.E.M. Janssen: J. Opt. Soc. Am. A **19**, 849 (2002)
33. J. Braat, P. Dirksen, A.J.E.M. Janssen: J. Opt. Soc. Am. A **19**, 858 (2002)
34. M. Abramowitz, I. Stegun: *Handbook of Mathematical Functions* (Dover, New York 1970)

4 Superresolution in Scanning Optical Systems

Roy Pike, Deeph Chana, Pelagia Neocleous, and Shi-hong Jiang

4.1 Introduction

In this contribution we describe some approaches which attempt to increase the resolution of optical imaging systems. In particular, we are interested in scanning systems, such as the confocal microscope, using both coherent and fluorescent light, and the optical disc. The basic optical imaging principles of both compact discs and current DVDs and those of a confocal scanning microscope using coherent illumination are very similar and thus some of our considerations can be applied to both. In so far as one might consider a disc system using fluorescent media, even more of the ground would be common.

We will first briefly consider several "direct" approaches to superresolution, namely, the Pendry "perfect lens" idea, Kino's solid immersion lens and Toraldo di Francia's apodising masks. We will then discuss our own approach by summarising work over recent years with the group of Professor Bertero at the University of Genoa and other colleagues which uses the methodolology of the burgeoning field of "inverse problems"[1]. However, this work comprises some fifty papers, including some very recent papers and reviews, [1–4], and we will not labour through the mathematics in any details here, but confine our remarks to general principles and results. We will, nevertheless, include some introductory and didactic analytical and numerical material and also present some general conclusions regarding potential gains in optical disc systems which we have reached in recent joint studies with the optical storage group at Philips laboratories Eindhoven, [5].

"Inverse problems" is now a wide interdisciplinary field in its own right. It provides a generalised form of Shannon's information theory and attempts to quantify the information content of experimental data. For example, apart from its application to optical imaging, we have used similar mathematical approaches to make progress in other fields of modern theoretical physics such as high-temperature superconductivity, early universe problems (quark-gluon plasma, baryogenisis), the design of radar and sonar antennas and speech recognition.

We will show how the application of inverse-problem theory to optics allows us to obtain improvements beyond the classical Rayleigh resolution limit for so-called type I systems and also beyond the performance of ideal so-called type II or confocal systems, both with coherent and incoherent light. We have applied these results

---

[1] see, for example, the well-established Journal *Inverse Problems* published by the Institute of Physics in the UK

Török/Kao (Eds.): Optical Imaging and Microscopy, Springer Series in Optical Sciences
Vol. 87 – © Springer-Verlag, Berlin Heidelberg 2003

to achieve robust increases in resolution in high-aperture confocal fluorescence microscopy, using specially calculated image-plane masks, which can reach nearly twice the conventional Rayleigh limit. By the use of such masks in optical disc systems we predict "point spread functions" to be twice as sharp as in the present CD/DVD systems when compared in the square-law detector electrical outputs. Due to the non-linearity of the detection process the resulting potential storage capacity gains are not clear as yet, as we shall discuss below.

4.2   Direct Methods

In proposing new methods for important practical applications it is clearly prudent to note advantages and disadvantages with other techniques aimed at the same end. Here we consider three such alternative schemes.

4.2.1   Pendry Lens

It has been suggested recently by Pendry, [6], that a thin metal slab can make a perfect lens. The refractive index of a metal can be negative. If it becomes -1 the metal supports surface plasmons. Then, according to Pendry, if the dimensions in the propagation direction of the imaging system are small compared to the wavelength of light we can use Maxwell's equations in the electrostatic limit and all diffraction blurring effects disappear. Indeed the lens can amplify the evanescent components of the electromagneic field back to their boundary values at the object. This is independent of the magnetic permeability, μ. The arrangement and the results for a two-slit object are shown in Fig. 4.1.

We have repeated Pendry's calculations for a silver slab of thickness 40 nm with the geometry of Fig. 4.1, taken from Pendry's paper. The refractive index ϵ is $5.7 - 9.0^2/\omega^2 - 0.4i$ which gives $\epsilon = -1 - 0.4i$ at 3.48 eV. The result is shown in Fig. 4.2 and, in fact, does not depend on where the slab is placed betweeen the object and image.

Although difficult to achieve in practice, this might look interesting and the use of an optical system with these dimensions would not be impossible to contemplate, for example, for an optical disc system. The question is, therefore, whether 80 nm is sufficiently small compared with an optical wavelength to approximate the electrostatic limit. Fortunately, the geometry of this system is sufficiently simple for an exact calculation to be carried out. We used the longer wavelength of 520 nm than that for silver to help the situation, with the same imaginary part of the refractive index. The results for the TE wave are

$$E_0(x, z) = \sum_{\sin \theta_0} u(\sin \theta_0)\, a_0\, e^{-k_0 z \sqrt{\sin^2 \theta_0 - 1}}\, e^{-ik_0 x \sin \theta_0} \tag{4.1}$$

$$E_2(x, z) = \sum_{\sin \theta_0} u(\sin \theta_0)\, T_\perp\, e^{k_0(h-z)\sqrt{\sin^2 \theta_0 - 1}}\, e^{-ik_0 x \sin \theta_0}\ , \tag{4.2}$$

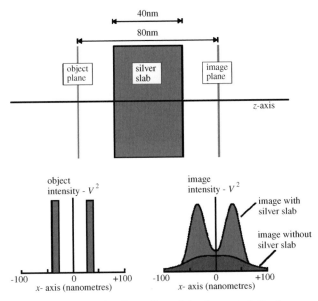

Fig. 4.1. Pendry perfect lens, silver slab, electrostatic limit

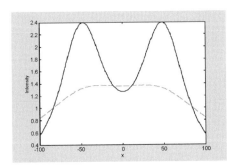

Fig. 4.2. Analytic calculations: silver slab, electrostatic limit. Diffracted image at 80 nm (*dashed line*), image with 40 nm silver-slab lens (*solid line*)

where $E_0(x, z)$ and $E_2(x, z)$ are the electric fields without and with the slab, respectively, x is the transverse coordinate, z is the propagation direction, θ_0 is the incident angle, k_0 is the incident wavevector, $u(\sin \theta_0)$ is the spatial-frequency amplitude of the incident field at the angle θ_0, $T_\perp$ is the transmission coefficient of the slab, a_0 is the amplitude of the incident wave and h is the slab thickness. The transmission coefficient of the slab is given by

$$T_\perp(\theta_0) = \frac{4n_0n_1}{(n_1 + n_0)(n_2 + n_1)} \frac{e^{-u_1h}}{1 - \frac{(n_0-n_1)(n_2-n_1)}{(n_0+n_1)(n_2+n_1)}e^{-2u_1h}}, \quad (4.3)$$

where $n_0 = n_2 = \cos(\theta)/\eta_0, n_1 = \sqrt{\epsilon_r/\mu_r}\cos(\theta_1)/\eta_0, \eta = \sqrt{i\mu\omega/(\sigma + i\epsilon\omega)}$, the characteristic frequency of the medium for plane-wave propagation, $u_1 = ik_0\sqrt{\mu_r\epsilon_r}\cos(\theta_1)$, $\sin(\theta_1) = \sin(\theta)/\sqrt{\mu_r\epsilon_r}$ and μ_r, ϵ_r and σ are the complex relative permeability, the complex relative susceptibility and the conductivity of the medium, respectively.

For a double slit with slit-width a and separation d

$$u(\sin(\theta_0)) = \frac{\sin(ak_0 \sin(\theta_0)/2)}{ak_0 \sin(\theta_0)/2} \cos(dk_0 \sin(\theta_0)/2) . \qquad (4.4)$$

Unfortunately for the perfect-lens idea, our results show that the answer to the question above is negative and, given that the magnetic permeability of such a lens would be 1 or greater it is not until we get down to a lens thickness of 20 nm that the image is satisfactory, by which time the diffraction losses themselves are not large. This is shown in Figs. 4.3, 4.4 and 4.5.

We have also used a simple finite-element numerical simulation to confirm these analytical results. We conclude that the use of a Pendry lens in optical-disc storage

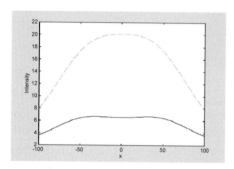

Fig. 4.3. Exact analytic calculations at 520 nm. Diffracted image at 80 nm (*dashed line*), image with 40 nm Pendry lens (*solid line*)

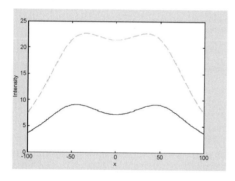

Fig. 4.4. Exact analytic calculations at 520 nm. Diffracted image at 60 nm (*dashed line*), image with 30 nm Pendry lens (*solid line*)

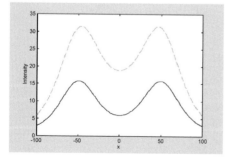

Fig. 4.5. Exact analytic calculations at 520 nm. Diffracted image at 40 nm (*dashed line*), image with 20 nm Pendry lens (*solid line*)

systems is not a promising way forward, unless materials with both surface plasmons and negative magnetic permeability in the visible region could be developed; this seems unlikely in the near future although it is not impossible in principle. See also the work by Shamonina et al. [7], in which Pendry's electrostatic approximation is not made, with similar results.

4.2.2 Kino's Solid Immersion Lens

The so-called solid immersion lens achieves effective apertures greater than are possible with conventional oil-immersion lenses by being operated in the near field. A schematic is shown in Fig. 4.6, [8]. The point spread function depends on the air gap as shown in Fig. 4.7

We illustrate the performance of such a lens for completeness in this chapter since it is a possible way forward for optical storage systems but since the air gaps required are so small, a sealed system would almost certainly be necessary to exclude dust and one then has the question of why use an optical disc rather than a hard-disc magnetic system which would have a much higher speed and capacity?

4.2.3 Toraldo di Francia's Apodising Masks

Making an analogy with superdirective antennas, it was suggested first by Toraldo di Francia, [9], that superresolution could be achieved by use of a pupil-plane mask. An annular ring system of amplitude- and/or phase-modulated pupil functions can

Fig. 4.6. Schematic of solid immersion lens

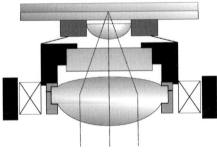

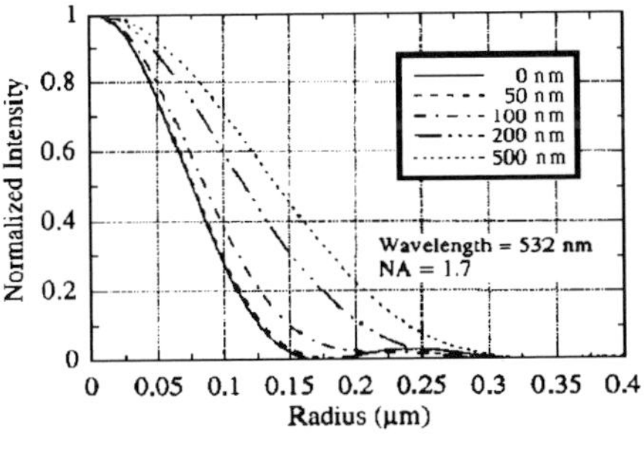

Fig. 4.7. Point spread functions with air gaps, stigmatic SIL, $N.A._{\text{eff}} = 1.7$

be used to modify the point spread function. With such a mask one could trade a resolution gain against a lower Strehl ratio and higher sidelobes.

Di Francia pointed out that there was no theoretical limit to the narrowness of the central peak and no limit either to the diameter of the surrounding dark region but, in practice, the size of the sidelobes increases prohibitively quickly and the Strehl ratio goes down to extremely small values as soon as one departs significantly from the uniform pupil and the normal Airy pattern.

A number of workers have investigated the performance of Toraldo masks using scalar theory (low numerical aperture) and numerical non-linear optimisation schemes. We show a recent example of a pair of conjugate phase-only pupil-plane filters, [10], in Fig. 4.8 which are designed to be used in the confocal transmission system of Fig. 4.9 to achieve axial superresolution.

The point spread functions are given in Fig. 4.10 which shows a 30% gain in axial resolution but with an increased sidelobe structure and an effective Strehl ratio for the combined illumination and collection optics of 0.0625 [11].

A similar pair of three-level, phase-only filters showed a small transverse resolution gain ($\approx 12\%$) with a Strehl ratio of 0.31 in one dimension, x, of the image plane at the expense of a similar loss of resolution in the orthogonal, y, direction.

A further attempt at tailoring axial resolution, [12], in the non-paraxial but still scalar Debye approximation shows similar results but the mask must now be used at a fixed aperture. Inversion of the imaging integral equation, which is discussed in

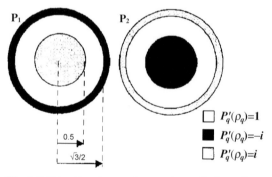

Fig. 4.8. Pair of conjugate phase-only pupil-plane filters

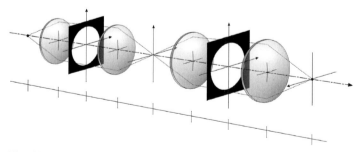

Fig. 4.9. Confocal transmission system with conjugate pupil-plane filters

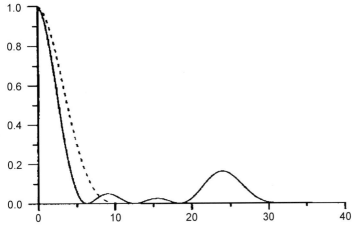

Fig. 4.10. Axial response funtion (normalised) of the conjugate filters of Kowalczyk et al.

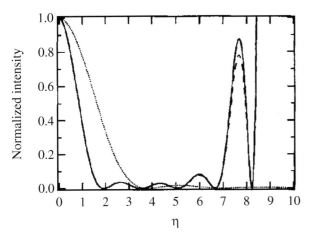

Fig. 4.11. Transverse response funtion (normalised) of the five-zone, pupil-plane, phase mask of Sales and Morris. The Airy pattern is also shown (*dotted line*)

detail below, is actually mentioned as a possibility in this paper but not considered further.

As another example we show in Fig. 4.11 the point spread function for the transverse response of a five-zone, pupil-plane, phase mask, [13]. One may see that the Toraldo principle is demonstrated with a 50% gain in transverse resolution but at the cost of enormous sidelobes (off the top of the graph) and a Strehl ratio of 2.10^{-7}.

Again in this subsection we have considered the Toraldo mask as a possible component for high-density storage systems but the rapid loss of energy in the central peak seems to preclude any realistic application. We should also say that the trend towards the use of higher numerical apertures for optical storage systems is not consistent with the limitations to the scalar case and circular symmetry of existing calculations of the Toraldo mask. We have also considered souce arrays but similar conclusions seem to apply.

4.3   Inverse Methods and Image-Plane Masks

In the application of the methods of the field of inverse problems to confocal scanning imaging systems one considers the image plane to contain "data" which must be used to determine as much information as possible about the object which generated it. It is also made clear that this information, in general, cannot be used to reconstruct the input completely since, as we shall see in a moment, the mathematical operator which transforms the input into the data is linear and compact and is thus a "smoothing" operator. Such an operator, in the presence of noise, irretrievably loses any components of the object which vary too rapidly. This gives rise to a defined "resolution" limit.

The theory shows that there is a "natural" basis for each such operator which provides a complete set of orthonormal components in which to expand the input and output (not necessarily the same for both) which are ordered in such a way that the components which are most robust to noise interference are those with the lower indices. A generalised, noise-dependent "Shannon number" can be defined, which is the index which limits the so-called "signal subspace". The noise prevents reliable recovery of the components with indices higher than this value, unless present in inordinate strength, which will live in the "noise subspace".

The process of reconstruction of the object in a scanning optical system using these concepts can be effected by placing a specially calculated, generally complex, optical mask in the image plane and integrating the transmitted complex amplitudes using suitable collection optics [14]. Alternatively, in the coherent case this can be calculated as a pupil-plane mask where it would act as a sophisticated type of Toraldo mask with uniaxial symmetry at high NA due to the effects of polarisation. In our experience the image-plane versions are simpler to fabricate and we do not discuss the pupil-plane versions any further.

We will give a brief outline of the definition and calculation of an image-plane mask for a generic imaging operator, A, defined by

$$(Af)(y) = \int A(x, y) f(x) \mathrm{d}x , \tag{4.5}$$

where x and y can be multidimensional. We call the object $f(x)$ and the image $g(y)$ and the imaging is described by the operator equation

$$g = Af. \tag{4.6}$$

We shall see a specific example of such an operator shortly in (4.16). (Of course you may peep now!)

The image-plane mask is calculated using the singular value decomposition (SVD) of the operator A, which we consider in the first place to act upon square integrable functions, i.e. technically it maps functions from one L^2 space into another. The SVD is given by the set of solutions $\{\alpha_k; u_k, v_k\}_{k=0}^{\infty}$ of the coupled integral equations

$$Au_k = \alpha_k v_k \tag{4.7}$$
$$A^* v_k = \alpha_k u_k , \tag{4.8}$$

where A^* denotes the adjoint of A. The u_k and v_k are called singular functions and provide orthonormal bases for f and g, respectively, (note that this notation is sometimes reversed). The α_k are called singular values and for a compact operator A they accumulate to zero with increasing k; they represent the strength with which each component is transferred into the image and hence how well it can withstand the addition of noise. Singular value decomposition can be accomplished using standard numerical packages found in many software libraries. Given this decomposition of the imaging operator, by expanding object and image in their respective bases and using their orthonormality it can be seen after a little algebra that the solution of (4.6) is

$$f(x) = \sum_{k=0}^{\infty} \frac{1}{\alpha_k} u_k(x) \int g(y) v_k(y) dy , \tag{4.9}$$

and specifically, on the optical axis

$$f(0) = \sum_{k=0}^{\infty} \frac{1}{\alpha_k} u_k(0) \int g(y) v_k(y) dy . \tag{4.10}$$

For a scanning system the determination of $f(0)$ at each point of the scan is sufficient to reconstruct the image and we utilise this fact for practical reasons, although this does entail some loss of information. Exchanging the sum in (4.10) with the integral we obtain

$$f(0) = \int g(y) M(y) dy , \tag{4.11}$$

where

$$M(y) = \sum_{k=0}^{\infty} \frac{1}{\alpha_k} u_k(0) v_k(y) . \tag{4.12}$$

The function $M(y)$ is called an image-plane optical mask and it can be seen that (4.11) explains the object reconstruction process described earlier. In practice the summation is truncated at a value of k which depends on the signal to noise ratio of the detected signal. Using the fact that (4.11) is the scalar product (M, g) in L^2 of M and g and that $g = Af$, we can see that

$$f(0) = (M, g) = (M, Af) = (A^* M, f) . \tag{4.13}$$

Thus we can write

$$f(0) = \int T(x)f(x)dx ,\tag{4.14}$$

where T is given by

$$T = A^*M .\tag{4.15}$$

Thus T is a smoothing operator or "mollifier".

4.4 Optical Systems for Scanning Imaging

We will now try to make some comparisons of the performance of three imaging systems used for scanning, namely, a conventional type I system, a confocal type II system and a system using an image-plane optical mask.

We consider an optical system with two identical aberration-free lenses each with numerical aperture (N.A.) for illumination and imaging, respectively, and uniform pupil function. The same lens serves for both in a reflective system. This is a special case of the general partially coherent optics described, for example, in early work of Hopkins [15]. In one dimension the action of each lens is described by the linear integral operator A, relating object $f(x)$ to image $g(y)$:

$$g(y) = (Af)(y) = \int_{-X/2}^{X/2} \frac{\sin \Omega(y - x)}{\pi(y - x)} f(x)dx ,\tag{4.16}$$

where the lens accepts light between the transverse wavenumbers $2\pi(N.A.)/\lambda$ of $-\Omega$ and Ω radians/metre (i.e. over a Fourier bandwidth of 2Ω) and the object lies between $-X/2$ and $X/2$. The support, y, of the image is, in theory, infinite but in practice will be defined by the rapid fall-off of the illumination or by a finite detector aperture. The optical system is shown in Fig. 4.12.

We will use optical units for x and y by multiplying actual units by Ω/π and the above equation then has the normalised form:

$$g(y) = \int_{-S/2}^{S/2} \frac{\sin \pi(y - x)}{\pi(y - x)} f(x)dx\tag{4.17}$$

$$= \int_{-S/2}^{S/2} \mathrm{sinc}\{\pi(y - x)\}f(x)dx\tag{4.18}$$

$$= \int_{-S/2}^{S/2} S(y - x)f(x)dx ,\tag{4.19}$$

where the script $S = X\Omega/\pi$ is the Shannon number and $S(\cdot)$ is the point spread function. In this form the numerical aperture is taken out of the further calculations and the band of the lens always lies between $-\pi$ and π. We use low-aperture scalar theory to make contact with the work of Braat [16] and Stallinga [5] and one-dimensional

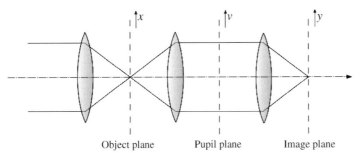

Fig. 4.12. Object, pupil and image planes

integrals for illustration. The one-dimensional results are didactic and more than academic in that they would apply (as a tensor product) to square pupils in two dimensions and they also provide some useful exact analytic formulae. The generalisation to circular pupils in two dimensions requires more lengthy computations but is straightforward (if not the generalisations for high-aperture vector theory, which are dealt with in [2]).

We fill the illumination lens with a plane wave, i.e. $f(x)$ of (4.17) is a point on axis at infinity; the image $g(y)$ (= $\text{sinc}(\pi y)$ in one dimension) then lies in its focal plane, the object plane, and provides the illumination profile of the scanning target (a gaussian beam can be dealt with by modifying the form of the source $f(x)$). We now redefine object and image for the action of the second lens, which again follows (4.17). Using s for the scanning variable, the objective lens sees as its object, $f(x)$, a (complex) field $S(x)R(x-s)$, where $R(x-s)$ is the scanned object transmittance (or reflectance). The field in the image plane of the objective (ignoring magnification) is thus defined by the linear integral operator, K, where

$$g(y, s) = (KR)(y, s) = \int_{-\infty}^{\infty} S(y - x)S(x)R(x - s)dx . \tag{4.20}$$

Here we have taken the limits of the integral to infinity since the support of the object will be defined by the illumination function, which will be effectively zero after a few diffraction rings. The three cases of interest to consider are:

 (i) An extended detector in the image plane
 (ii) A point detector in the image plane at $y = 0$
(iii) A mask in the image-plane followed by a detection system

A first suggestion as a way to compare these systems would be to use the concepts of Fourier spatial-frequency bandwidth and modulation transfer function and compare these between the systems. A second suggestion would be to calculate the response of each of the systems to an object consisting of a point "impulse" and compare the different object reconstructions. The first method should give just the Fourier transforms of the second. However, for our problem this will not be the case. This is because the above equivalence holds only for linear systems but the detectors respond to the modulus-squared of the field amplitudes g, in different ways and thus make the systems non-linear in a non-trivial fashion. We will discuss this point, which has led to some confusion in the field, further below but to see what causes the confusion we first go through the formal exercise from both frequency- and impulse-response points of view and then show how they differ and what practical relevence this has.

We shall calculate the spatial-frequency responses in these three cases by defining $R(x)$ as $\cos(lx)$ and the intensity impulse responses by defining $R(x)$ as $\delta(x)$. We first present the analysis for each case and then the results of numerical calculations. For reference it is well known (see, for example, [17]) that a one-dimensional (coherent) system in which the second lens is replaced by a small detector on the optical axis has a uniform frequency response up to a bandwidth l of π and is zero thereafter.

4.4.1   Analytical Results

(i) Extended Detector in the Image Plane This is known as a type I scanning system and its resolution is the same as that of a conventional type I microscope, see for example, Wilson and Sheppard's book [17] pp. 3 and 42. It is the system used in conventional optical disc readout [16,5].

Spatial-frequency response
Using (4.20), substituting $\cos\{l(x - s)\}$ for $R(x - s)$ and following the notation for detected intensity of Stallinga [5], in which the first subscript denotes [i]mage or [p]upil plane and the second a [p]oint or [e]xtended detector, we have

$$I_{i,e}(l, s) = \int_{\text{detector}} \left| \int_{-\infty}^{\infty} S(y - x)S(x) \cos\{l(x - s)\}\mathrm{d}x \right|^2 \mathrm{d}y . \tag{4.21}$$

Using our one-dimensional test-bed this becomes

$$I_{i,e}(s) = \int_{\text{detector}} \left| \int_{-\infty}^{\infty} \text{sinc}\{\pi(y - x)\}\text{sinc}(\pi x) \cos\{l(x - s)\}\mathrm{d}x \right|^2 \mathrm{d}y . \tag{4.22}$$

This can be written as

$$I_{i,e}(l, s) = \int_{\text{detector}} |\text{sinc}(\pi y) \otimes \{\text{sinc}(\pi y) \cos[l(y - s)]\}|^2 \, \mathrm{d}y , \tag{4.23}$$

where $\otimes$ denotes convolution. By Parseval's theorem this is equal to

$$\begin{aligned}
I_{p,e}(l, s) &= \int_{\text{pupil}} |\{C(v)\} \\
&\quad \{C(v) \otimes [\exp(ils)\delta(v - l) + \exp(-ils)\delta(v + l)]\}|^2 \mathrm{d}v \\
&= \int_{-\pi}^{\pi} |\{\exp(ils)C(v - l) + \exp(-ils)C(v + l)\}|^2 \mathrm{d}v ,
\end{aligned} \tag{4.24}$$

where $C(v) = \text{rect}(\pi v)$ is the characteristic function of the pupil, namely, unity when v is inside it and zero otherwise. This result shows that in this case the extended detector integrates the intensity of a double-slit diffraction pattern. The centres, widths and phases of the slits are piecewise functions of l depending on the overlap of the rect functions with the limits of the integral. We need to consider separately the regions $0 \leq l < \pi$ and $\pi \leq l \leq 2\pi$.

For $0 \leq l < \pi$

$$I_{p,e}(l, s) = \int_{-\pi}^{l-\pi} dv + 2 \int_{-\pi+l}^{\pi-l} |\{\exp(ils) + \exp(-ils)\}|^2 dv + \int_{\pi-l}^{\pi} dv$$

$$= 16(\pi - l)\cos^2(ls) + 2l , \tag{4.25}$$

and for $\pi \leq l \leq 2\pi$

$$I_{p,e}(l, s) = \int_{-\pi}^{\pi-l} dv + \int_{l-\pi}^{\pi} dv = 2(2\pi - l) . \tag{4.26}$$

This behaviour is shown in Fig. 4.13 which, in fact, is the result of a numerical simulation to be described below. The signal persists at frequencies l beyond π, decreasing to zero at $l = 2\pi$. However, for a pure sinusoidal input probe it carries no modulation beyond $l = \pi$.

Intensity impulse response
Using (4.20) the intensity impulse response is, as a function of scan position,

$$I_{i,e}(s) = \int_{detector} \left| \int_{-\infty}^{\infty} S(y - x)S(x)\delta(x - s)dx \right|^2 dy \tag{4.27}$$

$$= \int_{detector} |S(y - s)S(s)|^2 dy \tag{4.28}$$

$$= |S(s)|^2 \int_{detector} |S(y - s)|^2 dy . \tag{4.29}$$

For a detector which is larger than the effective size of $S(y)$ the integral will be a constant, independent of s, and thus

$$I_{i,e}(s) = |S(s)|^2 . \tag{4.30}$$

As pointed out by Stallinga, this response is the same as that given by integrating *intensity* over the pupil plane and it agrees with (3.7) of Ref. [17]. Note that for a point object the response is the same for both coherent and incoherent illumination.

(ii) Point Detector in the Image Plane This is known as a type II or "confocal" system. We put $y = 0$ in (4.20) and again follow the notation of Stallinga.

Spatial-frequency response
This is given by

$$I_{i,p}(l, s) = \left| \int \operatorname{sinc}^2(\pi x) \cos\{l(x - s)\}dx \right|^2 . \tag{4.31}$$

This function has the form shown in Fig. 4.14, obtained by numerical simulation below.

Intensity impulse response
The detected intensity as a function of scan position is

$$I_{i,p}(l, s) = \left| \iint_{-\infty}^{\infty} S(x)S(x)\delta(x - s)dx\,dy \right|^2 \tag{4.32}$$

$$= |S(s)|^4 . \tag{4.33}$$

This may be shown to be equivalent to integrating the *amplitude* over the pupil plane [17] p. 48 and in the one-dimensional case is 20% narrower at half height than $I_{i,e}$.

(iii) Mask in the Image Plane This is the new detection system developed at King's College, London, using inverse-problem techniques.

Spatial-frequency response
The effect of a mask in the image plane is shown in [14] to have the same bandwidth *for the amplitude image* as the confocal microscope (case (ii) above) but to fill it more effectively at higher frequencies. Explicitly, using (4.15) we find

$$T(x) = S(x) \int S(y - x)M(y)dy , \tag{4.34}$$

and by taking the Fourier transform of both sides of this equation we find that the transfer function $\hat{T}(\omega)$ is given by

$$\hat{T}(\omega) = \frac{1}{(2\pi)^n} \int \hat{S}(\omega - \omega')S(\omega')\hat{M}(\omega')d\omega' , \tag{4.35}$$

where $n = 1$ in one dimension and $n = 2$ in two dimensions.

In one dimension the infinite sum for the mask gives [14]

$$\hat{M}(\omega') = 4\pi\{\delta(\omega - \pi) + \delta(\omega + \pi)\} , \tag{4.36}$$

and

$$\hat{T}(\omega) = \frac{1}{(2\pi)} \int_{\omega-\pi}^{\pi} \hat{M}(\omega')d\omega' \qquad 0 < \omega < 2\pi , \tag{4.37}$$

so that

$$\hat{T}(\omega) = 1, \quad 0 < \omega < 2\pi . \tag{4.38}$$

Intensity impulse response
Application of (4.14) to a δ function gives an impulse response of $\frac{1}{2}\text{sinc}(2\pi s)$ and an intensity impulse response is found of

$$I_{i,m}(s) = \frac{1}{4}|\text{sinc}(2\pi s)|^2 , \tag{4.39}$$

where the extension of Stallinga's notation denotes a mask in the image plane. This result may also be calculated directly using (4.20).

4.4.2   Numerical Results

We have written a software package which calculates the images given by (4.20) and then implements the three detection schemes by numerical integration. The frequency-response functions are found by using a cosinusoidal grating of varying periods up to a bandwidth of 2π and following the modulations of the detector outputs as the grating is translated laterally in the object plane. Figs. 4.13, 4.14 and 4.15 show the output signals at a series of grating translations against the frequency of the grating for each of the three systems and Fig. 4.16 shows the normalised modulation depths at each frequency.

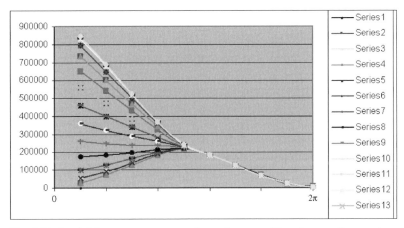

Fig. 4.13. Spatial frequency response of type I system. The plots are for a series of grating translations

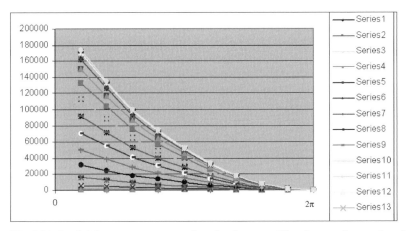

Fig. 4.14. Spatial frequency response of confocal system. The plots are for a series of grating translations

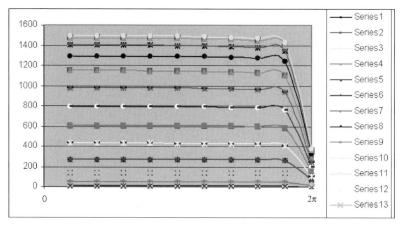

Fig. 4.15. Spatial frequency response of mask system. The plots are for a series of grating translations

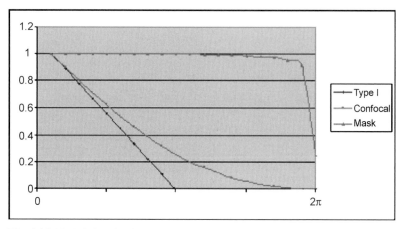

Fig. 4.16. Modulation depth versus spatial frequency (normalised)

It can be seen that the type I system loses the modulation of signal at a bandwidth of π although it has a DC response up to 2π. This is in complete agreement with (4.24), (4.25) and (4.26). There is a full filling of the 2π band in the mask case in agreement with (4.38) save for some numerical effect near the edge of the band. The response of the coherent detector is in line with the literature.

The impulse response functions were calculated by creating one dimensional "pits" of varying widths and scanning them in the object plane. We have also looked at the response to a pair of pits at different spacings with a mark-space ratio of unity. Fig. 4.17 shows the responses of the different detection schemes to a moving narrow pit, normalised to unity at the origin, together with the original sinc function which illuminates the pits, and Fig. 4.18 shows the pair responses. The two-point response function has also been calculated analytically and is shown in Fig. 4.19, it is seen

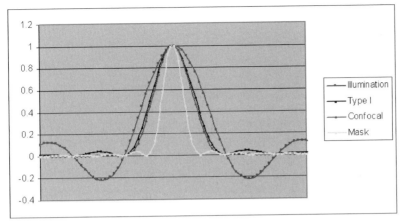

Fig. 4.17. Responses of the three detection systems to a moving narrow pit (normalised). The illumination profile is also shown

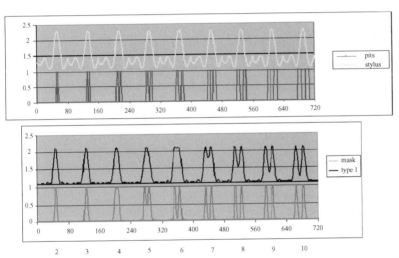

Fig. 4.18. Responses of the type I and mask detection systems to a pair of pits with mark-space ratio of unity at increasing spacings. The illumination profile is also shown above each spacing

that the results support the above numerical work. The absolute values of the response of the mask scheme in both frequency and point response is dependent upon the fact that it squares the integrated field rather than integrating the squares as in the other two cases. We have not had time yet to evaluate the exact relative efficiencies of these coherent systems for practical application but our recent experiments using image-plane masks in confocal scanning fluorescence microscopy, to be discussed below, give encouragingly good images at the same incident laser power.

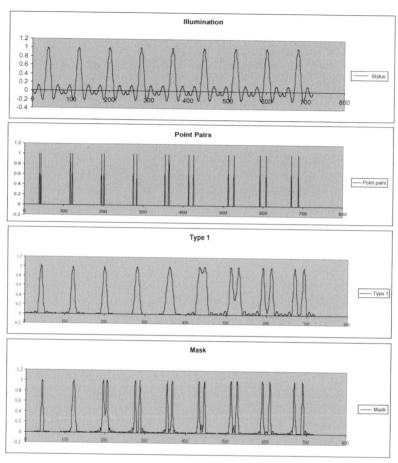

Fig. 4.19. Two-point response of the mask detection scheme. The illumination profile is also shown above each spacing. The spacings are as in Fig. 4.18

4.5   The Comparison of Non-linear Optical Scanning Systems

We seem to have agreement both analytically and numerically on the salient facts of scanning systems. We need to remember that this work only applies to square pupils; the case of circular pupils has yet to be treated in the same way (although the basic physics is unaltered and the results will be qualitatively similar). Our first aim was to resolve uncertainties about the respective (Fourier) band limits of the existing CD/DVD (type I) system and our proposed image-plane mask system but it now seems that this task was misconceived, in the sense that the use of a Fourier decomposition to discuss the performance of an optical (or any other) system is based implicitly upon the assumption that the system is linear. In fact, it is only as far as the optical fields are concerned that the scanning systems mentioned above are all linear.[2]

The images, $g(y, s)$, in all three cases, are given by (4.20); this is a Fredholm equation of the first kind and is manifestly linear. For a linear system the transfer function in Fourier space can be computed as the amplitude response of the system to sinusoidal input waves of increasing frequency, which will drop to zero at the edge of the "transmission band" or, alternatively, as the Fourier transform of the impulse-response function, which will give the same result. For the second method the input impulse can be decomposed into a uniformly filled band of frequencies (infinite in the theoretical limit but finite in practice) which will appear independently in the output with decreasing amplitudes, exactly the same as if they are presented one by one as in the first method.

However, the electrical signal which is used as output in a CD/DVD system is generated by a photon detector whose output is proportional to the modulus squared of the field amplitude and thus is no longer linearly dependent upon the input. For such a non-linear system these two measures of the "frequency response" are quite different, depending upon the exact nature of the non-linearities. Neither can be said to define an unambiguous "bandwidth" of the system and one sees that the concept of bandwidth loses its fundamental significance for the analysis of the coupled optical-electrical system.[3]

In our case we have seen that the optical-to-electrical impulse response of the type I system to a small single pit is

$$I_{i,e}(s) = |S(s)|^2 , \tag{4.40}$$

with a triangular filling of the band to 2π, while the optical-to-electrical impulse response of the mask system is

$$I_{i,m}(s) = |S(2s)|^2 , \tag{4.41}$$

with a triangular filling of the band to 4π which thus has double the bandwidth.

On the other hand if the calculation is done by injecting pure sinusoidal waves, the result is a triangular filling of the band to π for type I and a unit filling of the band to 2π for the mask system, which again has double the bandwidth but with different filling factors.

These results are shown in Figs. 4.20 and 4.21.

A third result is also possible for this non-linear system, *viz.*, by adding a DC offset to the sinusoidal input, i.e. imposing a weak signal modulation; we find that the frequency response using sinusoids is a triangle from $(0,1)$ to $(2\pi,0)$. This is the result found by Braat in "Principles of Optical Disc Systems" in 1985 [16] and according to Stallinga is used in the CD/DVD community.

---

[2] Note that in the case of a fluorescent (or incoherent) object we have the same linear equation (4.20) but with $S(*)$ denoting $\mathrm{sinc}^2(*)$ and g and R this time denoting the *intensities* in the object and image planes.

[3] Of course, the detector output, once generated, can be analysed into Fourier frequencies for discussion of onward linear electronic processing.

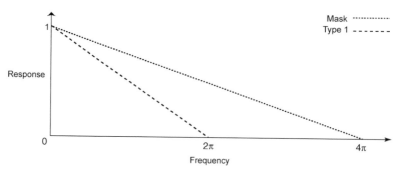

Fig. 4.20. Fourier spectrum of the point response function for type I and mask detection systems

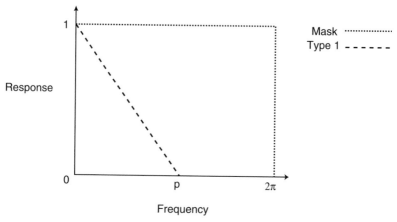

Fig. 4.21. Fourier spectrum of the sinusoidal grating response function for type I and mask detection systems

We should note that in current DVD systems the optical-to-electrical response given by (4.40) is not sufficient to overcome jitter due to adjacent-track interference and an additional "electronic equalisation" must be applied to enhance the tail of the detector response. This would not be necessary with the mask system but could also, in fact, be applied to the mask-detector response given by (4.41) to give a full 4π band in this case.

We have seen, therefore, that it is not a good idea to talk in terms of an optical-to-electrical "frequency response". Fourier analysis is not useful for non-linear systems since the components of the input are not additive in the output and any calculated output "frequency spectra" will be dependent on the particular non-linearities. This applies also to its use for the analysis of different impressed code patterns on optical storage discs since the results would depend in a complicated way on the actual signals. As we have seen above, three different "bandwidths" have already been obtained for three different forms of the input signal. However, whatever the input signal, an image-plane mask system uses all the information available above the noise level and is fully controlled by the number of terms which are used in the summation of (4.12).

The bottom line for the use of an image-plane mask is that it would seem to promise a near fourfold capacity increase over the present systems in the case where pits are resolved. The reduction in this factor will depend on the actual pattern being scanned and further calculations or simulations and/or experiments are needed to quantify this. For robust codes we would not expect much reduction but our software package is written to handle this problem and we hope to continue our calculations along these lines in the near future by simulating run-length-limited codes. The effects of various levels of noise can also be simulated. The whole exercise will then need to be repeated for circular pupils in two dimensions and we plan to move on to this as soon as possible in parallel with experimental work.

4.6 High-Aperture Image-Plane Masks

In the above discussions we have used simple low-aperture, one-dimensional calculations to illustrate the basic concepts. In reality, of course, this needs to be put into practice in real microscopes and optical storage systems where full vector analysis of high-aperture systems is required. As mentioned in the introduction, we have been making such calculations in recent years and the results can be found in references [1–3] mentioned above. We will just discuss here an example of a mask we have made for the Bio-Rad MRC 600 confocal scanning fluorescence microscope. In fluorescence microscopy we have no phase information and since the implementation of the image-plane mask detection system requires the subtraction of areas of the processed image from the rest we use a mask which both transmits and reflects. Two integrating detectors and an electronic subtractor process the transmitted and reflected components. For fluorescence microscopy circularly polarised light can be used and circular symmetry imposed. The image-plane mask may then be conveniently fabricated as a "binary-coded" version of the calculated continuous mask with aluminium rings deposited on silica flat. The binary ring pattern of the mask is devised to emulate the continuous profile of the calculated mask with sufficient accuracy to preserve the resolving power of the microscope while still being easy to manufacture [18]. The mask is made elliptical to be placed at 45 degrees to the optical axis and present a circular cross section to the incident light. A diagram of the mask is shown in Fig. 4.22.

An image which we have obtained recently of a field of 100 nm diameter fluorescent calibration spheres (Molecular Probes Ltd, Eugene, OR), is shown in Fig. 4.23 using the three detection schemes dicussed above, viz. type I, confocal and the KCL scheme, at an *N.A.* of 1.3 with 488 nm radiation. The inserts show profiles taken across the two spheres near the bottom left corner. The progressive increase in resolution is clearly evident. The leftmost sphere may be out of focus or at the edge of a scanning line. Absolute calibration is not yet available.

We have recently made calculations of high-aperture masks for coherent systems, [2], but have not yet implemented them experimentally. Here cylindrical sym-

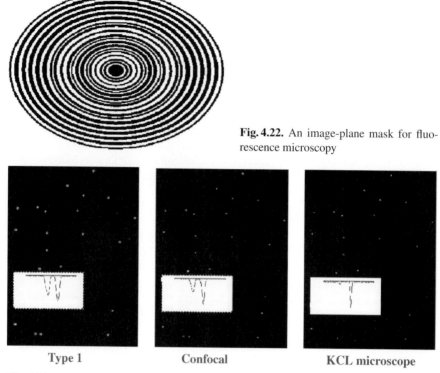

Fig. 4.22. An image-plane mask for fluorescence microscopy

| Type 1 | Confocal | KCL microscope |

Fig. 4.23. Images of 100 nm calibration spheres in fluorescence microscopy.

metry cannot be invoked and the masks are uniaxial along the direction of the incident polarisation, which must be linear in order for the mask calculation to reduce to
a linear inverse problem. An example is shown in Fig. 4.24. This work is in progress
and we hope to have results in the near future. It is also possible in the coherent case
to implement these masks in the pupil plane and this is also being investigated.

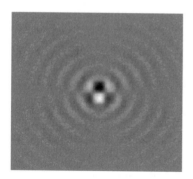

Fig. 4.24. A calculated image-plane mask for a coherent system of *N.A.* 0.6

Acknowledgements

The authors wish to thank the post-graduate students, post-doctoral fellows, academic visitors and staff who have contributed to this work over a number of years at King's College, London and who are now too many to list individually here, save perhaps for picking out Ulli Brand whose superb PhD thesis is devoted entirely to the design and construction of a superresolving confocal scanning fluorescence microscope.

Special thanks are due to Mario Bertero and his colleagues at the University of Genoa and Christine De Mol and her colleagues at the Université Libre de Bruxelles, who have been long-term collaborators in various inverse problems, including especially optical imaging. For help and joint work with us and special lens design in the early days we thank Fred Brakenhof and Hans van der Voort at the University of Amsterdam and Leitz, Wetzlar, respectively. More recently, Colin Sheppard at the University of Sydney and Peter Török at the Imperial College have engaged in helpful discussions on some of the more complex issues. As mentioned in the text, we have also benefitted from joint considerations of scanning systems with Sjoerd Stallinga, Ferry Zijp and Rob Hendriks and other members of the optical storage group at Philips Laboratories, Eindhoven; this has been within an FET programme "Super Laser Array Memory" of the European Union. Finally we acknowledge valuable collaboration with Mark Shipman and Ken Howard of University College, London in the final stages of applying our image-plane masks to biological fluorescence microscopy.

This work has been supported in its various phases by the NATO Scientific Affairs Division, the European Union, the US Army and the UK Engineering and Physical Sciences Research Council and presently, as just mentioned above, by the EU; we have also had help and encouragement from Andrew Dixon and generous support from his company, Bio-Rad Sciences Ltd.

References

1. U. Brand, G. Hester, J. Grochmalicki and E. R. Pike: J. Opt. A. **1**, 794 (1999)
2. J. Grochmalicki and E. R. Pike: Appl. Optics **39**, 6341 (2000)
3. E. R. Pike: 'Superresolution in fluorescence confocal microscopy and in DVD optical data storage'. In: *Confocal and Two-Photon Microscopy, Foundations, Applications and Advances*, ed. by A. Diaspro, (Wiley-Liss Inc., New York, 2002) pp. 499–524
4. E. R. Pike and S.-H. Jiang: J. Phys.: Condensed Matter, to appear (2002)
5. S. Stallinga: Private communications 29th Nov 2001 and 7th Jan 2002
6. J. B. Pendry: Phys. Rev. Lett. **85**, 3966 (2000)
7. E. Shamonina, V. A. Kalinin, K. H. Ringhofer and L. Solymar: Electronics Lett. **37**, 1243 (2001)
8. I. Ichimura, S. Hayashi and G. S. Kino: Appl. Optics **36**, 4339 (1997)
9. T. di Francia: Nuovo Cimento **9**, 426 (1952)
10. M. Kowalczyk, C. J. Zapata-Rodriguez and M. Martinez-Corral: Appl. Optics **37**, 8206 (1998)
11. M. Kowalczyk: private communication

12. M. Martinez-Corral, M. T. Caballero, E. H. K. Stelzer and J. Swoger: Opt. Express, **10**, 98 (2002)
13. T. R. M. Sales and G. M. Morris: J. Opt. Soc. Am. A **14**, 1637 (1997)
14. M. Bertero, P. Boccacci, R. E. Davies, F. Malfanti, E. R. Pike and J. G. Walker: Inverse Problems **8**, 1 (1992)
15. H. Hopkins: Proc. R. Soc. **A217**, 263 (1951)
16. J. Braat: 'Read-out of Optical Discs'. In: *Principles of Optical Disc Systems*, G. Bouwhuis, J. Braat. A. Huijser, J. Pasman, G. van Rosmalen and K. Shouhamer Immink (Adam Hilger, Bristol, 1985) Ch. 2
17. T. Wilson and C. Sheppard: *Theory and Practice of Scanning Optical Microscopy* (Academic Press, London 1984)
18. J. Grochmalicki, E. R. Pike, J. G. Walker, M. Bertero, P. Boccacci and R. E. Davies: J. Opt. Soc. Am. A **10**, 1074 (1993)

5 Depth of Field Control
in Incoherent Hybrid Imaging Systems

Sherif Sherif and Thomas Cathey

5.1 Introduction

A hybrid imaging system combines a modified optical imaging system and a digital post-processing step. We define a new metric to quantify the blurring of a defocused image that is more suitable than the defocus parameter for hybrid imaging systems.

We describe a spatial-domain method to design a pupil phase plate to extend the depth of field of an incoherent hybrid imaging system with a rectangular aperture. We use this method to obtain a pupil phase plate to extend the depth of field, which we refer to as the logarithmic phase plate. By introducing a logarithmic phase plate at the exit pupil and digitally processing the output of the detector, the depth of field is extended by an order of magnitude more than the Hopkins defocus criterion [1]. We compare the performance of the logarithmic phase plate with other extended-depth-of-field phase plates in extending the depth of field of incoherent hybrid imaging systems with rectangular and circular apertures.

We use our new metric for defocused image blurring to design a pupil phase plate to reduce the depth of field, thereby increasing the axial resolution, of an incoherent hybrid imaging systems with a rectangular aperture. By introducing this phase plate at the exit pupil and digitally processing the output of the detector output, the depth of field is reduced by more than a factor of two.

Finally, we examine the effect of using a charge-coupled device (CCD) optical detector, instead of an ideal optical detector, on the control of the depth of field.

5.2 Hybrid Imaging Systems

A hybrid imaging system is different from a system obtained by cascading a standard imaging system and a digital post-processing step. In a hybrid system, both the optical and digital modules are parts of a *single* system, and the imaging process is divided between them. Thus the final image in a hybrid system is obtained by digitally processing an intermediate optical image, as shown in Fig. 5.1. The additional digital degrees of freedom in a hybrid system can be used to improve its imaging performance beyond the best feasible performance of a similar standard system [2].

In this chapter, we control the depth of field of an incoherent hybrid imaging system by introducing a phase plate at its exit pupil and digitally processing the intermediate optical image. We use a phase plate instead of an amplitude plate or a complex plate to avoid any decrease of optical power at the image plane.

Török/Kao (Eds.): Optical Imaging and Microscopy, Springer Series in Optical Sciences
Vol. 87 – © Springer-Verlag, Berlin Heidelberg 2003

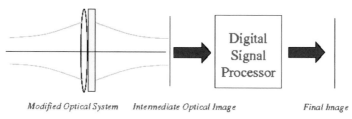

Modified Optical System   Intermediate Optical Image   Final Image

Fig. 5.1. Hybrid imaging system

5.2.1 Digital Post-Processing

The addition of a phase plate at the exit pupil of an optical system attenuates the magnitude of optical transfer function (OTF) of the original system, thereby attenuating most frequency components in the image. The addition of a phase plate at the exit pupil may also distort the phase of the original systems OTF. Thus in all the hybrid imaging systems described in this chapter, the digital post-processing steps involve a restoration digital filter, which amplifies the attenuated frequency components of the intermediate optical image and, if necessary corrects the phase of those frequency components.

For simplicity, we amplify the intermediate images attenuated frequency components and, if necessary correct their phase with a linear frequency-domain inverse filter whose frequency response is given by [3]

$$
H_{\text{inverse}}\left(f_x, f_y\right) = \begin{cases} \dfrac{H_{\text{clear-aperture}}\left(f_x, f_y\right)}{H_{\text{phase-plate}}\left(f_x, f_y\right)} & : H_{\text{phase-plate}}\left(f_x, f_y\right) \neq 0 \\ 0 & : H_{\text{phase-plate}}\left(f_x, f_y\right) = 0, \end{cases} \tag{5.1}
$$

where $H_{\text{clear-aperture}}$ is the in-focus OTF of the optical module with a clear aperture, without a phase plate at its exit pupil, and $H_{\text{phase-plate}}$ is the in-focus OTF of the optical module with a phase plate at its exit pupil. Since the inverse filter, H_{inverse}, is a high-pass filter, it will reduce the overall signal-to-noise ratio of the system. This reduction in the overall signal-to-noise ratio is one of the main drawbacks of hybrid imaging systems.

5.2.2 New Metric for Defocused Image Blurring

We define a new metric to quantify the blurring of a defocused image that is more suitable than the defocus parameter for hybrid imaging systems [4]. As mentioned in the previous section, a hybrid imaging system with a pupil phase plate has a blurred intermediate optical image. A specific restoration digital filter is used to amplify the attenuated frequency components of an intermediate optical image, and, if necessary correct the phase of those frequency components, thereby obtaining a final image. The degree of digital restoration of an out-of-focus object depends on the similarity between the in-focus digital filter used, and the out-of-focus digital filter required.

The angle in Hilbert space between any two functions is a measure of the similarity between these two functions. The smaller the angle between the two functions, the more similar are the two functions and vice versa. Thus as a new metric to quantify the defocused image blurring, we choose the angle in Hilbert space between a defocused point spread function (PSF) and the in-focus PSF. This angle, $0 \leq \theta \leq \frac{\pi}{2}$, is defined, for any defocus parameter value, ψ, as [5]

$$\cos \theta = \frac{\langle |h(u,0)|^2, |h(u,\psi)|^2 \rangle}{|||h(u,0)|^2|, |||h(u,\psi)|^2||}, \tag{5.2}$$

where the inner-product of the in-focus PSF, $|h(u,0)|^2$, and a defocused PSF, $|h(u,\psi)|^2$, is defined as

$$\langle |h(u,0)|^2, |h(u,\psi)|^2 \rangle = \int_{-\infty}^{\infty} |h(u,0)|^2 |h(u,\psi)|^2 du, \tag{5.3}$$

and the length in Hilbert space of the in-focus PSF is defined as

$$|||h(u,0)|^2|| = \left(\int_{-\infty}^{\infty} |h(u,0)|^2 |h(u,0)|^2 du \right)^{1/2}, \tag{5.4}$$

and the length in Hilbert space of the defocused PSF is defined as

$$|||h(u,\psi)|^2|| = \left(\int_{-\infty}^{\infty} |h(u,\psi)|^2 |h(u,\psi)|^2 du \right)^{1/2}. \tag{5.5}$$

Being a measure of the similarity between a defocused PSF and the in-focus PSF, the angle θ is a more general metric to quantify the blurring of a defocused image than the defocus parameter, ψ, which, for a given imaging system, is a measure of the defocus distance only.

5.3   Extended Depth of Field

Extended depth of field (EDF) in optical imaging systems has been the goal of many researchers over the last few decades. To solve this problem, most researchers used apodization techniques on standard imaging systems. Usually, an absorbing plate with a possible $\pm\pi$ phase plate was introduced at the exit pupil to extend the depth of field [6–10]. All of these apodization-based methods share two drawbacks: a decrease of optical power at the image plane and a possible decrease in image resolution. Another method for extending the depth of field of a standard imaging system without apodization was described in [11]. In this method, the focus must be varied during exposure; hence, it is not always practical. To extend the depth of

field of an incoherent hybrid imaging system, one method used an absorbing plate
at the exit pupil; hence, this method suffered from the above drawbacks [12].

In [13], a cubic phase plate was introduced at the exit pupil of an incoherent
hybrid imaging system, and a depth of field extension of an order of magnitude more
than the Hopkins defocus criterion [1] was achieved, without loss of optical power
or image resolution at the image plane. A less than ten times increase in the depth
of field of a nonparaxial hybrid imaging system was achieved using a logarithmic
asphere lens [14]. A logarithmic asphere is a lens that is divided into annular rings of
different focal lengths; hence, it has a continuous radial variation in its focal length.
This logarithmic asphere lens differs from the rectangular logarithmic phase plate
that we describe in this chapter in two fundamental ways: it is a circularly symmetric
lens and it is designed by applying Fermat's principle [15].

In this section, we describe a spatial-domain method to design a pupil phase
plate to extend the depth of field of an incoherent hybrid imaging system with a
rectangular aperture. We use this method to obtain a pupil phase plate to extend
the depth of field, which we refer to as the logarithmic phase plate. To verify that
a logarithmic phase plate can be used to extend the depth of field of an incoherent
diffraction-limited imaging system with a rectangular aperture, we show that the
PSF and the OTF of such system with a logarithmic phase plate at its exit pupil
are invariant with defocus. To demonstrate the extended depth of field, we compare
two sets of computer-simulated images of a chirp pattern for different defocus parameter values. The first set is obtained using a standard incoherent system and the
second set is obtained using a similar system, but with a logarithmic phase plate at
its exit pupil. Finally, we compare the performance of the logarithmic phase plate
with other extended-depth-of-field phase plates in extending the depth of field of
hybrid imaging systems with rectangular and circular apertures.

5.3.1 Design of a Rectangular EDF Phase Plate

Mathematical Separability of PSF The PSF of a defocused paraxial imaging system with a rectangular aperture, is given by [16]

$$\left| h\left(u, v, w_x, w_y\right) \right|^2 = \left| \kappa \int_{-y_{max}}^{y_{max}} \int_{-x_{max}}^{x_{max}} \exp\left\{ jk\left[\left(\frac{w_x x'^2}{x_{max}^2} + \frac{w_y y'^2}{y_{max}^2} \right) - \left(\frac{ux'}{z_i} + \frac{vy'}{z_i} \right) \right] \right\} dx' dy' \right|^2 ,$$

(5.6)

where (u, v) and (x', y') are rectangular coordinates in the image and exit pupil
planes, respectively, w_x and w_y are the defocus coefficients in the x' and y' directions, respectively, x_{max} and y_{max} are the half-widths of the aperture in the directions
of x' and y', respectively, k is the propagation constant, z_i is the image distance and
κ is a constant. Defining a normalized defocus coefficient, $w = \frac{w_x}{x_{max}^2} = \frac{w_y}{y_{max}^2}$, we can

write (5.6) as

$$|h(u, v, w)|^2 = \left| \kappa \int\limits_{-y_{max}}^{y_{max}} \int\limits_{-x_{max}}^{x_{max}} \exp\left\{ jk\left[(wx'^2 + wy'^2) - \left(\frac{ux'}{z_i} + \frac{vy'}{z_i} \right) \right] \right\} dx'dy' \right|^2 .$$

(5.7)

Since this defocused PSF is mathematically separable, we can restrict our analysis to a one-dimensional (1-D) defocused PSF,

$$|h(u, w)|^2 = \left| \sqrt{\kappa} \int\limits_{-x_{max}}^{x_{max}} \exp\left[jk\left(wx'^2 - \frac{ux'}{z_i} \right) \right] dx'dy' \right|^2 .$$

(5.8)

We introduce the normalized co-ordinates, $x = \frac{x'}{x_{max}}$, and we introduce a phase plate, $f(x, y)$, at the exit pupil and we drop all multiplicative constants, yielding

$$|h(u, w)|^2 = \left| \int\limits_{-1}^{1} \exp\left[jk\left(wx_{max}^2 x^2 - f(x) - \frac{ux_{max}x}{z_i} \right) \right] dx \right|^2 .$$

(5.9)

Axial Symmetry Condition of PSF For an extended depth of field, the PSF of an imaging system must be invariant with defocus. Any defocus-invariant PSF is also symmetric about the image plane; hence, any phase plate, $f(x)$, that extends the depth of field must satisfy the PSF axial symmetry condition,

$$|h(u, w)|^2 = |h(u, -w)|^2 .$$

(5.10)

It can be shown [20] that (5.10) is satisfied if and only if the phase plate, $f(x)$, is an odd function, which implies that it must not have any focusing power. If $f(x)$ has any focusing power, it will shift the location of the image plane.

Asymptotic Approximation of a Defocused PSF To simplify the mathematical analysis, we obtain the asymptotic approximation of the integral in a defocused PSF, (5.9), as $k \to \infty$. We use the stationary phase method [15] and [17–19], which is an excellent approximation for any large value of k (small wavelength λ). Optical wavelengths are small enough to justify the stationary phase approximation, which yields results similar to geometrical optics since both assume infinitely small wavelength.

The stationary phase method states that the only significant contributions to the integral in (5.9) occur at the points (stationary points) where the gradient of the phase term vanishes,

$$\frac{d}{dx}\left(wx_{max}^2 x^2 - f(x) - \frac{ux_{max}x}{z_i} \right)\bigg|_{x=x_s} = 0 .$$

(5.11)

A stationary point, $x_s(w, u)$, is defined by the stationary point equation,

$$2wx_{max}^2 x_s - f'(x_s) - \frac{ux_{max}}{z_i} = 0 . \tag{5.12}$$

It can be shown [20] that the asymptotic approximation of a defocused PSF of a system with a rectangular aperture, (5.9), using the stationary phase method, is given by

$$|h(u, w)|^2 \simeq \left| \frac{\lambda}{2wx_{max}^2 - f''(x_s)} \right| . \tag{5.13}$$

Differentiating the stationary point equation, (5.12), with respect to w, we obtain

$$2x_{max}^2 x_s + 2wx_{max}^2 \frac{\partial x_s}{\partial w} - f''(x_s) \frac{\partial x_s}{\partial w} = 0. \tag{5.14}$$

We substitute (5.14) in (5.13) to obtain a more useful expression for the approximate defocused PSF,

$$|h(u, w)|^2 \simeq \left| \frac{\lambda \left(\frac{\partial x_s}{\partial w} \right)}{2x_{max}^2 x_s} \right| . \tag{5.15}$$

Condition for Extended Depth of Field For an extended depth of field, we seek the stationary point, $x_s(w, u')$, from which we can obtain $f(x)$, such that the approximate defocused PSF, (5.15), does not change with respect to w at an arbitrary point in the image plane, u'

$$\frac{d}{dw} \left| \frac{\lambda \left(\frac{\partial x_s}{\partial w} \right)}{2x_{max}^2 x_s} \right| = 0 . \tag{5.16}$$

Evaluating, we get an ordinary differential equation,

$$\frac{d}{dw} \left| \frac{\lambda \left(\frac{\partial x_s}{\partial w} \right)}{2x_{max}^2 x_s} \right| = 0 , \tag{5.17}$$

which has a solution,

$$x_s(w, u') = \pm \frac{1}{\beta''} \exp\left(\pm \frac{w}{\alpha''} \right) \qquad\qquad \alpha'', \beta'' > 0 \tag{5.18}$$

Obtaining the inverse function of this solution, we have

$$w(x_s) = \pm \alpha'' \log(\beta'' x_s) \qquad\qquad x_s > 0 \tag{5.19}$$

and

$$w(x_s) = \pm \alpha'' \log(-\beta'' x_s) \qquad\qquad x_s < 0 \tag{5.20}$$

From (5.12), for an arbitrary point in the image plane, u', we have

$$w(x_s) = \frac{1}{2x_{max}^2 x_s} \left[f'(x_s) + \frac{u' x_{max}}{z_i} \right] \tag{5.21}$$

We equate the right hand side of (5.19) and the right hand side of (5.21), yielding, for $x_s > 0$,

$$\pm\alpha'' \log(\beta'' x_s) = \frac{1}{2x_{max}^2 x_s} \left[f'(x_s) + \frac{u' x_{max}}{z_i} \right]. \tag{5.22}$$

We integrate both sides, yielding

$$f(x_s) = \pm\alpha' x_{max}^2 \int x_s \log(\beta'' x_s) \mathrm{d}x_s - \frac{u' x_{max} x_s}{z_i} + c_1 , \tag{5.23}$$

where $\alpha' = 2\alpha''$ and c_1 is an integration constant. Since $\log(ab) = \log(a) + \log(b)$, we have

$$f(x_s) = \pm\alpha' x_{max}^2 \int [x_s \log(x_s) + x_s \log(\beta'')] \mathrm{d}x_s - \frac{u' x_{max} x_s}{z_i} + c_1. \tag{5.24}$$

Evaluating the integral, we obtain

$$f(x_s) = \pm\alpha' x_{max}^2 \left[\frac{x_s}{2} \left(\log(x_s) - \frac{1}{2} \right) + \beta' \frac{x_s^2}{2} \right] - \frac{u' x_{max} x_s}{z_i} + c_2 \quad \alpha' > 0 \tag{5.25}$$

where $\beta' = \log(\beta'')$ and c_2 is a new integration constant. Simplifying, we obtain

$$f(x_s) = \pm\alpha x_{max}^2 x_s^2 (\log(x_s) + \beta) - \frac{u' x_{max} x_s}{z_i} + c_2 \qquad \alpha > 0 , \tag{5.26}$$

where $\alpha = \frac{\alpha'}{2}$ and $\beta = \beta' - \frac{1}{2}$.

Similarly for $x_s < 0$,

$$f(x_s) = \pm\alpha x_{max}^2 x_s^2 (\log(-x_s) + \beta) - \frac{u' x_{max} x_s}{z_i} + c_2 \qquad \alpha > 0 . \tag{5.27}$$

We combine (5.26) and (5.27) after choosing the signs that make $f(x)$ an odd function, and we ignore the constant c_2 (a constant delay), yielding

$$f(x) = \mathrm{sgn}(x)\alpha x_{max}^2 x^2 (\log |x| + \beta) - \frac{u' x_{max} x}{z_i} . \tag{5.28}$$

Because of mathematical separability, our desired two-dimensional phase plate to extend the depth of field, $f(x, y)$, is given by

$$\boxed{\begin{aligned} f(x, y) = &\mathrm{sgn}(x)\alpha x_{max}^2 x^2 (\log |x| + \beta) - \frac{u' x_{max} x}{z_i} \\ &+ \mathrm{sgn}(y)\alpha y_{max}^2 y^2 (\log |y| + \beta) - \frac{u' y_{max} y}{z_i} . \end{aligned}} \tag{5.29}$$

We refer to $f(x, y)$ as the logarithmic phase plate.

Optimum Logarithmic Phase Plate We obtain the optimum values of the logarithmic phase plate parameters, α, β and u', for an arbitrary paraxial imaging system, using Zemax: an optical design program [21]. Using Zemax, we specify an incoherent imaging system, whose parameters are shown in Table 5.1, and with a logarithmic phase plate at its exit pupil. We use the numerical optimization routine

Table 5.1. Imaging system parameters

λ (nm)	f(mm)	$F\#$	z_i (mm)	M
587.56	20.0	4.0	40.0	-1

of Zemax to obtain the optimum values of α, β and u' such that an objective function is minimized.

The first part of our objective function is defined as the difference between various points of the in-focus OTF and their corresponding points of different defocused OTFs. If this part of our objective function is minimized, the systems OTF will be as defocus invariant as possible. The second part of our objective function is defined as the difference between various points of the in-focus OTF and their corresponding points of the in-focus diffraction-limited OTF. If this part of the objective function is minimized, the systems OTF will have the highest values possible. In essence, the minimization of our objective function yields the highest value of the OTF for a system that is as invariant with defocus as possible.

The optimum values of the logarithmic phase plate parameters, which correspond to the imaging system parameters shown in Table 5.1, are shown in Table 5.2. The initial value of our objective function, which corresponds to $\alpha = 0, \beta = 0$ and

Table 5.2. Logarithmic phase plate optimum parameters

α (mm)	β	u' (mm)
4.23×10^{-4}	0.57	-2.18×10^{-4}

$u' = 0$ is 0.51221. The optimum value of our objective function, which corresponds to the optimum parameters in Table 5.2 is 0.04000. The value of our optimum objective function increases from 0.0400 to 0.04031, which is a change of only 0.06% of the initial value of our objective function, when we set $u' = 0$. Thus u' has a negligible effect in extending the depth of field and will be ignored. Thus the practical values of the logarithmic phase plate parameters, which correspond to the imaging system parameters shown in Table 5.1, are shown in Table 5.3.

The profile of the logarithmic phase plate, whose parameters are shown in Table 5.3, is shown in Fig. 5.2.

Table 5.3. Logarithmic phase plate practical parameters

α (mm)	β	u' (mm)
4.23×10^{-4}	0.57	0.0

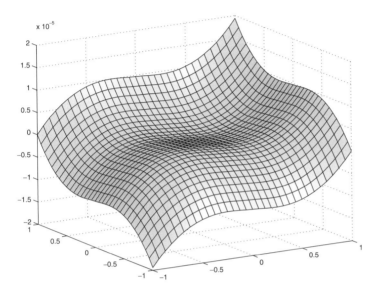

Fig. 5.2. Profile of a logarithmic phase plate

5.3.2   Performance of a Logarithmic Phase Plate

Defocused PSF Using a Logarithmic Phase Plate The PSF of a diffraction-limited imaging system, whose parameters are shown in Table 5.1, and with a logarithmic phase plate, whose parameters are shown in Table 5.3, at its exit pupil is shown in Fig. 5.3 for different values of the defocus parameter ψ. We note that, apart from a slight lateral shift with defocus, the shape and the intensity of the PSF shown in Fig. 5.3 are invariant with defocus over a wide range of defocus parameter values, compared to the PSF of a similar defocused standard diffraction-limited imaging system, shown in Fig. 5.4.

A quantitative way to show that the PSF of a diffraction-limited system with a logarithmic phase plate at its exit pupil is invariant with defocus, compared to the PSF of a similar standard imaging system, is to evaluate the angle in Hilbert space between the in-focus PSF and different defocused PSFs, (5.2).

In Fig. 5.5, we show the angle in Hilbert space between the in-focus PSF and defocused PSFs of a diffraction-limited standard imaging system, whose parameters are shown in Table 5.1. On the same figure, Fig. 5.5, we also show the angle in Hilbert space between the in-focus PSF and defocused PSFs of the same imaging

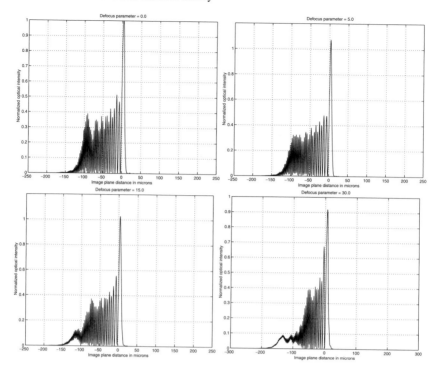

Fig. 5.3. Defocused diffraction-limited PSF using a logarithmic phase plate

system, but with a logarithmic phase plate, whose parameters are shown in Table 5.3, at its exit pupil.

From Fig. 5.5, we note that, for all shown defocus parameter values, the angle in Hilbert space between the in-focus PSF and defocused PSFs of a system with a logarithmic phase plate at its exit pupil has a smaller value than the corresponding angle in Hilbert space between the in-focus PSF and defocused PSFs of a standard system. Thus the PSF of a diffraction-limited imaging system with a logarithmic phase plate at its exit pupil is invariant with defocus, compared to the PSF of a similar standard diffraction-limited imaging system.

Defocused OTF Using a Logarithmic Phase Plate The OTF of a diffraction-limited imaging system, whose parameters are shown in Table 5.1, and with a logarithmic phase plate, whose parameters are shown in Table 5.3, at its exit pupil is shown in Fig. 5.6 for different values of the defocus parameter ψ. We note that the OTF shown in Fig. 5.6 is invariant with defocus over a wide range of defocus parameter values, compared to the OTF of a similar defocused imaging system, shown in Fig. 5.7. We also note that for all defocus parameter values, the OTF shown in Fig. 5.6 has no nulls; hence, there is no loss of spatial frequencies in the image.

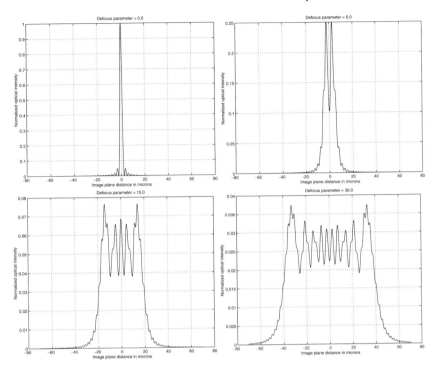

Fig. 5.4. Defocused diffraction-limited PSF using a clear rectangular aperture

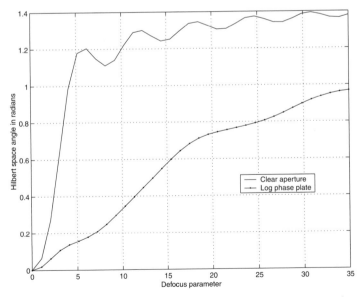

Fig. 5.5. Diffraction-limited Hilbert space angles using a clear rectangular aperture and using a logarithmic phase plate

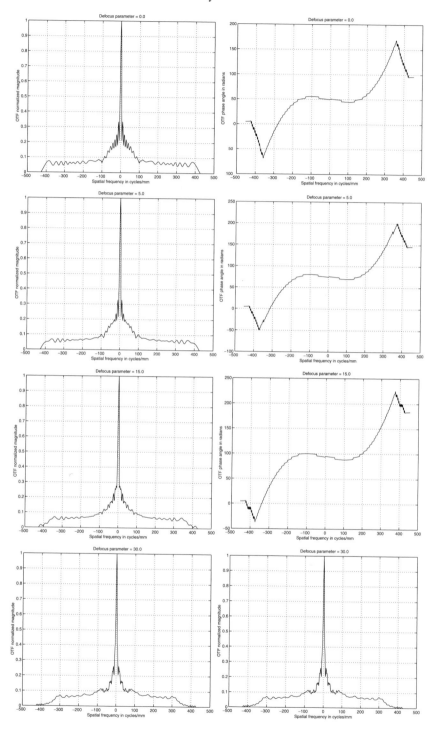

Fig. 5.6. Defocused diffraction-limited OTF using a logarithmic phase plate

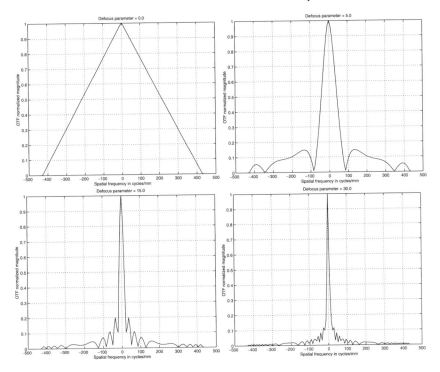

Fig. 5.7. Defocused diffraction-limited OTF using a clear rectangular aperture

The Woodward function of the pupil function of an imaging system, with a rectangular aperture, represents a polar display of the systems OTFs for different values of defocus [22]. The Woodward function of the pupil function of a diffraction-limited imaging system with a logarithmic phase plate at its exit pupil is shown in Fig. 5.8.

On comparing the Woodward function shown in Fig. 5.8 with the Woodward function of the pupil function of a standard imaging system, shown in Fig. 5.9, we note that radial lines through the origin of the Woodward function shown in Fig. 5.8 have nearly the same values as a function of angle, for a broad range of angles. We also note that the Woodward function shown in Fig. 5.8 has non-zero values uniformly distributed along the normalized frequency axis. Thus the OTF of an imaging system with a logarithmic phase plate at its exit pupil is invariant with defocus over a wide range of defocus parameter values compared to the OTF of a defocused standard imaging system.

Simulated Imaging Example To demonstrate the extended depth of field, we compare two sets of computer-simulated images of a chirp pattern for different defocus parameter values. On the left column of Fig. 5.10, we show computer-simulated images of a chirp pattern, for different defocus parameter values, that we obtained

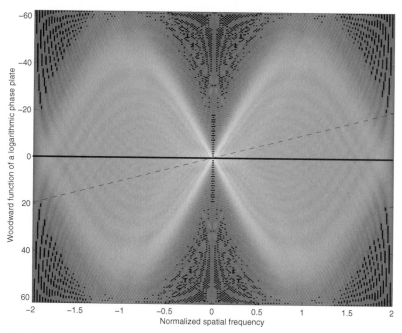

Fig. 5.8. Woodward function of a logarithmic phase plate

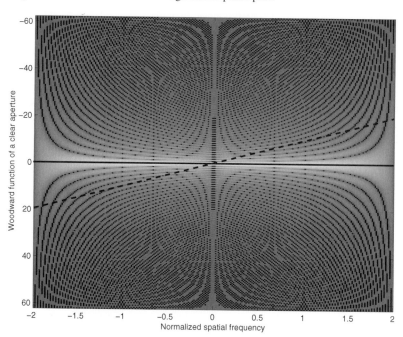

Fig. 5.9. Woodward function of a clear rectangular aperture

using an incoherent standard diffraction-limited imaging system, whose parameters are shown in Table 5.1. On the right column of Fig. 5.10, we show computer-simulated images of the same chirp pattern, for different defocus parameter values, that we obtained using a similar imaging system with a logarithmic phase plate, whose parameters are shown in Table 5.3, at its exit pupil. In each column, the value of the defocus parameter is changed from 0 (in-focus) to 30.

We obtained these images by using an inverse filter whose frequency response is given by [3]

$$H_{inverse}\left(f_x, f_y\right) = \begin{cases} \frac{H_{clear-aperture}\left(f_x, f_y\right)}{H_{log-plate}\left(f_x, f_y\right)} & : H_{log-plate}\left(f_x, f_y\right) \neq 0 \\ 0 & : H_{log-plate}\left(f_x, f_y\right) = 0, \end{cases} \tag{5.30}$$

where $H_{clear-aperture}$ is the in-focus OTF of the diffraction-limited imaging system with a clear aperture, without a phase plate at its exit pupil, and $H_{log-plate}$ is the in-focus OTF of the diffraction-limited imaging system with the logarithmic phase plate at its exit pupil.

From Fig. 5.10, we note that the imaging system with a logarithmic phase plate at its exit pupil has a depth of field that is an order of magnitude more than the Hopkins defocus criterion.

5.3.3   Performance Comparison of Different EDF Phase Plates

Imaging System with a Rectangular Aperture   In this section, we compare the performance of a logarithmic phase plate with a cubic phase plate [13] in extending the depth of field of a diffraction-limited imaging system with a rectangular aperture. A cubic phase plate is mathematically written as [13]

$$f(x, y) = \alpha(x^3 + y^3). \tag{5.31}$$

We obtain the optimum value of the cubic phase plate parameter, α, using Zemax. We use the numerical optimization routine of Zemax to minimize the same objective function that we used to obtain the optimum values of the logarithmic phase plate parameters shown in Table 5.3. For the diffraction-limited imaging system whose parameters are shown in Table 5.1, the optimum value of the cubic phase plate parameter, α, is 0.005 mm.

In Fig. 5.11, we show the angle in Hilbert space between the in-focus PSF and defocused PSFs of a diffraction-limited standard imaging system, whose parameters are shown in Table 5.1, with a cubic phase plate with $\alpha = 0.005$ mm. On the same figure, Fig. 5.11, we also show the angle in Hilbert space between the in-focus PSF and defocused PSFs of the same imaging system, with a logarithmic phase plate, whose parameters are shown in Table 5.3, at its exit pupil.

From Fig. 5.11, we note that, for all defocus parameter values less than 15, the angle in Hilbert space between the in-focus PSF and defocused PSFs of a system with a cubic phase plate at its exit pupil has an approximately equal value to the corresponding angle in Hilbert space between the in-focus PSF and defocused PSFs

126   Sherif Sherif and Thomas Cathey

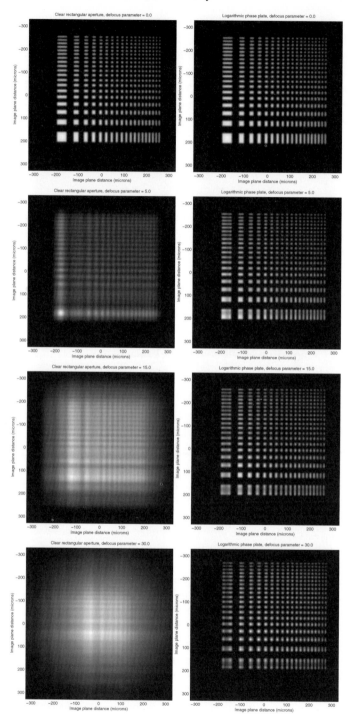

Fig. 5.10. Defocused diffraction-limited images using a clear rectangular aperture and using a logarithmic phase plate

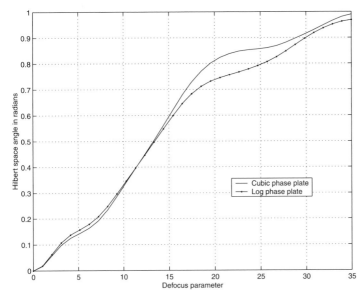

Fig. 5.11. Diffraction-limited Hilbert space angles using a cubic phase plate and using a logarithmic phase plate

of a system with a logarithmic phase plate at its exit pupil. For all defocus parameter values between 15 and 30, the angle in Hilbert space between the in-focus PSF and defocused PSFs of a system with a cubic phase plate at its exit pupil is slightly larger than the corresponding angle in Hilbert space between the in-focus PSF and defocused PSFs of a system with a logarithmic phase plate at its exit pupil.

Thus the performances of the logarithmic phase plate and the cubic phase plate in extending the depth of field of an imaging system with a rectangular aperture are very similar, with a slight advantage of the logarithmic phase plate for higher defocus parameter values. However, it is important to note that the values of the parameters of a plate are very critical to the its performance.

Imaging System with a Circular Aperture In this section, we compare the performance of a rectangular logarithmic phase plate, a rectangular cubic phase plate, a logarithmic asphere [14], and an EDF circular phase plate [23] in extending the depth of field of a diffraction-limited hybrid imaging system with a circular aperture. A logarithmic asphere, $f(r)$, is a circularly symmetric, stand-alone phase imaging element and is given by [14]

$$f(r) = \left\{ \left[\sqrt{r^2 + z_i^2} - z_i \right] + \frac{r_{max}^2}{2(z_2 - z_1)} \times \left[\log\left(2\frac{z_2 - z_1}{r_{max}^2} \left\{ \sqrt{r^2 + \left(z_1 + \frac{z_2 - z_1}{r_{max}^2} \right)^2} \right. \right. \right. \right.$$
$$\left. \left. \left. \left. + \left(z_1 + \frac{z_2 - z_1}{r_{max}^2} \right) \right\} + 1 \right) - \log\left(4\frac{z_2 - z_1}{r_{max}^2} z_1 + 1 \right) \right] \right\}$$

$$(5.32)$$

where z_i is the in-focus image distance, z_1 and z_2 are defocused image distances and r_{max} is the radius of the aperture. An EDF circular plate, $f(r, \theta)$, is a non-circularly symmetric pupil phase plate and is given by [23]

$$f(r, \theta) = \begin{cases} f_R(r, \theta) & : & -\frac{\pi}{2} \leq \theta \leq \frac{\pi}{2} \\ f_L(r, \theta) & : & -\frac{\pi}{2} \geq \theta \geq \frac{\pi}{2}, \end{cases} \tag{5.33}$$

where

$$f_L(r, \theta) = \alpha r^{5.33} \theta^3 \tag{5.34}$$

and

$$f_L(r, \theta) = -f_R(r, \theta - \pi) \tag{5.35}$$

In Fig. 5.12, we show the angle in Hilbert space between the in-focus PSF and defocused PSFs of a diffraction-limited standard imaging system, whose parameters are shown in Table 5.1, with different EDF phase plates at its exit pupil. The first EDF phase plate is a logarithmic phase plate, whose parameters are shown in Table 5.3. The second EDF phase plate is a cubic phase plate with $\alpha = 0.005$ mm. The third EDF phase plate is an EDF circular plate with $\alpha = 0.0031$ mm. On the same figure, Fig. 5.12, we show the angle in Hilbert space between the in-focus PSF and defocused PSFs of a diffraction-limited imaging system, whose parameters are shown in Table 5.1, which uses a logarithmic asphere, (5.32), instead of a lens. We choose $z_1 = 39.64$ mm and $z_2 = 40.36$ mm which, for the given imaging system, are the distances equivalent to defocus parameters $\psi = -30$ and $\psi = 30$, respectively.

From Fig. 5.12, we note that the angle in Hilbert space between the in-focus and defocused PSFs of an imaging system that uses a logarithmic asphere is not symmetric about the image plane. We also note that, in general, the angle of a system with a cubic phase plate or a logarithmic phase plate at its exit pupil is lower than the corresponding angle of a system with an EDF circular phase plate at its exit pupil or the corresponding angle of a diffraction-limited imaging system which uses a logarithmic asphere.

Thus both performances of the logarithmic phase plate and the cubic phase plate are better than both performances of the EDF circular phase plate and the logarithmic asphere in extending the depth of field of an imaging system with a circular aperture. The performance of the cubic phase plate has a slight advantage over the logarithmic phase plate for higher defocus parameter values. However, it is important to note that the values of the parameters of a plate are very critical to the its performance.

5.4   Reduced Depth of Field

Reduced depth of field (RDF), thereby increased axial resolution, in optical imaging systems is of great importance in three-dimensional (3-D) imaging. If the magnitudes of the defocused optical transfer functions (OTFs) of a hybrid imaging system

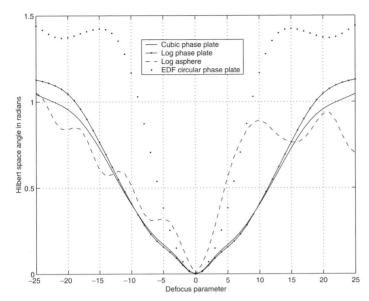

Fig. 5.12. Diffraction-limited Hilbert space angles using different EDF phase plates

have no nulls within the pass-band of the system, there will be no irrecoverable loss of information in the axial direction. When there is no irrecoverable loss of information in the axial direction, the lower the depth of field, the higher the optical-sectioning capacity of an imaging system [24] and [25].

However, compared to extended depth of field, not too much attention has been given to this problem. Continuously varying amplitude pupil plates [26], and annular binary pupil plates, [27] and [28], were used to reduce the length of the central lobe of the axial intensity PSF of a standard imaging systems. Similar to the EDF problem, these apodization-based methods to reduce the depth of field share two drawbacks: a decrease of optical power at the image plane and a possible decrease in image resolution. In [29], a phase-only pupil plate was used to reduce the axial spot size of a confocal scanning microscope. Another method to reduce the depth of field of a standard optical system used structured illumination, [30]. In this method, three images taken at different positions are processed to produce a single reduced depth of field image; hence, it is not always efficient.

In this section, we use our new metric for defocused image blurring to design a pupil phase plate to reduce the depth of field, thereby increasing the axial resolution, of an incoherent hybrid imaging system with a rectangular aperture. By introducing this phase plate at the exit pupil and digitally processing the output of the detector, the depth of field is reduced by more than a factor of two.

5.4.1 Design of a Rectangular RDF Phase Plate

1-D Defocused PSF Similar to our approach to extend the depth of field, we reduce the depth of field by introducing a phase plate at the exit pupil of the imaging system and digitally processing the intermediate optical image. Following our analysis in Sect. 5.3.1, the 1-D PSF of a defocused paraxial imaging system with a phase plate, $f(x)$, at its exit pupil can be written as

$$|h(u, w)|^2 = \left| \int_{-1}^{1} \exp\left[jk\left(\psi_x x^2 - f(x) - \frac{ux_{max}x}{z_i} \right) \right] dx \right|^2 , \qquad (5.36)$$

where $\psi_x = kw_x$ is the defocus parameter in the x direction.

Condition for Reduced Depth of Field For a reduced depth of field, we seek a phase plate which results in maximum image blurring at a slightly defocused plane that is specified by a relatively small defocus parameter value. To obtain our desired phase plate, $f(x)$, which reduces the depth of field, we substitute (5.36) into (5.2) and solve the optimization problem,

$$\min_{f} \frac{\left\langle |h(u, 0)|^2, |h(u, \psi)|^2 \right\rangle}{|||h(u, 0)|^2||, |||h(u, \psi)|^2||} , \qquad (5.37)$$

for a relatively small value of the defocus parameter ψ_x.

Rectangular RDF phase plate We solve the optimization problem (5.37) for $\psi_x = 1$ by assuming that our desired phase plate, $f(x)$, is a periodic phase grating [4]. We assume $f(x)$ to be a periodic grating so that the PSF of the imaging system is an array of narrowly spaced spots. An array of narrowly spaced spots would change its overall shape considerably, as its spots spread out and overlap due to propagation. As mentioned in Sect. 5.3.1, to avoid any shifting in the location of the image plane, $f(x)$ must not have any focusing power. Thus we represent $f(x)$ by an odd and finite Fourier series,

$$f(x) = \sum_{n=1}^{N} b_n \sin(2\pi n v x). \qquad (5.38)$$

After substituting (5.38) into (5.37), we obtain the optimum values of the fundamental spatial frequency, v, and the coefficients, b_n, numerically by using the method of steepest descent, [31].

The initial value of θ which corresponds to a system with a clear aperture is 0.0601 radians. The optimum value of θ which corresponds to a system with a periodic phase grating, (5.38), and with a number of coefficients $N = 5$ is 0.1355 radians. We note that an increase in the number of coefficients from $N = 4$ to $N = 5$ results in a negligible 0.15%, in the optimal value of θ. Thus we restrict the number of phase grating coefficients to $N = 5$. The optimum values of the fundamental frequency, v, and the coefficients, b_n, which correspond to an F/4 imaging system are shown in Table 5.4.

Table 5.4. Rectangular RDF phase grating optimum parameters

$v(cycles/x_{max})$	$b_1(\mu m)$	$b_2(\mu m)$	$b_3(\mu m)$	$b_4(\mu m)$	$b_5(\mu m)$
1.04	1.1705	-0.0437	0.0271	-0.0325	-0.007

Because of mathematical separability, our desired 2-D phase plate to reduce the depth of field, $f(x, y)$, is given by

$$f(x, y) = \sum_{n=1}^{N} b_n \sin(2\pi nvx) + \sum_{n=1}^{N} b_n \sin(2\pi nvy) . \tag{5.39}$$

We refer to $f(x, y)$, whose coefficients are shown in Table 5.4, as the rectangular RDF phase grating. The profile of the rectangular RDF phase grating is shown in Fig. 5.13.

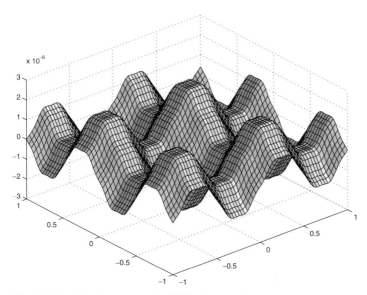

Fig. 5.13. Profile of a rectangular RDF phase grating

5.4.2   Performance of a Rectangular RDF Phase Grating

Defocused PSF using a rectangular RDF phase grating The PSF of a diffraction-limited imaging system, whose parameters are shown in Table 5.1, and with an RDF phase grating, whose parameters are shown in Table 5.4, at its exit pupil is shown in Fig. 5.14, for different values of the defocus parameter ψ. We note that the variation

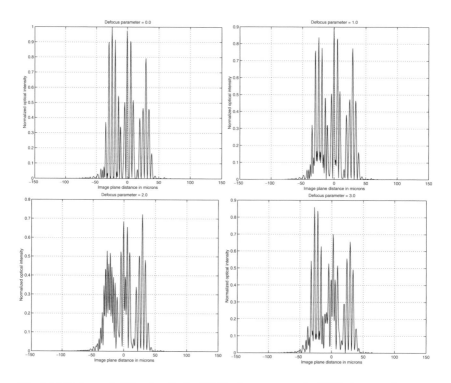

Fig. 5.14. Defocused diffraction-limited PSF using a rectangular RDF phase grating

with defocus in the shape of the PSF shown in Fig. 5.14 is greater than the variation with defocus in the shape of the PSF of a similar standard diffraction-limited system, shown in Fig. 5.15.

In Fig. 5.16, we show the angle in Hilbert space between the in-focus PSF and defocused PSFs of a diffraction-limited standard imaging system, whose parameters are shown in Table 5.1. On the same figure, Fig. 5.16, we also show the angle in Hilbert space between the in-focus PSF and defocused PSFs of the same imaging system, but with an RDF phase plate, whose parameters are shown in Table 5.4, at its exit pupil.

From Fig. 5.16, we note that, for all shown defocus parameter values, the angle in Hilbert space between the in-focus PSF and defocused PSFs of a system with a rectangular RDF phase grating at its exit pupil, has a greater value than the corresponding angle in Hilbert space between the in-focus PSF and defocused PSFs

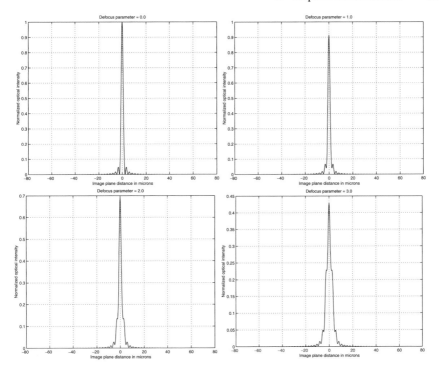

Fig. 5.15. Defocused diffraction-limited PSF using a clear rectangular aperture

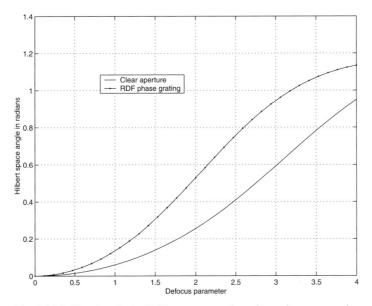

Fig. 5.16. Diffraction-limited Hilbert space angles using a clear rectangular aperture and using a rectangular RDF phase grating

of a standard system. Thus the shape of the PSF of a diffraction-limited imaging system, with a rectangular RDF phase grating at its exit pupil, varies with defocus more than the shape of the PSF of a similar standard diffraction-limited imaging system. Furthermore, we note that, for lower defocus parameter values, the angle in Hilbert space is larger than the corresponding angle between the in-focus PSF and defocused PSFs of a standard system by more than a factor of two. Thus a diffraction-limited system with a rectangular RDF phase grating at its exit pupil has less than half the depth of field of a similar diffraction-limited system.

Defocused OTF Using a Rectangular RDF Phase Grating The OTF of a diffraction-limited imaging system, whose parameters are shown in Table 5.1, and with an RDF phase grating, whose parameters are shown in Table 5.4, at its exit pupil is shown in Fig. 5.17 for different values of the defocus parameter ψ.

We note that there is rapid variation with defocus in the phase of the OTF shown in Fig. 5.17. Thus the variation with defocus in this OTF is greater than the variation with defocus in the OTF of a similar standard diffraction-limited system, shown in Fig. 5.18. We also note that the in-focus OTF shown in Fig. 5.17 has no nulls; hence, there is no loss of spatial frequencies in the image.

Simulated Imaging Example To demonstrate the reduced depth of field, we compare two sets of computer-simulated images of a chirp pattern for different defocus parameter values. On the left column of Fig. 5.19, we show computer-simulated images of a chirp pattern, for different defocus parameter values, that we obtained using an incoherent standard diffraction-limited imaging system, whose parameters are shown in Table 5.1. On the right column of Fig. 5.19, we show computer-simulated images of the same chirp pattern, for different defocus parameter values, that we obtained using a similar imaging system with an RDF phase grating, whose parameters are shown in Table 5.4, at its exit pupil. In each column, the value of the defocus parameter is changed from 0.0 (in-focus) to 3.0.

We obtained these images by using an inverse filter whose frequency response is given by [3]

$$
H_{\text{inverse}}\left(f_x, f_y\right) = \begin{cases} \frac{H_{\text{clear-aperture}}\left(f_x, f_y\right)}{H_{\text{RDF-grt}}\left(f_x, f_y\right)} & : H_{\text{RDF-grt}}\left(f_x, f_y\right) \neq 0 \\ 0 & : H_{\text{RDF-grt}}\left(f_x, f_y\right) = 0 , \end{cases}
\tag{5.40}
$$

where $H_{\text{clear-aperture}}$ is the in-focus OTF of the diffraction-limited imaging system with a clear aperture, without a phase plate at its exit pupil, and $H_{\text{RDF-grt}}$ is the in-focus OTF of the diffraction-limited imaging system with the EDF phase grating at its exit pupil.

From Fig. 5.19, we note that the depth of field of the imaging system with a rectangular RDF phase grating at its exit pupil is reduced, compared to the depth of field of the standard system.

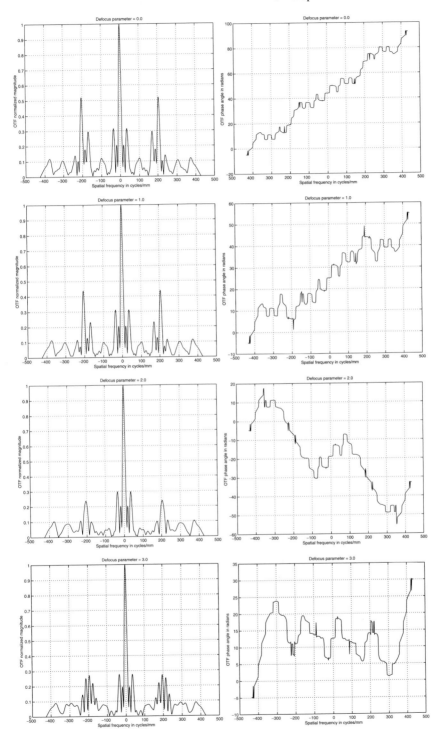

Fig. 5.17. Defocused diffraction-limited OTF using a rectangular RDF phase grating

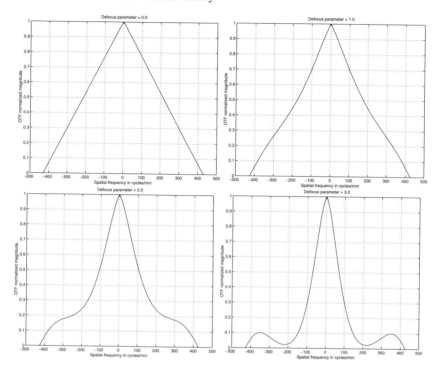

Fig. 5.18. Defocused diffraction-limited OTF using a clear rectangular aperture

5.5   CCD Effect on Depth of Field Control

5.5.1   Charge-Coupled Device-Limited PSF

A charge-coupled device (CCD) detector can be modeled as an array of rectangular pixels. Assuming uniform responsivity across each and every pixel, the 1-D PSF of a defocused CCD-limited imaging system can be written as [32],

$$\left[|h\,(u,w)|^2\right]_{\text{CCD–limited}} = \left\{|h\,(u,w)|^2 * \text{rect}\left(\frac{u}{a}\right)\right\} \text{comb}\left(\frac{u}{u_s}\right) \tag{5.41}$$

where a is the 1-D pixel size, $\text{comb}\left(\frac{u}{u_s}\right)$ is a train of Dirac delta functions with spacing u_s and $*$ is a 1-D convolution operator.

5.5.2   CCD Effect on Depth of Field Extension

In Fig. 5.20, we show the angle in Hilbert space between the in-focus PSF and defocused PSFs of a diffraction-limited imaging system, whose parameters are shown in Table 5.1, with a logarithmic phase plate, whose parameters are shown in Table 5.3, at its exit pupil. On the same figure, Fig. 5.20, we also show the angle in Hilbert

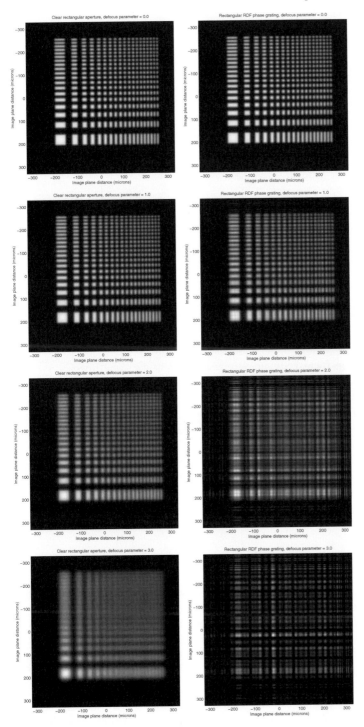

Fig. 5.19. Defocused diffraction-limited images using a clear rectangular aperture and using a rectangular RDF phase grating

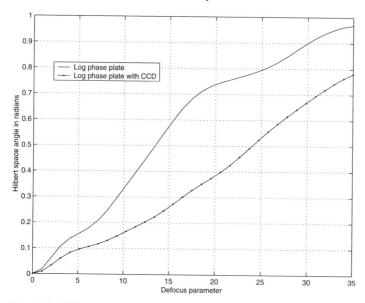

Fig. 5.20. Diffraction-limited and CCD-limited Hilbert space angles using a logarithmic phase plate

space between the in-focus PSF and defocused PSFs of the same imaging system, but with a CCD of pixel size $a = 2.0$ μm.

From Fig. 5.20, we note that, for all shown defocus parameter values, the angle in Hilbert space between the in-focus PSF and defocused PSFs of a CCD-limited system with a logarithmic phase plate at its exit pupil has a smaller value than the corresponding angle in Hilbert space of a similar diffraction-limited system. Thus the PSF of a CCD-limited imaging system varies even less with defocus, compared to the PSF of a similar diffraction-limited imaging system.

In general, the use of a CCD optical detector with a diffraction-limited system, which has a logarithmic phase plate at its exit pupil, helps in extending the depth of field.

5.5.3 CCD Effect on Depth of Field Reduction

In Fig. 5.21, we show the angle in Hilbert space between the in-focus PSF and defocused PSFs of a diffraction-limited imaging system, whose parameters are shown in Table 5.1, with an RDF phase grating, whose parameters are shown in Table 5.4, at its exit pupil. On the same figure, Fig. 5.22, we also show the angle in Hilbert space between the in-focus PSF and defocused PSFs of the same imaging system, but with a CCD of pixel size $a = 1.0$ μm.

From Fig. 5.21, we note that, for all shown defocus parameter values, the angle in Hilbert space between the in-focus PSF and defocused PSFs of a CCD-limited system with an RDF phase grating at its exit pupil has a smaller value than the

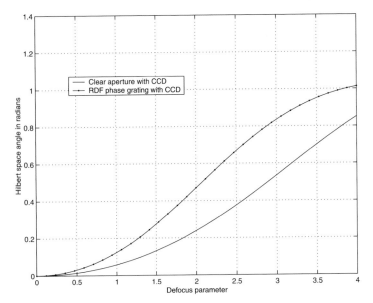

Fig. 5.21. CCD-limited Hilbert space angles using a clear rectangular aperture and a rectangular RDF phase grating

corresponding angle in Hilbert space of a similar diffraction-limited system. Thus the PSF of a CCD-limited imaging system does not vary as much with defocus, compared to the PSF of a similar diffraction-limited imaging system.

In Fig. 5.22, we show the angle in Hilbert space between the in-focus PSF and defocused PSFs of a diffraction-limited imaging system, whose parameters are shown in Table 5.1, with an RDF phase grating, whose parameters are shown in Table 5.4, at its exit pupil. On the same figure, Fig. 5.22, we also show the angle in Hilbert space between the in-focus PSF and defocused PSFs of the same imaging system, but with a CCD of pixel size $a = 1.0$ μm.

From Fig. 5.22, we note that, for lower defocus parameter values, the angle in Hilbert space between the in-focus PSF and defocused PSFs of a system with a rectangular RDF phase grating at its exit pupil, is larger than the corresponding angle in Hilbert space of a standard system by more than a factor of two. Thus a CCD-limited system with a rectangular RDF phase grating at its exit pupil also has less than half the depth of field of a similar standard CCD-limited system.

In general, the use of a CCD optical detector with a diffraction-limited system, which has an RDF phase grating at its exit pupil, has a negative effect on the reduction of the depth of field. However, the use of an RDF phase grating to reduce the depth of field of a CCD-limited imaging system by a factor of 2 is possible.

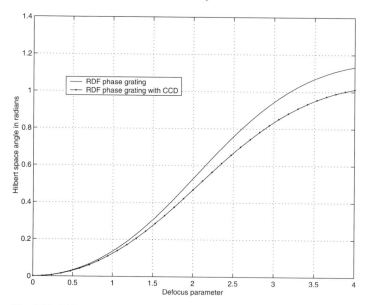

Fig. 5.22. Diffraction-limited and CCD-limited Hilbert space angles using a rectangular RDF phase grating

5.6   Conclusions

We defined a new metric to quantify the blurring of a defocused image that is more suitable than the defocus parameter for hybrid imaging systems.

We described a spatial-domain method to design a pupil phase plate to extend the depth of field of an incoherent hybrid imaging system with a rectangular aperture. We used this method to obtain a pupil phase plate to extend the depth of field, which we referred to as the logarithmic phase plate. By introducing a logarithmic phase plate at the exit pupil and digitally processing the output of the detector, the depth of field was extended by an order of magnitude more than the Hopkins defocus criterion. We compared the performance of the logarithmic phase plate with other extended-depth-of-field phase plates in extending the depth of field of incoherent hybrid imaging systems with rectangular and circular apertures.

We used our new metric for defocused image blurring to design a pupil phase plate to reduce the depth of field, thereby increasing the axial resolution, of an incoherent hybrid imaging systems with a rectangular aperture. By introducing this phase plate at the exit pupil and digitally processing the output of the detector output, the depth of field was reduced by more than a factor of two.

Finally, we examined the effect of using a CCD optical detector, instead of an ideal optical detector, on the control of the depth of field. We found that the use of a CCD with a diffraction-limited system helps in extending the depth of field and the use of a CCD has a negative effect on the reduction of the depth of field. However, the use of an RDF phase grating to reduce the depth of field of a CCD-limited imaging system by a factor of 2 is possible.

Acknowledgment

This material is based upon work supported by, or in part by, the U.S. Army Research Laboratory and the U.S. Army Research Office under contract/grant number DAAD 19-00-1-0514.

References

1. H. H. Hopkins : Proc. Roy. Soc. A **231**, 91 (1955)
2. W. T. Cathey, B. R. Frieden, W. T. Rhodes and C. K. Rushford: J. Opt. Soc. Amer. A **1**, 241 (1984)
3. A. K. Jain: *Fundamentals of Digital Image Processing* (Prentice Hall, New Jersey 1989)
4. S. S. Sherif and W. T. Cathey: Appl. Opt. **41**, 6062 (2002)
5. L. E. Franks: *Signal Theory*, Revised edn. (Dowden and Culver, Stroudsburg 1981)
6. W. T. Welford: J. Opt. Soc. Amer. **50**, 794 (1960)
7. M. Mino and Y. Okano: Appl. Opt. **10**, 2219 (1971)
8. J. Ojeda-Castenada, P. Andres and A. Diaz: Opt. Letters **11**, 478 (1986)
9. J. Ojeda-Castenada, E. Tepichin and A. Diaz: Appl. Opt. **28**, 266 (1989)
10. J. Ojeda-Castenada and L. R. Berriel-Valdos: Appl. Opt. **29**, 994 (1990)
11. G. Hausler: Opt. Comm. **6**, 994 (1972)
12. J. Ojeda-Castenada, R. Ramos and A. Noyola-Isgleas: Appl. Opt. **27**, 2583 (1988)
13. E. R. Dowski and W. T. Cathey: Appl. Opt. **34**, 1859 (1995)
14. W. Chi and N. George: Opt. Letters **26**, 875 (2001)
15. M. Born and E. Wolf: *Principles of Optics*, 6th edn. (Cambridge University Press, Cambridge 1997)
16. J. W. Goodman: *Introduction to Fourier Optics*, 2nd edn. (McGraw-Hill, New York 1996)
17. E. T. Copson: *Asymptotic Expansions* (Cambridge University Press, Cambridge 1967)
18. E. L. Key, E. N. Fowle and R. D. Haggarty: IRE Int. Conv. Rec. **4**, 146 (1961)
19. E. N. Fowle: IEEE Trans. Inf. Theory **10**, 61 (1964)
20. S. S. Sherif, E. R. Dowski and W. T. Cathey: to appear in J. Opt. Soc. Amer. A (2003)
21. Focus Software, Inc.: *Zemax: Optical Design Program, User's Guide* (Tucson, Arizona 2000)
22. K. Brenner, A. Lohmann and J. Ojeda-Castenada: Opt. Comm. **44**, 323 (1983)
23. S. S. Sherif: Depth of Field Control in Incoherent Hybrid Imaging Systems. Ph.D. Dissertation, University of Colorado, Boulder (2002)
24. J. R. Swedlow, J. W. Sedat and D. A. Agard: 'Deconvolution in Optical Microscopy'. In: *Deconvolution of Images and Spectra*, 2nd edn. ed. by P. A. Janssen (Academic Press, San Diego 1997) pp. 284–309
25. A. Erhardt, G. Zinser, D. Komitowski and J. Bille: Appl. Opt. **24**, 194 (1985)
26. C. J. Sheppard and Z. S. Hegedus: J. Opt. Soc. Amer. A **5**, 643 (1988)
27. M. Martinez-Corral, P. Andres, J. Ojeda-Castenada and G. Saavedra: Opt. Comm. **119**, 491 (1995)

28. M. Martinez-Corral, P. Andres, C. Zapata-Rodriguez and M. Kowalczyk: Opt. Comm. **165**, 267 (1999)
29. T. R. M. Sales and G. M. Morris: Opt. Comm. **156**, 227 (1998)
30. M. Neil, R. Juskaitis and T. Wilson: Opt. Lett. **22**, 1905 (1997)
31. G. S. Beveridge and R. S. Schechter: *Optimization: Theory and Practice* (McGraw-Hill, New York 1970)
32. J. E. Greivenkamp and A. E. Lowman: Appl. Opt. **33**, 5029 (1994)

6 Wavefront Coding Fluorescence Microscopy Using High Aperture Lenses

Matthew R. Arnison, Carol J. Cogswell, Colin J. R. Sheppard, Peter Török

6.1 Extended Depth of Field Microscopy

In recent years live cell fluorescence microscopy has become increasingly important in biological and medical studies. This is largely due to new genetic engineering techniques which allow cell features to grow their own fluorescent markers. A popular example is green fluorescent protein. This avoids the need to stain, and thereby kill, a cell specimen before taking fluorescence images, and thus provides a major new method for observing live cell dynamics.

With this new opportunity come new challenges. Because in earlier days the process of staining killed the cells, microscopists could do little additional harm by squashing the preparation to make it flat, thereby making it easier to image with a high resolution, shallow depth of field lens. In modern live cell fluorescence imaging, the specimen may be quite thick (in optical terms). Yet a single 2D image per time-step may still be sufficient for many studies, as long as there is a large depth of field as well as high resolution.

Light is a scarce resource for live cell fluorescence microscopy. To image rapidly changing specimens the microscopist needs to capture images quickly. One of the chief constraints on imaging speed is the light intensity. Increasing the illumination will result in faster acquisition, but can affect specimen behaviour through heating, or reduce fluorescent intensity through photobleaching.

Another major constraint is the depth of field. Working at high resolution gives a very thin plane of focus, leading to the need to constantly "hunt" with the focus knob while viewing thick specimens with rapidly moving or changing features. When recording data, such situations require the time-consuming capture of multiple focal planes, thus making it nearly impossible to perform many live cell studies.

Ideally we would like to achieve the following goals:

- use all available light to acquire images quickly,
- achieve maximum lateral resolution,
- and yet have a large depth of field.

However, such goals are contradictory in a normal microscope.

For a high aperture aplanatic lens, the depth of field is [1]

$$\Delta z = 1.77\lambda \ / \ \left[4\sin^2 \frac{\alpha}{2} \left(1 - \frac{1}{3}\tan^4 \frac{\alpha}{2} \right) \right], \tag{6.1}$$

Török/Kao (Eds.): Optical Imaging and Microscopy, Springer Series in Optical Sciences
Vol. 87 − © Springer-Verlag, Berlin Heidelberg 2003

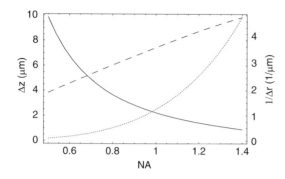

Fig. 6.1. Depth of field (*solid line*), lateral resolution (*dashed line*) and peak intensity at focus (*dotted line* – arbitrary units) for an oil immersion (n_{oil} = 1.518) aplanatic microscope objective with a typical range of NA and λ_0 = 0.53 μm as the vacuum wavelength

where Δz is defined as the distance along the optical axis for which the intensity is more than half the maximum. Here the focal region wavelength is λ and the aperture half-angle is α. A high aperture value for the lateral resolution can be approximated from the full-width at half-maximum (FWHM) of the unpolarised intensity point spread function (PSF) [2]. We can use the same PSF to find the peak intensity at focus, as a rough indication of the high aperture light collection efficiency,

$$I_{focus} \propto \left[1 - \frac{5}{8}(\cos^{\frac{3}{2}} \alpha)(1 + \frac{3}{5} \cos \alpha) \right]^2 . \tag{6.2}$$

These relationships are plotted in Fig. 6.1 for a range of numerical apertures (NA),

$$NA = n_1 \sin \alpha \tag{6.3}$$

where n_1 is the refractive index of the immersion medium. Clearly maximising the depth of field conflicts with the goals of high resolution and light efficiency.

6.1.1   Methods for Extending the Depth of Field

A number of methods have been proposed to work around these limitations and produce an extended depth of field (EDF) microscope.

Before the advent of charge-coupled device (CCD) cameras, Häusler [3] proposed a two step method to extend the depth of focus for incoherent microscopy. First, an axially integrated photographic image is acquired by leaving the camera shutter open while the focus is smoothly changed. The second step is to deconvolve the image with the integration system transfer function. Husler showed that as long as the focus change is more than twice the thickness of the object, the transfer function does not change for parts of the object at different depths – effectively the transfer function is invariant with defocus. The transfer function also has no zeros, providing for easy single-step deconvolution.

This method could be performed easily with a modern microscope, as demonstrated recently by Juškaitis et al. [4]. However, the need to smoothly vary the focus is a time-consuming task requiring some sort of optical displacement within the microscope. This is in conflict with our goal of rapid image acquisition.

A similar approach is to simply image each plane of the specimen, stepping through focus, then construct an EDF image by taking the axial average of the 3D image stack, or some other more sophisticated operation which selects the best focused pixel for each transverse specimen point. This has been described in application to confocal microscopy [5], where the optical sectioning makes the EDF post-processing straightforward. Widefield deconvolution images could also be used. In both cases the focal scanning and multiple plane image capture are major limitations on overall acquisition speed.

Potuluri et al. [6] have demonstrated the use of rotational shear interferometry with a conventional widefield transmission microscope. This technique, using incoherent light, adds significant complexity, and reduces the signal-to-noise ratio (SNR). However the authors claim an effectively infinite depth of field. The main practical limit on the depth of field is the change in magnification with depth (perspective projection) and the rapid drop in image contrast away from the imaging lens focal plane.

Another approach is to use a pupil mask to increase the depth of field, combined with digital image restoration. This creates a digital–optical microscope system. Designing with such a combination in mind allows additional capabilities not possible with a purely optical system. We can think of the pupil as encoding the optical wavefront, so that digital restoration can decode a final image, which gives us the term *wavefront coding*.

In general a pupil mask will be some complex function of amplitude and phase. The function might be smoothly varying, and therefore usable over a range of wavelengths. Or it might be discontinuous in step sizes that depend on the wavelength, such as a binary phase mask.

Many articles have explored the use of amplitude pupil masks [7–9], including for high aperture systems [10]. These can be effective at increasing the depth of field, but they do tend to reduce dramatically the light throughput of the pupil. This poses a major problem for low light fluorescence microscopy.

Wilson et al. [11] have designed a system which combines an annulus with a binary phase mask. The phase mask places most of the input beam power into the transmitting part of the annular pupil, which gives a large boost in light throughput compared to using the annulus alone. This combination gives a ten times increase in depth of field. The EDF image is laterally scanned in x and y, and then deconvolution is applied as a post-processing step.

Binary phase masks are popular in lithography where the wavelength can be fixed. However, in widefield microscopy any optical component that depends on a certain wavelength imposes serious restrictions. In epi-fluorescence, the incident and excited light both pass through the same lens. Since the incident and excited light are at different wavelengths, any wavelength dependent pupil masks would need to be imaged onto the lens pupil from beyond the beam splitter that separates the incoming and outgoing light paths. This adds significant complexity to the optical design of a widefield microscope.

The system proposed by Wilson et al. [11] is designed for two-photon confocal microscopy. Optical complexity, monochromatic light, and scanning are issues that confocal microscopy needs to deal with anyway, so this method of PSF engineering adds relatively little overhead.

Wavefront coding is an incoherent imaging technique that relies on the use of a smoothly varying phase-only pupil mask, along with digital processing. Two specific functions that have been successful are the cubic [12,13] and logarithmic [14] phase masks, where the phase is a cubic or logarithmic function of distance from the centre of the pupil, in either radial or rectangular co-ordinates. The logarithmic design is investigated in detail in Chap. 5.

The cubic phase mask (CPM) was part of the first generation wavefront coding systems, designed for general EDF imaging. The CPM has since been investigated for use in standard (low aperture) microscopy [15]. The mask can give a ten times increase in the depth of field without loss of transverse resolution.

Converting a standard widefield microscope to a wavefront coding system is straightforward. The phase mask is simply placed in the back pupil of the microscope objective. The digital restoration is a simple single-step deconvolution, which can operate at video rates. Once a phase mask is chosen to match a lens and application, an appropriate digital inverse filter can be designed by measuring the PSF. The resulting optical–digital system is specimen independent.

The main trade off is a lowering of the SNR as compared with normal widefield imaging. The CPM also introduces an imaging artefact where specimen features away from best focus are slightly laterally shifted in the image. This is in addition to a perspective projection due to the imaging geometry, since an EDF image is obtained from a lens at a single position on the optical axis. Finally, as the CPM is a rectangular design, it strongly emphasises spatial frequencies that are aligned with the CCD pixel axes.

High aperture imaging does produce the best lateral resolution, but it also requires more complex theory to model accurately. Yet nearly all of the investigations of EDF techniques reviewed above are low aperture. In this chapter we choose a particular EDF method, wavefront coding with a cubic phase plate, and investigate its experimental and theoretical performance for high aperture microscopy.

6.2   High Aperture Fluorescence Microscopy Imaging

A wavefront coding microscope is a relatively simple modification of a modern microscope. A system overview is shown in Fig. 6.2.

The key optical element in a wavefront coding system is the waveplate. This is a transparent molded plastic disc with a precise aspheric height variation. Placing the waveplate in the back focal plane of a lens introduces a phase aberration designed to create invariance in the optical system against some chosen imaging parameter.

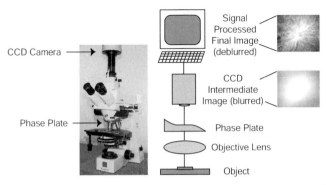

Fig. 6.2. An overview of a wavefront coding microscope system. The image-forming light from the object passes through the objective lens and phase plate and produces an intermediate encoded image on the CCD camera. This blurred image is then digitally filtered (decoded) to produce the extended depth of field result. Examples at right show the fluorescing cell image of Fig. 6.7(c) at each stage of the two-step process. At lower left an arrow shows where the phase plate is inserted into the microscope

A cubic phase function on the waveplate is useful for microscopy, as it makes the low aperture optical transfer function (OTF) insensitive to defocus.

While the optical image produced is quite blurry, it is uniformly blurred over a large range along the optical axis through the specimen (Fig. 6.3). From this blurred intermediate image, we can digitally reconstruct a sharp EDF image, using a measured PSF of the system and a single step deconvolution. The waveplate and digital filter are chosen to match a particular objective lens and imaging mode, with the digital filter further calibrated by the measured PSF. Once these steps are carried out, wavefront coding works well for any typical specimen.

The EDF behaviour relies on modifying the light collection optics only, which is why it can be used in other imaging systems such as photographic cameras, without needing precise control over the illumination light. In epi-fluorescence both the illumination light and the fluorescent light pass through the waveplate. The CPM provides a beneficial effect on the illumination side, by spreading out the axial range of stimulation in the specimen, which will improve the SNR for planes away from best focus.

6.2.1   Experimental Method

The experimental setup followed the system outline shown in Fig. 6.2. We used a Zeiss Axioplan microscope with a Plan Neofluar 40× 1.3 NA oil immersion objective. The wavefront coding plate was a rectangular cubic phase function design (CPM 127-R60 Phase Mask from CDM Optics, Boulder, CO, USA) with a peak to valley phase change of 56.6 waves at 546 nm across a 13 mm diameter optical surface. This plate was placed in a custom mount and inserted into the differential interference contrast slider slot immediately above the objective, and aligned so that it was centred with the optical axis, covering the back pupil.

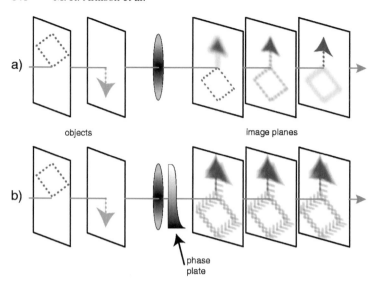

objects   image planes

phase
plate

Fig. 6.3. How points are imaged in standard versus wavefront coding systems: (**a**) Conventional (small depth of field) system with two axially-separated objects to the left of a lens. Because each object obtains best focus at a different image plane, the arrow object points decrease in diameter toward their plane of best focus (far right), while the object points of the diamond are increasingly blurred. (**b**) Inserting a CPM phase plate causes points from both objects to be equivalently blurred over the same range of image planes. Signal processing can be applied to any one of these images to remove the constant blur and produce a sharply-focused EDF image

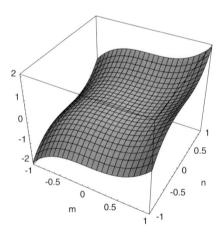

Fig. 6.4. Height variation across the cubic phase mask given in (6.4), for $A = 1$

A custom square aperture mask was inserted into an auxiliary slot 22 mm above the lens, with the square mask cut to fit inside the 10 mm circular pupil of the objective lens. This square mask is needed due to the rectangular nature of the CPM function,

$$\varphi(m, n) = A(m^3 + n^3) \,, \tag{6.4}$$

where m and n are the Cartesian co-ordinates across the pupil and A is the strength of the phase mask (see Fig. 6.4). The square mask was rotated to match the mn axes of the CPM. For comparison, standard widefield fluorescence imaging was performed without the CPM or the square aperture mask in place.

Images were taken in epi-fluorescence mode with a mercury lamp (HBO 50 W) and fluorescein isothiocyanate (FITC) fluorescence filters in place. Images were recorded with a Photometrics cooled camera (CH250) with a Thomson TH 7895 CCD at 12-bit precision. To ensure we were sampling at the maximum resolution of the 1.3 NA lens, a 2.5× eyepiece was inserted just before the camera inside a custom camera mount tube. This tube also allowed precise rotational alignment of the camera, in order to match the CCD pixel array axes with the CPM mn axes.

With 100× total magnification and 19 μm square CCD pixels, this setup gave a resolution of 0.19 μm per pixel. This is just below the theoretical maximum resolution of 0.22 μm for a 1.3 NA lens (see Fig. 6.1), for which critical sampling would be 0.11 μm per pixel, so the results are slightly under sampled.

The PSF was measured using a 1 μm diameter polystyrene bead stained with FITC dye. The peak emission wavelength for FITC is 530 nm. Two dimensional PSF images were taken over a focal range of 10 μm in 1 μm steps. This PSF measurement was used to design an inverse filter to restore the EDF image. The OTF was obtained from the Fourier transform of the 2D PSF.

Each wavefront coding intermediate image was a single exposure on the CCD camera. A least squares filter was incorporated into the inverse filter to suppress noise beyond the spatial frequency cutoff of the optical system. A final wavefront coding image was obtained by applying the inverse filter to a single intermediate image.

6.2.2   PSF and OTF Results

The measured PSFs and derived OTFs for the focused and 4 μm defocused cases are shown in Fig. 6.5, comparing standard widefield microscopy with wavefront coding using a CPM.

The widefield PSF shows dramatic change with defocus as expected for a high aperture image of a 1 μm bead. But the wavefront coding PSF shows very little change after being defocused by the same amount.

The OTF measurements emphasise this focus independence for the wavefront coding system. While the in-focus OTF for the widefield system has the best overall response, the OTF quickly drops after defocusing. The widefield defocused OTF also has many nulls before the spatial frequency cutoff, indicated in these results by a downward spike. These nulls make it impossible in widefield to use the most straightforward method of deconvolution – division of the image by the system OTF in Fourier space. Time consuming iterative solutions must be used instead.

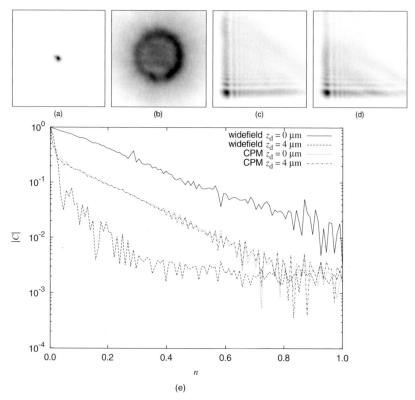

(a)   (b)   (c)   (d)

(e)

Fig. 6.5. Experimental PSFs and OTFs for the widefield and wavefront coding systems as measured using a 1 μm fluorescent bead and a NA = 1.3 oil objective. For each type of microscope, a PSF from the plane of best focus is followed by one with 4 μm defocus. The upper images (**a-d**) show the intensity of a central region of the PSF whilst the lower graph (**e**) gives the magnitude of the OTF for a line $m = 0$ through the OTF for each case: (**a**) widefield $z_d = 0$ μm (*solid line*), (**b**) widefield defocused $z_d = 4$ μm (*dashed line*), (**c**) CPM $z_d = 0$ μm (*dotted line*), (**d**) CPM defocused $z_d = 4$ μm (*dash-dotted line*). The spatial frequency n has been been normalised so that $n = 1$ lies at the CCD camera spatial frequency cutoff. The PSFs have area 13 μm × 13 μm

The wavefront coding system OTF shows a reduced SNR compared with the in-focus widefield OTF. Yet the same SNR is maintained through a wide change in focus, indicating a depth of field at least 8 times higher than the widefield system. The CPM frequency response extends to 80% of the spatial frequency cutoff of the widefield case before descending into the noise floor. This indicates that the wavefront coding system has maintained much of the transverse resolution expected from the high aperture lens used. Because there are no nulls in the CPM OTF at spatial frequencies below the SNR imposed cutoff, deconvolution can be performed using a single-pass inverse filter based on the reciprocal of the system OTF.

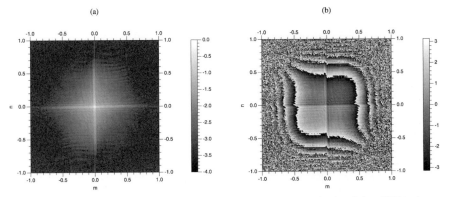

Fig. 6.6. The measured CPM in-focus 2D OTF: (**a**) is the magnitude of the OTF in $\log_{10}$ scale, and (**b**) is the wrapped OTF phase in radians. The spatial frequencies m and n have been been normalised so that $|m|, |n| = 1$ lies at the CCD camera spatial frequency cutoff

A limiting factor on the SNR, and therefore the wavefront coding system resolution, is the CCD camera dynamic range of 12 bits, giving a noise floor of at least 2.4×10^{-4}. From Fig. 6.5(e) the effective noise floor seems to be a bit higher at 10^{-3}. This has a greater impact on the off-axis spatial frequencies, where a higher SNR is required to maintain high spatial frequency response, an effect which is clearly seen in the measured 2D OTF in Fig. 6.6.

6.2.3   Biological Imaging Results

In order to experimentally test high resolution biological imaging using the CPM wavefront coding system in epi-fluorescence, we imaged an anti-tubulin / FITC-labeled HeLa cell. For comparison, we also imaged the same mitotic nucleus in both a standard widefield fluorescence microscope and a confocal laser scanning system (Fig. 6.7). The first widefield image, Fig. 6.7(a), shows a mitotic nucleus with one centriole in sharp focus, while a second centriole higher in the specimen is blurred. This feature became sharp when the focus was altered by 6 μm, as shown in Fig. 6.7(b). The wavefront coding system image in Fig. 6.7(c) shows a much greater depth of field, with both centrioles in focus in the same image. We observed a depth of field increase of at least 6 times compared with the widefield system, giving a 6 μm depth of field for the wavefront coding system for the NA = 1.3 oil objective.

For further comparison, we imaged the same specimen using a confocal microscope. A simulated EDF image is shown in Fig. 6.7(d), obtained by averaging 24 planes of focus. This gives an image of similar quality to the wavefront coding image. However, the confocal system took over 20 times longer to acquire the data for this image, due to the need to scan the image point in all three dimensions. There is also a change in projection geometry between the two systems. The confocal EDF image has orthogonal projection, whereas the wavefront coding EDF image has perspective projection.

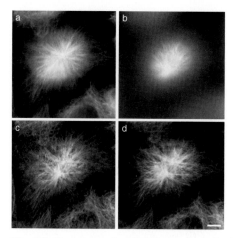

Fig. 6.7. Comparison images of an antitubulin / FITC-labeled HeLa cell nucleus obtained using three kinds of microscope. (**a-b**) Conventional widefield fluorescence images of the same mitotic nucleus acquired at two different focal planes, 6 μm apart in depth. Misfocus blurring is prevalent, with only one of the two centrioles in focus in each image. (**c**) A CPM wavefront coding image of this nucleus greatly increases focal depth so that now both centrioles in the mitotic spindle are sharply focused. (**d**) An equivalent confocal fluorescence EDF image obtained by averaging 24 separate planes of focus, spaced 0.5 μm apart. The resolutions of the wavefront coding and confocal images are comparable but the confocal image took over 20 times longer to produce. Note that wavefront coding gives a perspective projection and confocal gives an isometric projection, which chiefly accounts for their slight difference in appearance. Objective NA=1.3 oil, scale bar: 6 μm

6.3   Wavefront Coding Theory

In this section we will investigate theoretical models for wavefront coding microscopy. We present a summary of the development of the cubic phase function and the paraxial theory initially used to model it. We then analyse the system using vectorial high aperture theory, as is normally required for accuracy with a 1.3 NA lens.

High aperture vectorial models of the PSF for a fluorescence microscope are well developed [16,17]. The Fourier space equivalent, the OTF, also has a long history [18–20]. However, the CPM defined in (6.4) is an unusual microscope element:

1. Microscope optics usually have radial symmetry around the optical axis, which the CPM does not.
2. The CPM gives a very large phase aberration of up to 60 waves, whilst most aberration models are oriented towards phase strengths on the order of a wave at most.
3. In addition, the CPM spreads the light over a very long focal range, whilst most PSF calculations can assume the energy drops off very rapidly away from focus.

These peculiarities have meant we needed to take particular care with numerical computation in order to ensure accuracy, and in the case of the OTF modeling the radial asymmetry has motivated a reformulation of previous symmetric OTF theory.

6.3.1   Derivation of the Cubic Phase Function

There are various methods that may be used to derive a pupil phase function which has the desired characteristics for EDF imaging. The general form of a phase function in Cartesian co-ordinates is

$$T(m,n) = \exp\left[ik\varphi(m,n)\right] , \tag{6.5}$$

where m, n are the lateral pupil co-ordinates and $k = 2\pi/\lambda$ is the wave-number. The cubic phase function was found by Dowski and Cathey [13] using paraxial optics theory by assuming the desired phase function is a simple 1D function of the form

$$\varphi(m) = Am^\gamma, \ \gamma \neq \{0, 1\}, \ A \neq 0 . \tag{6.6}$$

By searching for the values of A and γ which give an OTF which does not change through focus, they found, using the stationary phase approximation and the ambiguity function, that the best solution was for $A \gg 20/k$ and $\gamma = 3$. Multiplying out to 2D, this gives the cubic phase function in (6.4).

6.3.2   Paraxial Model

Using the Fraunhofer approximation, as suitable for low NA, we can write down a 1D pupil transmission function encompassing the effects of cubic phase (6.4) and defocus,

$$T(m) = \exp\left[ik\varphi(m)\right] \exp\left(im^2\psi\right) , \tag{6.7}$$

where ψ is a defocus parameter. We then find the 1D PSF is

$$E(x) = \int_{-1}^{1} T(m) \exp(ixm) \, dm , \tag{6.8}$$

where x is the lateral co-ordinate in the PSF . The 1D OTF is

$$C(m) = \int_{-1}^{1} T\left(m' + m/2\right) T^*\left(m' - m/2\right) dm' . \tag{6.9}$$

The 2D PSF is simply $E(x)E(y)$.

Naturally this 1D CPM gives behaviour in which, in low aperture systems at least, the lateral x and y imaging axes are independent of each other. This gives significant speed boosts in digital post-processing. Another important property of the CPM is that the OTF does not reach zero below the spatial frequency cutoff which means that deconvolution can be carried out in a single step. The lengthy

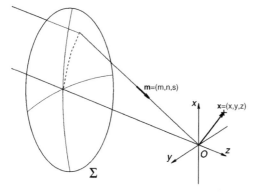

Fig. 6.8. Diagram of the light focusing geometry used in calculating the high NA PSF, indicating the focal region co-ordinate *x* and the pupil co-ordinate *m*, the latter of which may also be thought of as a unit vector aligned with a ray from the pupil to the focal point O

iterative processing of widefield deconvolution is largely due to the many zeros in the conventional defocused OTF. Another important feature of Fraunhofer optics is that PSF changes with defocus are limited to scaling changes. Structural changes in the PSF pattern are not possible.

This paraxial model for the cubic phase mask has been thoroughly verified experimentally for low NA systems [12,15].

6.3.3   High Aperture PSF Model

We now explore the theoretical behaviour for a high NA cubic phase system. Normally we need high aperture theory for accurate modeling of lenses with NA > 0.5. However large aberrations like our cubic phase mask can sometimes overwhelm the high NA aspects of focusing. By comparing the paraxial and high NA model results we can determine the accuracy of the paraxial approximation for particular wavefront coding systems.

The theory of Richards and Wolf [2] describes how to determine the electric field near to the focus of a lens which is illuminated by a plane polarised quasi-monochromatic light wave. Their analysis assumes very large values of the Fresnel number, equivalent to the Debye approximation. We can then write the equation for the vectorial amplitude PSF $E(x)$ of a high NA lens illuminated with a plane polarised wave as the Fourier transform of the complex vectorial pupil function $Q(m)$ [19],

$$E(x) = -\frac{ik}{2\pi} \iiint Q(m) \exp{(ikm \cdot x)} \, dm \; . \tag{6.10}$$

Here $m = (m, n, s)$ is the Cartesian pupil co-ordinate, and $x = (x, y, z)$ is the focal region co-ordinate. The z axis is aligned with the optical axis, and s is the corresponding pupil co-ordinate, as shown in Fig. 6.8. The vectorial pupil function $Q(m)$ describes the effect of a lens on the polarisation of the incident field, the complex value of any amplitude or phase filters across the aperture, and any additional aberration in the lens focusing behaviour from that which produces a perfect spherical wavefront converging on the focal point.

From the Helmholtz equation for a homogeneous medium, assuming constant refractive index in the focal region, we know that the pupil function is only non-zero on the surface of a sphere with radius k,

$$Q(m) = P(m)\,\delta\left(|m| - k^2\right) . \tag{6.11}$$

Because the pupil function only exists on the surface of a sphere, we can slice it along the $s = 0$ plane into a pair of functions

$$Q(m) = Q(m)\,\frac{k}{s}\,\delta\left(s - \sqrt{k^2 - l^2}\right) + Q(m)\,\frac{k}{s}\,\delta\left(s + \sqrt{k^2 - l^2}\right) , \tag{6.12}$$

representing forward and backward propagation [21,22]. Here we have introduced a radial co-ordinate $l = \sqrt{m^2 + n^2}$. Now we take the axial projection $P_+(m, n)$ of the forward propagating component of the pupil function,

$$P_+(m, n) = \int_0^\infty Q(m)\,\frac{k}{s}\,\delta\left(s - \sqrt{k^2 - l^2}\right) ds \tag{6.13}$$

$$= Q(m, n, s_+)\,\frac{1}{s_+} , \tag{6.14}$$

where we have normalised the radius to $k = 1$ and indicated the constraint on s to the surface of the sphere with

$$s_+ = \sqrt{1 - l^2} . \tag{6.15}$$

For incident light which is plane-polarised along the x axis, we can derive a vectorial strength function $a(m, n)$, from the strength factors used in the vectorial point spread function integrals [2,22,23]

$$a(m, n) = \begin{pmatrix} (m^2 s_+ + n^2)/l^2 \\ -mn(1 - s_+)/l^2 \\ -m \end{pmatrix} \tag{6.16}$$

where we have converted from the spherical polar representation in Richards and Wolf to Cartesian co-ordinates.

We can now model polarisation, apodisation and aperture filtering as amplitude and phase functions over the projected pupil,

$$P_+(m, n) = \frac{1}{s_+}\,a(m, n)S(m, n)T(m, n) \tag{6.17}$$

representing forward propagation only ($\alpha \leq \pi/2$), where $S(m, n)$ is the apodisation function, and $T(m, n)$ is any complex transmission filter applied across the aperture of the lens. T can also be used to model aberrations.

Microscope objectives are usually designed to obey the sine condition, giving aplanatic imaging [24], for which we write the apodisation as

$$S(m, n) = \sqrt{s_+} . \tag{6.18}$$

By applying low angle and scalar approximations, we can derive from (6.17) a paraxial pupil function,

$$P_+(m, n) \cong T(m, n) . \tag{6.19}$$

Returning to the PSF, we have

$$E(x) = -\frac{ik}{2\pi} \iint_\Sigma P_+(m, n) \exp(ikm_+ \cdot x) \, dm \, dn , \tag{6.20}$$

integrated over the projected pupil area Σ. The geometry is shown in Fig. 6.8. We use $m_+ = (m, n, s_+)$ to indicate that m is constrained to the pupil sphere surface.

For a clear circular pupil of aperture half-angle α, the integration area Σ_{circ} is defined by

$$0 \leq l \leq \sin\alpha , \tag{6.21}$$

while for a square pupil which fits inside that circle, the limits on Σ_{sq} are

$$\begin{aligned} |m| &\leq \sin\alpha / \sqrt{2} \\ |n| &\leq \sin\alpha / \sqrt{2} \end{aligned} . \tag{6.22}$$

The transmission function T is unity for a standard widefield system with no aberrations, while for a cubic phase system (6.4) and (6.5) give

$$T_c(m, n) = \exp\left[ikA\left(m^3 + n^3\right)\right] . \tag{6.23}$$

6.3.4   High Aperture OTF Model

A high aperture analysis of the OTF is important, because the OTF has proven to be more useful than the PSF for design and analysis of low aperture wavefront coding systems. For full investigation of the spatial frequency response of a high aperture microscope, we would normally look to the 3D OTF [18–20,25]. We have recently published a method for calculating the 3D OTF suitable for arbitrary pupil filters [21] which can be applied directly to find the OTF for a cubic phase plate. But since an EDF system involves recording a single image at one focal depth, a frequency analysis of the 2D PSF at that focal plane is more appropriate. This can be performed efficiently using a high NA vectorial adaptation of 2D Fourier optics [22].

This adaptation relies on the Fourier projection–slice theorem [26], which states that a slice through real space is equivalent to a projection in Fourier space:

$$f(x, y, 0) \Longleftrightarrow \int F(m, n, s) \, ds \tag{6.24}$$

where $F(m, n, s)$ is the Fourier transform of $f(x, y, z)$. We have already obtained the projected pupil function $P_+(m, n)$ in (6.17). Taking the 2D Fourier transform and applying (6.24) gives the PSF in the focal plane

$$E(x, y, 0) \Longleftrightarrow P_+(m, n) \ . \tag{6.25}$$

Since fluorescence microscopy is incoherent, we then take the intensity and 2D Fourier transform once more to obtain the OTF of that slice of the PSF

$$C(m, n) \Longleftrightarrow |E(x, y, 0)|^2 \ . \tag{6.26}$$

We can implement this approach using 2D fast Fourier transforms to quickly calculate the high aperture vectorial OTF for the focal plane.

6.3.5   Defocused OTF and PSF

To investigate the EDF performance, we need to calculate the defocused OTF. Defocus is an axial shift z_d of the point source being imaged relative to the focal point. By the Fourier shift theorem, a translation z_d of the PSF is equivalent to a linear phase shift in the 3D pupil function,

$$E(x, y, 0 + z_d) \Longleftrightarrow \exp(iksz_d) Q(m, n, s) \ . \tag{6.27}$$

Applying the projection-slice theorem as before gives a modified version of (6.25)

$$E(x, y, z_d) \Longleftrightarrow \int \exp(iksz_d) Q(m, n, s) \, ds \ . \tag{6.28}$$

allowing us to isolate a pupil transmission function that corresponds to a given defocus z_d,

$$T_d(m, n, z_d) = \exp(iks_+ z_d) \ , \tag{6.29}$$

which we incorporate into the projected pupil function $P_+(m, n)$ from (6.17), giving

$$P_+(m, n, z_d) = \frac{1}{s_+} a(m, n) S(m, n) T_d(m, n, z_d) T_c(m, n) \ . \tag{6.30}$$

If we assume a low aperture pupil, we can approximate (6.15) to second order, giving the well known paraxial aberration function for defocus

$$T_d(m, n, z_d) \cong \exp\left(-ikz_d \frac{l^2}{2}\right) \ . \tag{6.31}$$

Finally, using $\mathcal{F}$ to denote a Fourier transform, we write down the full algorithm for calculating the OTF of a transverse slice through the PSF:

$$C(m, n, z_d) = \mathcal{F}^{-1} \left\{ |\mathcal{F}[P_+(m, n, z_d)]|^2 \right\} \ . \tag{6.32}$$

Table 6.1. Optical parameters used for PSF and OTF simulations

Optical parameter	Simulation value
Wavelength	530 nm
Numerical aperture	NA = 1.3 oil
Oil refractive index	$n_1 = 1.518$
Aperture half angle	$\alpha = \pi/3$
Pupil shape	Square
Pupil width	7.1 mm
Cubic phase strength	25.8 waves peak to valley

It is convenient to calculate the defocused PSF using the first step of the same approach:

$$E(x, y, z_d) = \mathcal{F}\left[P_+(m, n, z_d)\right] . \tag{6.33}$$

6.3.6   Simulation Results

We have applied this theoretical model to simulate the wavefront coding experiments described earlier, using the parameters given in Table 6.1. The theoretical assumption that the incident light is plane polarised corresponds to the placement of an analyser in the microscope beam path. This polarisation explains some xy asymmetry in the simulation results.

Due to the large phase variation across the pupil, together with the large defocus distances under investigation, a large number of samples of the cubic phase function were required to ensure accuracy and prevent aliasing. We created a 2D array with 1024^2 samples of the pupil function $P+$ from (6.30) using (6.22) for the aperture cutoff. We then padded this array out to 4096^2 to allow for sufficient sampling of the resulting PSF, before employing the algorithms in (6.33) and (6.32) to calculate the PSF and OTF respectively. Using fast Fourier transforms, each execution of (6.32) took about 8 minutes on a Linux Athlon 1.4 GHz computer with 1 GB of RAM.

The wavefront coding inverse filter for our experiments was derived from the theoretical widefield (no CPM) OTF and the measured CPM OTF. The discrepancy in the focal plane theoretical widefield OTF between the paraxial approximation and our vectorial high aperture calculation is shown in Fig. 6.9(a). We show a similar comparison of the defocused widefield OTF in Fig. 6.9(b). We can see there is a major difference in the predictions of the two models, especially at high frequencies. The discrepancy between the models increases markedly with defocus. This implies that the best deconvolution accuracy will be obtained by using the vectorial widefield OTF when constructing the digital inverse filter for a high aperture system.

We now investigate the simulated behaviour of a CPM system according to our vectorial theory. Figures 6.10 and 6.11 show the vectorial high aperture PSF and OTF for the focal plane with a strong CPM. The defocused $z_d = 4$ µm vectorial CPM OTF (not shown) and the paraxial in-focus and defocused $z_d = 4$ µm CPM PSFs and OTFs (not shown) are all qualitatively similar to the vectorial CPM results shown in Figs. 6.10 and 6.11.

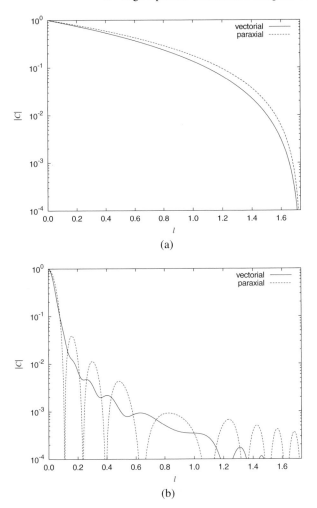

Fig. 6.9. A comparison of widefield (no CPM) OTFs using our vectorial (*solid line*) and paraxial (*dashed line*) simulations: (**a**) in-focus at $z_d = 0$ μm and (**b**) defocused to $z_d = 4$ μm. For a diagonal line through the OTF along $m = n$, we have plotted the value of the 2D projected OTF for each case. While the structure of the in-focus OTF curves is similar for the two models, the relative difference between them increases with spatial frequency, reaching over 130% at the cutoff. Once defocus is applied, the two models predict markedly different frequency response in both structure and amplitude

However, if we perform a quantitative comparison we see that there are marked differences. Figure 6.12 shows the relative strength of the CPM OTF for a diagonal cross section. The differences between the models are similar to the widefield OTF in Fig. 6.9(a) for the in-focus case, with up to 100% difference at high spatial fre-

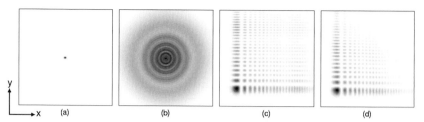

y
x   (a)   (b)   (c)   (d)

Fig. 6.10. The simulated vectorial high aperture PSF for widefield and wavefront coding, showing the effect of defocus: (**a**) widefield in-focus $z_d = 0$ μm, (**b**) widefield defocused $z_d = 4$ μm, (**c**) CPM in-focus $z_d = 0$ μm, (**d**) CPM defocused $z_d = 4$ μm. This amount of defocus introduces very little discernible difference between the CPM PSFs. Indeed paraxial CPM simulations (not shown here) are also similar in structure. The PSFs shown have the same area as Fig. 6.5 (13 μm × 13 μm). The incident polarisation is in the x direction. The images are normalised to the peak intensity of each case. Naturally the peak intensity decreases with defocus, but much less rapidly in the CPM system

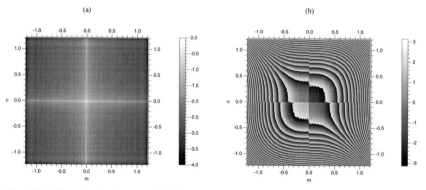

Fig. 6.11. The simulated vectorial high aperture in-focus CPM OTF: (**a**) is the magnitude of the OTF in $\log_{10}$ scale, and (**b**) is the wrapped phase in radians. While the frequency response is much stronger along the m and n axes, the magnitude remains above 10^{-3} throughout the spatial frequency cutoff. The phase of the OTF is very similar to the cubic phase in the pupil. Compensating for the OTF phase is important in digital restoration. The $z_d = 4$ μm defocused OTF (not shown) has a similar appearance to this in-focus case. See Fig. 6.6 to compare with the measured OTFs

quencies. However, as the defocus increases, the structure of the vectorial CPM OTF begins to diverge from the paraxial model, as well as the point where the strength drops below 10^{-4}. This is still a much lower discrepancy than the widefield model for similar amounts of defocus, as is clear by comparison with Fig. 6.9.

These plots allow us to assess the SNR requirements for recording images with maximum spatial frequency response. For both widefield and CPM systems, the experimental dynamic range will place an upper limit on the spatial frequency response. In widefield a 10^3 SNR of will capture nearly all spatial frequencies up to the cutoff (see Fig. 6.9(a)), allowing for good contrast throughout. Further increases

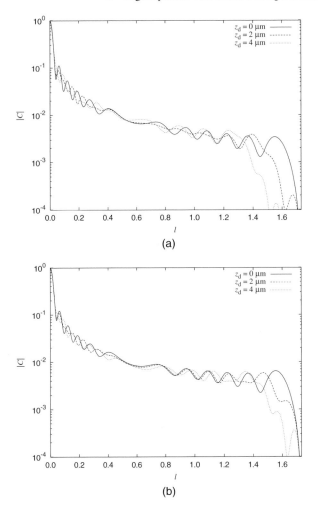

Fig. 6.12. The magnitude of the wavefront coding OTF for the (**a**) vectorial and (**b**) paraxial models, plotted along a diagonal line $m = n$ through the OTF, with different values of defocus: in-focus $z_d = 0$ μm (*solid line*), defocused $z_d = 2$ μm (*dashed line*), defocused $z_d = 4$ μm (*dotted line*). In common with the the widefield system, the models differ the most at high spatial frequencies, up to 300% for the in-focus case. As defocus increases, the differences become more extreme, with the vectorial simulation predicting a quicker reduction in effective cutoff

in SNR will bring rapidly diminishing returns, only gradually increasing the maximum spatial frequency response.

For CPM imaging the same 10^3 SNR will produce good contrast only for low spatial frequencies, with the middle frequencies lying less than a factor of ten above the noise floor, and the upper frequencies dipping below it. However, a SNR of 10^4

will allow a more reasonable contrast level across the entire OTF. For this reason, a 16-bit camera, together with other noise control measures, is needed for a CPM system to achieve the full resolution potential of high aperture lenses. This need for high dynamic range creates a trade off for rapid imaging of living specimens – faster exposure times will reduce the SNR and lower the resolution.

Arguably the most important OTF characteristic used in the EDF digital deconvolution is the phase. As can be seen from Fig. 6.11 the CPM OTF phase oscillates heavily due to the strong cubic phase. This corresponds to the numerous contrast reversals in the PSF. The restoration filter is derived from the OTF, and therefore accurate phase in the OTF is needed to ensure that any contrast reversals are correctly restored.

A comparison of the amount of OTF phase difference between focal planes for the vectorial and paraxial models is shown in Fig. 6.13. We calculated this using the unwrapped phase, obtained by taking samples of the OTF phase along a line $m = n$, then applying a 1D phase unwrapping algorithm to those samples. After finding the unwrapped phases for different focal planes, $z_d = 2$ μm and $z_d = 4$ μm, we then subtracted them from the in focus case at $z_d = 0$ μm.

Ideally the OTF phase difference between planes within the EDF range should be very small. It is clear however that there are some notable changes with defocus. Both paraxial and vectorial models show a linear phase ramp, with oscillations.

This linear phase ramp is predicted by the stationary phase approximation to the 1D CPM OTF, Eq. (A12) in Dowski and Cathey [13]. Since the Fourier transform of a phase ramp is a lateral displacement, this gives a lateral motion of the PSF for different focal planes. In practice this has the effect of giving a slightly warped projection. A mismatch between the microscope OTF and the inverse filter of this sort will simply result in a corresponding offset of image features from that focal plane. Otherwise spatial frequencies should be recovered normally.

The oscillations will have a small effect; they are rapid and not overly large in amplitude: peaking at $\pi/2$ for both vectorial and paraxial models. This will effectively introduce a source of noise between the object and the final recovered image. Whilst these oscillations are not predicted by the stationary phase approximation, they are still evident for the paraxial model.

The most dramatic difference between the two models is in the curvature of the vectorial case, which is particularly striking in the $z_d = 4$ μm plane, and not discernible at all in the paraxial case (Fig.6.13). The primary effect of this curvature will be to introduce some additional blurring of specimen features in the $z_d = 4$ μm plane, which the inverse filter will not be able to correct. The total strength of this curvature at $z_d = 4$ μm is about 2π across the complete $m = n$ line, or one wave, which is a significant aberration.

6.3.7   Discussion

The CPM acts as a strong aberration which appears to dominate both the effects of defocus and of vectorial high aperture focusing. The paraxial approximation certainly loses accuracy for larger values of defocus, but not nearly so much as in the

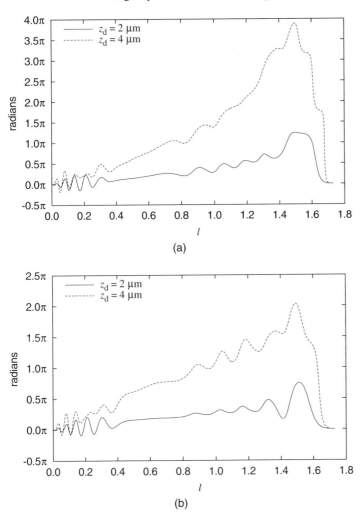

Fig. 6.13. The relative OTF phase angle between focal planes, along a diagonal line $m = n$ through the CPM OTF, for (**a**) the vectorial model, and (**b**) the paraxial model. For both (a) and (b) we show two cases, the unwrapped phase difference between the $z_d = 0$ μm and $z_d = 2$ μm OTF (*solid line*) and the unwrapped phase difference between $z_d = 0$ μm and $z_d = 4$ μm (*dashed line*). All cases show a linear phase ramp with an oscillation of up to $\pi/2$. This phase ramp corresponds to a lateral shift of the PSF. The vectorial case shows an additional curvature and larger overall phase differences of up to 4π radians (or 2 waves) across the spectrum

defocused widefield case. Yet significant differences remain between the two models, notably a one wave curvature aberration in the vectorial case, and this suggests that vectorial high aperture theory will be important in the future design of high aperture wavefront coding systems.

We can also look at the two models as providing an indication of the difference in performance of CPM wavefront coding between low aperture and high aperture systems. The curvature aberration in the high aperture case varies with defocus, which means that it cannot be incorporated into any single-pass 2D digital deconvolution scheme. This effectively introduces an additional blurring of specimen features in planes away from focus, lowering the depth of field boost achieved with the same CPM strength in a low aperture wavefront coding system.

In general the CPM performs a little better at low apertures for EDF applications. But the high aperture CPM system still maintains useful frequency response across the full range of an equivalent widefield system, especially for on-axis frequencies.

6.4   Conclusion

Wavefront coding is a new approach to microscopy. Instead of avoiding aberrations, we deliberately create and exploit them. The aperture of the imaging lens still places fundamental limits on performance. However wavefront coding allows us to trade off those limits between the different parameters we need for a given imaging task. Focal range, signal to noise, mechanical focus scanning speed and maximum frequency response are all negotiable using this digital–optical approach to microscopy.

The high aperture experimental results presented here point to the significant promise of wavefront coding. The theoretical simulations predict an altered behaviour for high apertures, which will become more important with higher SNR imaging systems. For large values of defocus, these results predict a tighter limit on the focal range of EDF imaging than is the case for paraxial systems, as well as additional potential for image artefacts due to aberrations.

The fundamental EDF behaviour remains in force at high apertures, as demonstrated by both experiment and theory. This gives a solid foundation to build on. The CPM was part of the first generation wavefront coding design. Using simulations, new phase mask designs can be tested for performance at high apertures before fabrication. With this knowledge, further development of wavefront coding techniques may be carried out, enhancing its use at high apertures.

Acknowledgments

We would like to thank W. Thomas Cathey and Edward R. Dowski Jr. of CDM Optics Inc, Boulder, CO, USA. Experimental assistance was provided by Eleanor Kable, Theresa Dibbayawan, David Philp and Janey Lin at the University of Sydney and Claude Rosignol at Colorado University. Peter Török acknowledges the partial support from the European Union within the framework of the Future and Emerging Technologies-SLAM program and the EPSRC, UK.

References

1. C. J. R. Sheppard: J. Microsc. **149**, 73 (1988)
2. B. Richards, E. Wolf: Proc. Roy. Soc. A **253**, 358 (1959)
3. G. Häusler: Opt. Commun. **6**, 38 (1972)
4. R. Juškaitis, M. A. A. Neil, F. Massoumian, T. Wilson: 'Strategies for wide-field extended focus microscopy'. In: *Focus on microscopy conference* (Amsterdam, 2001)
5. C. J. R. Sheppard, D. K. Hamilton, I. J. Cox: Proc. R. Soc. Lond. A **A387**, 171 (1983)
6. P. Potuluri, M. Fetterman, D. Brady: Opt. Express **8**, 624 (2001)
7. J. Ojeda-Castañeda, R. Ramos, A. Noyola-Isgleas: Appl. Opt. **27**, 2583 (1988)
8. J. Ojeda-Castañeda, E. Tepichin, A. Diaz: Appl. Opt. **28**, 2666 (1989)
9. W. T. Welford: J. Opt. Soc. Am. **50**, 749 (1960)
10. J. Campos, J. C. Escalera, C. J. R. Sheppard, M. J. Yzuel: J. Mod. Optics **47**, 57 (2000)
11. T. Wilson, M. A. A. Neil, F. Massoumian: 'Point spread functions with extended depth of focus'. In: *Proc. SPIE, volume 4621* (San Jose, CA, 2002) pp. 28–31
12. S. Bradburn, W. T. Cathey, E. R. Dowski, Jr.: App. Opt. **36**, 9157 (1997)
13. E. R. Dowski, Jr., W. T. Cathey: App. Opt. **34**, 1859 (1995)
14. W. Chi, N. George: Opt. Lett. **26**, 875 (2001)
15. S. C. Tucker, W. T. Cathey, E. R. Dowski, Jr.: Optics Express **4**, 467 (1999)
16. P. D. Higdon, P. Török, T. Wilson: J. Microsc. **193**, 127 (1999)
17. P. Török, P. Varga, Z. Laczik, G. R. Booker: J. Opt. Soc. Am. A **12**, (325) (1995)
18. B. R. Frieden: J. Opt. Soc. Am. **57**, 56 (1967)
19. C. W. McCutchen: J. Opt. Soc. Am. **54**, 240 (1964)
20. C. J. R. Sheppard, M. Gu, Y. Kawata, S. Kawata: J. Opt. Soc. Am. A **11**, 593 (1994)
21. M. R. Arnison, C. J. R. Sheppard: Opt. Commun. **211**, 53 (2002)
22. C. J. R. Sheppard, K. G. Larkin: Optik **107** 79 (1997)
23. M. Mansuripur: J. Opt. Soc. Am. A **3**, 2086 (1986)
24. H. H. Hopkins: Proc. Phys. Soc. **55**, 116 (1943)
25. C. J. R. Sheppard, C. J. Cogswell: 'Three-dimensional imaging in confocal microscopy'. In: *Confocal microscopy* (Academic Press, London 1990) pp. 143–169.
26. R. N. Bracewell: *Two-dimensional imaging* (Prentice Hall, Englewood Cliffs, NJ 1995)

Part II

Nonlinear Techniques in Optical Imaging

7 Nonlinear Optical Microscopy

François Lagugné Labarthet and Yuen Ron Shen

7.1 Introduction

The constant evolution of optical microscopy over the past century has been driven by the desire to improve the spatial resolution and image contrast with the goal to achieve a better characterization of smaller specimens. Numerous techniques such as confocal, dark-field, phase-contrast, Brewster angle and polarization microscopies have emerged as improvement of conventional optical microscopy. Being a pure imaging tool, conventional optical microscopy suffers from its low physical and chemical specificity. This can be remedied by combining it with spectroscopic technique like fluorescence, infrared or Raman spectroscopy. Such microscopes have been successfully applied to the study of a wide range of materials with good spectral resolution. However their spatial resolution is restricted by the diffraction limit imposed by the wavelength of the probe light. In infrared microscopy, for instance, the lateral resolution is a few microns which is insufficient to resolve sub-micron structures. Conventional microscopy also does not provide microscopic information about the real surface structure of a specimen. Even in reflection geometry, they can only probe the structure of a surface layer averaged over a thickness of a reduced wavelength. Furthermore, they are insensitive to the polar organization of molecules in the layer although this could be important. In biophysics, for example, it is interesting to know the polar orientation of molecules adsorbed on a membrane and its influence on the membrane physiology.

In this context, nonlinear optical measurements used in conjunction with microscopy observation have created new opportunities [1]. Second-order nonlinear processes such as second harmonic (SH) or sum frequency generation (SFG) are highly surface-specific in media with inversion symmetry and uniquely suited for *in-situ* real-time investigation of surfaces and interfaces [2–4]. With their sub-monolayer surface sensitivity, SHG and SFG microscopies can be used to characterize inhomogeneities, impurities, formation of domains on surfaces or buried interfaces by mapping out the spatial variation of nonlinear susceptibilities at the interfaces. They are also sensitive to the orientation and distribution of molecules useful for evaluation of the structure and reactivity of a surface [5,6]. Third-order processes such as third harmonic generation (THG), coherent anti-Stokes Raman scattering (CARS) and two-photon excited fluorescence (TPEF) microscopies are of interest for the study of buried structures [7–9]. Because the output of second- and third- order processes scales, respectively, with the square and the cube of the excitation intensity, the focal excitation volume is greatly reduced, enhancing the depth resolution and

Török/Kao (Eds.): Optical Imaging and Microscopy, Springer Series in Optical Sciences
Vol. 87 – © Springer-Verlag, Berlin Heidelberg 2003

reducing the out-of-focus background noise. This simple idea has stimulated the development of THG, CARS and TPEF microscopies in an effort to image buried structures in transparent materials or in biological tissues with a three dimensional sectioning capability. Using ultra-short pulses (10^{-12}–10^{-15} s) CARS has also been demonstrated to be a possible technique to probe the interior of living cells in real time [10].

Vibrational spectra are known as fingerprints of molecules. Nonlinear optical microspectrometry finds more interest in the mapping of localized domains via vibrational resonances. Microscopy with infrared (IR)–visible (Vis) sum frequency vibrational spectroscopy allows imaging of molecule-specific domains at a surface. Using a conventional far-field microscope, it can have a spatial resolution one order of magnitude better than Fourier-transform infrared microspectrometry. CARS microscopy also provides vibrational identification of chemical species and is an alternative way to conventional Raman confocal microscopy. Microspectrometry in the electronic transition region is also useful. Multiphoton fluorescence microscopy is becoming a standard technique for biological research [11]. In contrast to one-photon-excited fluorescence, multiphoton-excited fluorescence allows the use of input radiation in the transparent region of a sample and is capable of probing the interior structure of the sample with little laser-induced damages.

This brief introduction to nonlinear optical microscopy would be incomplete without mentioning the recent development in combining nonlinear optical measurements with near-field scanning optical microscopy (NSOM) techniques. In the past several years a growing number of studies of NSOM-SHG/SFG, NSOM-THG and NSOM-TPEF have been reported [12–15]. The spatial resolution of such optical microscopy is limited by the tip apex radius of the optic fiber; it varies from 20 to 200 nm depending on the techniques. In such an experiment, the sample-tip distance is maintained constant by using a feedback mechanism, which then also yields the surface topographical information. The latter can be correlated with the nonlinear NSOM result on specific molecular properties of the sample with a high spatial resolution. With short laser pulses, transient properties of nanostructures can also be probed.

In this chapter, we review a number of nonlinear optical techniques combined with microscope measurements that have been developed in recent years. We will describe separately, SHG, SFG, THG, CARS and multiphoton fluorescence microscopies. For each technique, we will first recall the underlying principle and describe the typical optical setup. We will then focus our interest on several chosen examples taken from the literature. Finally, the resolution limit as well as possible improvements of the various techniques will be discussed.

7.2   Second Harmonic Nonlinear Microscopy

7.2.1   Basic Principle of SHG

We describe here briefly the basic principle of SHG and the extension to the case of a strongly focused fundamental input beam. Details can be found elsewhere [16,17].

SHG originates from a nonlinear polarization $\boldsymbol{P}^{(2)}(2\omega)$ induced in a medium by an incoming field $\boldsymbol{E}(\omega)$:

$$\boldsymbol{P}^{(2)}(2\omega) = \varepsilon_0 \overset{\leftrightarrow}{\chi}^{(2)} : \boldsymbol{E}(\omega)\boldsymbol{E}(\omega), \tag{7.1}$$

where ε_0 is the vacuum permittivity and $\overset{\leftrightarrow}{\chi}^{(2)}$ denotes the nonlinear susceptibility of the medium. If the medium has inversion symmetry, then $\overset{\leftrightarrow}{\chi}^{(2)}$ for the bulk vanishes under the electric dipole approximation. At the surface or interface, however, the inversion symmetry is necessarily broken, and the corresponding $\overset{\leftrightarrow}{\chi}^{(2)}_S$ is non-zero. This indicates that SHG can be highly surface-specific . As a result, SHG has been developed into a useful tool for many surface studies [18,19].

For SHG in reflected direction from a surface or interface, it has been shown that the output signal is given by

$$S(2\omega) = \frac{(2\omega)^2}{8\varepsilon_0 c^3 \cos^2\beta} \left|\chi^{(2)}_{\text{eff}}\right|^2 [I(\omega)]^2 AT, \tag{7.2}$$

where β is the exit angle of the SH output, $I(\omega)$ is the beam intensity at ω, A is the beam area at the surface, T is the input laser pulse-width, and

$$\chi^{(2)}_{\text{eff}} = \left(\overset{\leftrightarrow}{L}_{2\omega} \cdot \hat{\boldsymbol{e}}_{2\omega}\right) \cdot \overset{\leftrightarrow}{\chi}^{(2)}_S : \left(\overset{\leftrightarrow}{L}_\omega \cdot \hat{\boldsymbol{e}}_\omega\right)\left(\overset{\leftrightarrow}{L}_\omega \cdot \hat{\boldsymbol{e}}_\omega\right), \tag{7.3}$$

with $\hat{\boldsymbol{e}}_\omega$ being the unit polarization vector and $\overset{\leftrightarrow}{L}_\omega$ the Fresnel factor at frequency ω. The surface nonlinear susceptibility $\overset{\leftrightarrow}{\chi}^{(2)}_S$ is related to the molecular hyperpolarizability $\overset{\leftrightarrow}{\alpha}^{(2)}$ by a coordinate transformation

$$\left(\chi^{(2)}_S\right)_{ijk} = N_S \sum_{i'j'k'} \left[\alpha^{(2)}_{i'j'k'} \left\langle (\hat{\imath} \cdot \hat{\imath}')(\hat{\jmath} \cdot \hat{\jmath}')(\hat{k} \cdot \hat{k}') \right\rangle \right], \tag{7.4}$$

where i, j, k and i', j', k' stand for the laboratory and molecular coordinate axes, the angular brackets refer to an average over the molecular orientational distribution, and N_S is the surface density of molecules. At a surface of high symmetry, only a few $(\chi^{(2)}_S)_{ijk}$ elements are independent and nonvanishing. They can be determined from SHG measurements with different input/output polarizations. From $(\chi^{(2)}_S)_{ijk}$ together with knowledge of $\alpha^{(2)}_{i'j'k'}$, an approximate orientational distribution of the molecules could be deduced.

From (7.4), it is seen that spatial variation of molecular species, molecular density, and molecular orientation and arrangement can lead to spatial variation of $\overset{\leftrightarrow}{\chi}^{(2)}_S$. Such variation on the micrometer or nanometer scale could be probed by SHG microscopy.

The above theoretical description of SHG assumes the usual plane-wave approximation. However, SHG microscopy often requires a strongly focused input laser beam. Obviously, (7.2) for SHG output has to be modified by a geometric factor to take into account the focusing geometry. Fortunately, this geometric factor should remain unchanged as the focal spot scans over a flat surface, and therefore the spatial variation of the SHG output should still reflect the spatial variation of $|\chi^{(2)}_{\text{eff}}|^2$ over the surface, yielding a surface microscope image.

Geometries for Second Harmonic Far-Field Microscopy The first experiments of second harmonic microscopy were performed in the middle of the seventies by Hellwarth and Christensen [20]. In these experiments, grain structures and defects localized at the surface of thin films of nonlinear materials (ZnSe, GaAs, CdTe) were imaged by the second harmonic signal ($\lambda = 532$ nm) generated in the transmitted direction using a nanosecond pulsed Nd:YAG laser. Boyd and coworkers [21] first demonstrated the possibility of SH microscopy imaging of a surface monolayer. Even with a crude focusing and scanning setup, they were able to observe SHG in reflection, with a spatial resolution of several microns, from a hole burned in a Rhodamine 6G monolayer deposited on a silica substrate. Later, Schultz et al. [22,23] used SH microscopy to study surface diffusion of Sb adsorbates on Ge. They adopted the parallel imaging scheme with a photodiode array. The spatial resolution limited by the pixel size of the 2-D detector was estimated to be $\sim 5\,\mu m$. With the use of a 5 ns and 10 Hz pulsed Nd:YAG laser beam as the pump source, an acquisition time of 5–10 minutes was required.

In recent years, tremendous progress on laser technology, imaging systems, photodetectors and scanning devices has been made. This facilitates the development of SH microscopy. On the other hand, growing interest in surfaces, surface monolayers and thin films has also fostered the interest in SH microscopy [24]. The technique has been applied to the study of a wide range of surfaces and interfaces including Langmuir monolayers at the air-water interface [25], multilayers assemblies [26], self assembled monolayers [27], metallic [28] and semiconductors surfaces [29–31], nonlinear optical materials [32–34], and biomembranes [35–38].

Using a nsec Q-switched Nd:YAG laser and a CCD detector array, Flörsheimer and coworkers has developed SH microscopy in various geometries (Fig. 7.1a–d) and applied it to Langmuir monolayers [5,39–42]. Variation of the polarizations of the input and output beams allows the determination of polar orientation of molecules in microdomains. As examples, the SH microscopy images of a Langmuir monolayer of a 2-docosylamino-5-nitropyridine (DCANP) on water are displayed in Fig. 7.1e and f. They were obtained with the beam geometry described in Fig. 7.1d and the input/output polarization combinations depicted in Fig. 7.1g and h, respectively. A domain size of a few μm can be easily resolved.

In simple cases like 2-docosylamino-5-nitropyridine (DCANP), the nonlinear polarizability of the molecules can be approximated by a single tensor element $\alpha_{zzz}^{(2)}$ along the molecular axis $\hat{z}$. If the molecules are well aligned in a monolayer domain, then the nonlinear susceptibility of the domain is also dominated by a single tensor element $\chi_{Z'Z'Z'}^{(2)}$ along the alignment direction $\hat{Z}'$. In this case, the s-in/s-out polarization combination of SHG probes an effective nonlinear susceptibility

$$\left|\chi_{\text{eff}}^{(2)}\right|_{ss} = \left|F_{ZZ}\chi_{ZZZ}^{(2)}\sin^3\phi\right| = \left|F_{ZZ}\chi_{Z'Z'Z'}^{(2)}\cos^3\theta\sin^3\phi\right|, \tag{7.5}$$

where θ is the tilt angle of $\hat{Z}'$ from the surface plane, ϕ is the angle between the incidence plane and $\hat{Z}$, which is the projection of $\hat{Z}'$ on the surface, and F_{ZZ} is the product of the three relevant Fresnel factors (see (7.3)). The beam geometry of Fig. 7.1g corresponds to such a case.

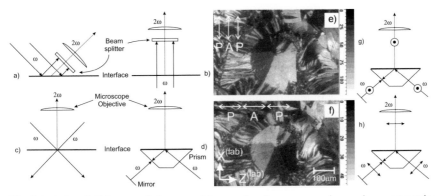

Fig. 7.1. (**a,b,c,d**) Schemes of the second harmonic microscopes using various geometries. (**e,f**) SH micrographs of a DCANP Langmuir film at the surface of water. Image (**e**) is obtained using the *ss* polarization configuration (**g**) while image (**f**) is obtained using the *pp* polarization configuration (**h**) [5]. (Copyright (1999) Wiley-VCH)

For the beam geometry of Fig. 7.1h, the effective nonlinear susceptibility deduced is

$$\left|\chi_{\mathrm{eff}}^{(2)}\right|_{pp} = \left|F_{ZZ}\cos^2\beta_1\cos^3\phi\chi_{ZZZ}^{(2)} + F_{ZY}\sin^2\beta_1\cos\phi\chi_{ZYY}^{(2)}\right|$$
$$= \left|\left(F_{ZZ}\cos^2\beta_1\cos^3\theta\cos^3\phi + F_{ZY}\sin^2\beta_1\cos\theta\sin^2\theta\cos\phi\right)\chi_{Z'Z'Z'}^{(2)}\right|,$$

$$(7.6)$$

where β_1 is the incidence angle of the fundamental input, F_{ZZ} and F_{ZY} are the respective products of Fresnel factors, and $z - y$ defines the incidence plane. It is seen from (7.5) that the Z direction can be determined from the variation of SHG with ϕ with the *s*-in/*s*-out polarization. Then from (7.5) and (7.6), $\chi_{Z'Z'Z'}^{(2)}$ and θ can be deduced if β_1 is known and $(\chi_{\mathrm{eff}}^{(2)})_{ss}$ and $(\chi_{\mathrm{eff}}^{(2)})_{pp}$ are measured. Thus the various domains in Fig. 7.1e and f clearly correspond to domains of different molecular alignments.

SH microscopy has also been used to probe three dimensional structures of biological tissues and defects in materials. In this case, SHG originates from the focal region in the bulk. Using a reflection geometry and femtosecond pulses from a mode-locked dye laser, Guo et al. [43] have obtained SH images of skin tissues. In their setup, a 27× microscope objective was used for both excitation and collection of SHG, and the sample was mounted on a *XYZ* stage for lateral and axial movement, allowing 3-D imaging of a fascia membrane and muscle attached to the biological tissue. Interpretation of the images in term of molecular organization could be improved by analyzing the polarization dependence of the collected signal. In the case of the muscles tissues, *in-vivo* investigations could also be of interest to monitor the physiological change of the fibrils membranes under stress.

Gauderon et al. [44] have used femtosecond pulses from a Ti:sapphire laser to obtain by reflection 3-D SH microscopy images of lithium triborate (LBO) crystal fragments embedded in agar. The experimental arrangement in shown in Fig. 7.2a. The sample on the scanning stage was scanned by the focused laser beam both laterally and axially. A typical image is presented in Fig. 7.2b describing a scanned volume of $70\,\mu m \times 70\,\mu m \times 30\,\mu m$. Terraces, isolated microcrystallites and columnar arrangements of the crystal fragments are clearly observed. The spatial resolution of the image depends on the dimensions of the focal point. Since SHG is proportional to the square of the local fundamental input intensity, the effective point-spread function of the SH output is smaller than that of the input by a $\sqrt{2}$ factor allowing a resolution of $0.61\lambda/(\sqrt{2}NA)$, where λ is the input wavelength and NA denotes the numerical aperture of the focusing lens. With $\lambda=790\,nm$ and NA = 0.8 this lead to a spatial resolution of $0.4\,\mu m$, that is comparable to the maximum resolution obtained in confocal linear optical microscopy. It could be further improved using a confocal pinhole in the detection optics.

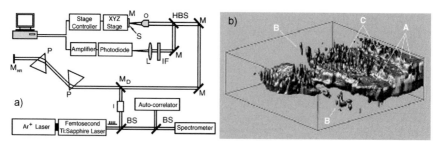

Fig. 7.2. (a) Optical setup used for SH tomographic imaging in the reflectance mode. BS: beam splitter, HBS: harmonic dichroic filter, L: lens, M: mirrors, O: objective, P: prisms, S: sample. **(b)** 3-D second harmonic generation($70\,\mu m \times 70\,\mu m \times 30\,\mu m$) of lithium triborate crystal fragments. Terraces (A), isolated crystallites (B), columnar stacking (C), can be identified [44]. (Copyright (1998) Optical Society of America)

7.2.2   Coherence Effects in SH Microscopy

In SH microscopy, interferences from various features on the surface and/or background can lead to severe distortion of the SH image although, in most studies, such effects have not been taken into account. Parallel (non-scanning) SHG imaging using a 2-D detector array, for example, could suffer from interferences between contributions originating from different sites of the sampled area. On the other hand, despite longer acquisition time, the local nature of a scanning focused fundamental beam reduces interference effects and the morphology interpretation of images is generally more straightforward [45].

However, even if the fundamental beam is tightly focused, the coherent nature of SHG implies that the sample signal can still be distorted by background contribution from sources other than the sample [46]. To cancel a uniform background contribution, one can insert a NLO crystal in the optical pathway [45,47]. By adjusting the crystal orientation properly, it is possible to generate a SH field from the crystal with proper phase and amplitude to cancel the background SH field. Figure

7.3 shows an example. The original SH microscopy image of a damaged area of Si–SiO$_2$ interface has two minima and a central peak. After the background suppression, only the central peak with a slightly larger width is observed. The same background correction scheme may not work with the parallel imaging method if the background contribution is non-uniform over the probed area.

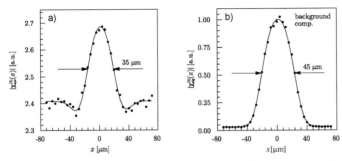

Fig. 7.3. (a) SH signal profile of a damaged area before correction of the background contribution. The step size is 4 μm and the dwell time at each sample spot was 5 s. (b) Damaged area after background compensation [47]

One can utilize the interference effect to obtain phase-contrast microscope images. In a recent experiment, Flörsheimer et al. [48] coherently mix the SH wave generated from a sample (interface) with the reference SH wave from a crystalline quartz plate. The resultant SH wave has a spatial intensity variation depending on $\cos[(\phi_S(r) - \phi_R]$ where ϕ_S and ϕ_R are the phases of the SH waves from the sample and the reference, respectively. The microscope image then reflects the spatial variation of $\phi_S(r)$, which can provide additional information concerning the sample.

7.2.3 Scanning Near-Field Nonlinear Second Harmonic Generation

Far-field SH microscopy still has a spatial resolution limited to μm^2. Incorporation of near-field optics in SH microscopy overcomes the diffraction limit and allows studies of nanosructures. Near-field SHG measurements on rough metallic surfaces [49], ferroelectric domains [50], piezoceramics [51], and LB films [12] have been reported.

The first near-field SHG images were obtained by Bozhevolnyi et al. [12,52]. Their experimental setup (Fig. 7.4a) is a NSOM with a shear-force-based feedback system. The femtosecond fundamental input pulses from a mode-locked Ti:sapphire laser irradiate the sample through a single-mode fiber with an uncoated sharp tip, and both the fundamental and the SH output in the transmitted direction are collected by photodetectors. The fiber tip has its axial position controlled by the tuning fork sensor technique [53]. The scheme then allows the mapping of surface topography with a few nm resolution like that in atomic force microscopy when the sample surface is scanned. The topography can be correlated with the observed spatial variations of the fundamental and SH outputs from the sample. As an example, Fig. 7.4b presents the three images of a LiNbO$_3$ crystal with three small scatterers on it. All the three images display the presence of the unidentified particles at the surface. In the fundamental and SH images, the particles appear as dark spots. The decrease of signals over the particles is presumably due to scattering loss as the tip moves over the particles. The observed lateral resolution of SH-NSOM was around 100–150 nm in the experiments that map the domain edges of a Langmuir–Blodgett film [12,52] and a quasi-phase matching crystal [54]. Similar resolution was obtained by Smolyaninov et al. [51].

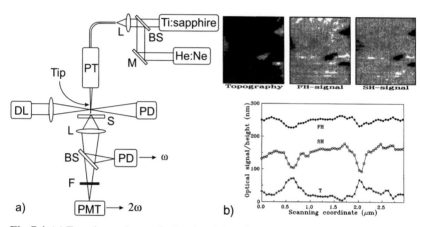

Fig. 7.4. (a) Experimental setup for local probing of second-order susceptibilities with a near-field optical microscope. BS: beam splitters, M: mirror, L: micro-objectives, S: sample, PT: piezo translators, DL: diode laser, PD: photodiodes, F: filter and PMT: photomultiplier tube. The detected second harmonic power versus the pump power squared is shown in the inset. **(b)** Topography, fundamental and SH near-field images ($2.5 \times 3 \ \mu m^2$) of a LiNbO$_3$ crystal with corresponding cross-sections of these images. The images were recorded simultaneously with a total acquisition time of 9 minutes. Reprinted from [52], Copyright (1998), with permission from Elsevier Science

There are many variations of beam geometry in SHG-NSOM. With the fundamental input incident on a sample through a fiber tip, the SH output can be detected either in the transmitted and reflected direction. It is also possible to use a fiber tip to collect near-field SHG from a sample which is broadly illuminated by the fundamental input. One can also have SHG in an evanescent mode and use a metal tip close to the sample surface to induce a SH leak from the surface.

As is true for all NSOM, the theoretical description of SHG-NSOM is difficult. The attempts to model SHG-NSOM have been reported recently [55–57]. The SH signal depends in a complicated way on the polarization of the fundamental field, and multiple scattering of both the fundamental and the SH waves in the medium must be taken into account. Thus, unlike in far-field SHG, little information can be deduced from the polarization dependence of SHG-NSOM. Near-field SHG from mesoscopic structures using either the tip illumination mode [58] or the tip collection mode [59] has been described by a self-consistent integral equation formalism. The calculation shows a strong localization of the SH field around the mesostructures in both lateral and axial dimensions. This strong 3-D localization of SHG is of particular interest to imaging of nanostructures with a NSOM.

7.3 Sum Frequency Generation Microscopy

7.3.1 Basic Principle of Sum Frequency Generation

As a second-order nonlinear optical process, sum frequency generation (SFG) is also highly surface-specific in media with inversion symmetry, and can be used as a surface probe. The IR-Vis SFG has already been developed into a powerful surface vibrational spectroscopic technique [60,61]. In IR-Vis SFG, a visible laser pulse (ω_{Vis}) overlaps with a infrared pulse (ω_{IR}) at a surface and induces a nonlinear polarization $P^{(2)}$ at the sum frequency $\omega_{\mathrm{SF}} = \omega_{\mathrm{Vis}} + \omega_{\mathrm{IR}}$:

$$P^{(2)}(\omega_{\mathrm{SF}} = \omega_{\mathrm{Vis}} + \omega_{\mathrm{IR}}) = \varepsilon_0 \overset{\leftrightarrow}{\chi}{}^{(2)} : E(\omega_{\mathrm{Vis}})E(\omega_{\mathrm{IR}}), \tag{7.7}$$

which radiates and generates the SF output. As in SHG (7.2), the SF output in the reflected direction is given by:

$$S(\omega_{\mathrm{SF}}) \propto \left| \chi_{\mathrm{eff}}^{(2)}(\phi) \right|^2 I(\omega_{\mathrm{Vis}})I(\omega_{\mathrm{IR}})AT, \tag{7.8}$$

with

$$\chi_{\mathrm{eff}}^{(2)}(\phi) = \left(\overset{\leftrightarrow}{L}_{\omega_{\mathrm{SF}}} \cdot \hat{e}_{\omega_{\mathrm{SF}}} \right) \cdot \overset{\leftrightarrow}{\chi}{}_{\mathrm{S}}^{(2)} : \left(\overset{\leftrightarrow}{L}_{\omega_{\mathrm{Vis}}} \cdot \hat{e}_{\omega_{\mathrm{Vis}}} \right)\left(\overset{\leftrightarrow}{L}_{\omega_{\mathrm{IR}}} \cdot \hat{e}_{\omega_{\mathrm{IR}}} \right) \tag{7.9}$$

and $\overset{\leftrightarrow}{\chi}{}_{\mathrm{S}}^{(2)}$ is related to the molecular hyperpolarizability $\overset{\leftrightarrow}{\alpha}{}^{(2)}$ by (7.4). If ω_{IR} is tuned over vibrational resonances, $\overset{\leftrightarrow}{\alpha}{}^{(2)}$ must exhibit corresponding resonance enhancement and can be expressed in the form

$$\overset{\leftrightarrow}{\alpha}{}^{(2)} = \overset{\leftrightarrow}{\alpha}{}_{\mathrm{NR}}^{(2)} + \sum_q \frac{\overset{\leftrightarrow}{a}_q}{(\omega_{\mathrm{IR}} - \omega_q) + i\Gamma_q}, \tag{7.10}$$

where $\overset{\leftrightarrow}{\alpha}{}_{\mathrm{NR}}^{(2)}$ is the non-resonant contribution and $\overset{\leftrightarrow}{a}_q$, ω_q and Γ_q denote the strength, resonant frequency and damping constant of the qth vibrational mode, respectively. Correspondingly, we have

$$\overset{\leftrightarrow}{\chi}{}_{\mathrm{S}}^{(2)} = \overset{\leftrightarrow}{\chi}{}_{\mathrm{NR}}^{(2)} + \sum_q \frac{\overset{\leftrightarrow}{A}_q}{(\omega_{\mathrm{IR}} - \omega_q) + i\Gamma_q}. \tag{7.11}$$

If the infrared input is tunable and scans over vibrational resonances, the resonant enhancement of the SF output naturally yields a surface vibrational spectrum. Such spectra with different input/output polarization combinations can yield information about surface composition and structure in a local surface region. This technique has found many applications in many areas of surface science. However, only a few studies combining SFG vibrational spectroscopy and optical microscope techniques have been reported.

7.3.2   Far-Field SFG Microscopy

Combination of SF vibrational spectroscopy with optical microscopy allows detection or mapping of selective molecular species at a surface or interface. It also has the advantage of an improved spatial resolution in comparison to conventional FTIR microscopy because the resolution is now limited by the SF wavelength instead of the IR wavelength. This is similar to Raman microscopy, but the latter is not surface specific.

Flörsheimer has extended his development of SH microscopy to SF microscopy [5,42]. The experimental arrangement with the input beams in total reflection geometry is depicted in Fig. 7.5a. A Langmuir–Blodgett monolayer film to be imaged is adsorbed on the base surface of a prism. Two input beams enter by the two hypotenuse faces of the prism and the SF output exits from the film surface at an oblique angle.

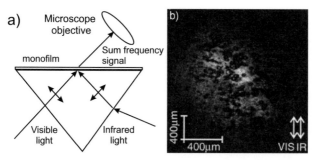

Fig. 7.5. (a) The thin film is deposited on the prism and is illuminated with two p-polarized pulsed beams, ω_{Vis} and ω_{IR}, impinging on the film in the total reflection geometry. The visible frequency, ω_{Vis}, is fixed at 18800 cm^{-1} and the infrared frequency, ω_{IR}, is tunable. No output analyzer is used and the SF signal is found to be mainly p polarized. (b) Sum frequency image of a LB monolayer of arachidic acid. The infrared frequency (2962 cm^{-1}) is tuned to the anti-symmetric stretching mode of methyl groups. Note that the scale bars are not of equal size due to the oblique incidence of the collecting objective. Figure 7.5b is reprinted with permission from [42]. Copyright (1999) American Chemical Society

Figure 7.5b shows an SFG image of a monolayer of arachidic acid molecules obtained with parallel imaging with the IR frequency set at 2962 cm^{-1} in resonance with the CH$_3$ stretching mode. The image exhibits dark holes and clear non-uniformity in the illumination region. Similar experiment with the IR frequency at 2850 cm^{-1} in resonance with the methylene stretch yielded little SF signal. These

results indicate that the alkyl chains of the monolayer are in nearly all-trans configuration. The bright areas in the image in Fig. 7.5b represent regions with a densely
packed molecules in trans conformation. The dark holes likely originate from a lack
of molecules in the areas. They were not observed in freshly prepared LB films, but
appeared and grew with time due to a dewetting or desorption.

In the experimental geometry of Fig. 7.5a, emission of SF output at an oblique
angle could lead to image distortion in the parallel imaging process and deterio-
ration of spatial resolution. The problem could be solved by having the SF output
emitted normal to the surface. Unfortunately this may not be possible with the total-
reflection geometry described in Fig. 7.5a. Matching of wave-vector components
along the surface would require that the visible input beam be transmissive at the
surface.

7.3.3 Near-Field SFG Imaging

Spatial resolution of SF microscopy can be improved to much below micron scale
by near-field detection of the SF output, as demonstrated independently by Shen et
al. [13], Schaller and Saykally [62] and Humbert et al. [63]. However, so far, only
studies of SFG-allowed media have been reported.

The experimental setup of Shen et al. is described in Fig. 7.6a. It allows near-
field imaging of SHG and SFG as well as topography of a sample with a lateral
resolution of about 120 nm. The wavelengths of the two input beams from a nsec
laser are fixed at 1.047 μm and 0.524 μm in the experiment. SHG and SFG from the
sample deposited on the base surface of the prism are generated in the total reflection
mode. The spatial variations of the evanescent waves above the sample surface are
probed via optical tunneling by an Al-coated fiber with a 120 nm aperture attached to
an XYZ piezoelectric scanner. With a shear force feedback to keep the probe-sample
distance constant, the fiber tip also yields the surface topography as the probe scans
the sample surface. Figure 7.6b shows the surface topography as well as the SH
and SF microscopy images of three isolated NNP (N-(4-nitrophenyl)-(L)-prolinol)
nanocrystals of ~360 nm in size that were deposited on the prism surface. The three
images appear to be well correlated.

Schaller and Saykally [62] have performed near-field SFG microscopy with a
tunable IR input. They adopted a commercial NSOM head for collection of the SF
output (Fig. 7.7a). A femtosecond Ti:sapphire laser was used to provide the visible
input and to pump an optical parametric amplifier/difference frequency generator
system to generate the tunable IR input. The microscopy was applied to a thin film
of ZnSe with the IR wavelength varied from 3.1 to 4.4 μm. Figure 7.7b shows the
surface topography and SF images at two IR wavelengths for a 10 μm^2 area of the
sample. The SF images reveals strain patterns that are not observable in the surface
topography. The lateral spatial resolution is estimated to be about 1/20 of the IR
wavelength.

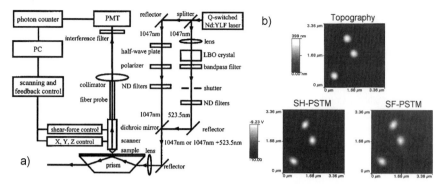

Fig. 7.6. (a) Optical set up used for both SHG and SFG photon scanning tunneling microscopy. (b) Topography, SH and SF images of NNP nanocrystals [13]

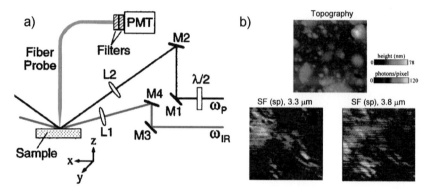

Fig. 7.7. (a) Setup for SFG-NSOM. The uncoated NSOM fiber probe tip (apex ∼50 nm) operates in the collection mode. Infrared light is tunable from 2.8 to 10 μm. The two input beams overlap over a surface of 100 μm². (b) Topographic and SFG images (10 μm²) of ZnSe thin film deposited on a substrate by chemical vapor deposition. Topographic and SH images are simultaneously acquired in 30 min. The spatial resolution is estimated to be 190 nm. Reprinted with permission from [62]. Copyright (2001) American Chemical Society

7.4   Third Harmonic Generation Microscopy

While second-order nonlinear optical processes require a medium without inversion symmetry, third-order processes such as third-harmonic generation are allowed in any medium. The nonlinear polarization induced in a medium responsible for THG is

$$P^{(3)}(3\omega) = \varepsilon_0 \overleftrightarrow{\chi}^{(3)} : E(\omega)E(\omega)E(\omega), \qquad (7.12)$$

where $\overleftrightarrow{\chi}^{(3)}$ is the third-order nonlinear susceptibility for the process. In the plane wave approximation, the TH output is given by

$$S(3\omega) \propto \left| \chi_{\text{eff}}^{(3)}(\phi) \right|^2 I(\omega)I(\omega)I(\omega), \qquad (7.13)$$

with,

$$\chi_{\text{eff}}^{(3)}(\phi) = \left(\overleftrightarrow{L}_{3\omega} \cdot \hat{e}_{3\omega}\right) \cdot \overleftrightarrow{\chi}^{(3)} : \left(\overleftrightarrow{L}_\omega \cdot \hat{e}_\omega\right)\left(\overleftrightarrow{L}_\omega \cdot \hat{e}_\omega\right)\left(\overleftrightarrow{L}_\omega \cdot \hat{e}_\omega\right). \tag{7.14}$$

If the fundamental input beam is strongly focused in the bulk of the medium, then the TH output becomes [64,17]

$$S_{3\omega} \propto P_\omega^3 \left| \int_0^l \frac{\chi^{(3)}(z')\exp(i\Delta k z')dz'}{(1 + 2iz'/b)^2} \right|^2, \tag{7.15}$$

where the integration extends over the length of the medium, P_ω is the fundamental beam power, k_ω and $k_{3\omega}$ are the wave vectors at ω and 3ω, respectively, $b = k_\omega . w_0^2$ is the confocal parameter of the fundamental beam with a waist radius of w_0 and $\Delta k = 3k_\omega - k_{3\omega}$ is the phase mismatch. The integration in (7.15) nearly vanishes if $\Delta k \leq 0$ and $l \gg 1/\Delta k$ but is finite for $\Delta k > 0$ [17]. However, in a medium with normal dispersion, we expect $\Delta k < 0$ and hence a vanishing THG. Since third order nonlinearity away from resonance is generally very weak and tight focusing for THG is often required, this is obviously a shortcoming for using THG as a probe. Nevertheless, Tsang has pointed out that with focusing at an interface, the symmetry of the focusing geometry is broken and (7.15) needs to be modified [65]. Then even with $\Delta k < 0$, THG in the forward direction can be appreciable.

In combination with a transmission optical microscope, THG can be used to image a transparent sample with a 3-D microscopy capability. Indeed, liquid crystal phase transitions [66], semiconductor microstructures [67], optical fiber structures [68,69] and biological samples [7,70] have been investigated with this technique. Recently, Schaller et al. have imaged red blood cells using THG-NSOM [14].

To illustrate the 3-D imaging capability of THG microscopy, we present in Fig. 7.8 the experimental setup and result of Squier et al. [71]. A regeneratively amplified femtosecond Ti:sapphire laser is used to pump an optical parametric amplifier, whose output is focused on the sample for TH imaging. The sample is mounted on an XYZ scanning stage. Figure 8.6 in Chap. 8 shows an example of such an image. The 3-D array of letters inscribed in a glass can be recognized with micron resolution.

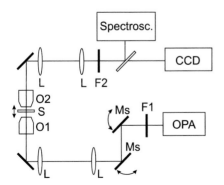

Fig. 7.8. Optical setup used by Squier *et al.* for both writing and reading microdamage patterns in optical glass: OPA: optical parametric amplifier, F1: long pass filter, Ms: scanning mirrors, O1 and O2: objective lenses, F2: blocking filter, M1: insertable mirror [71]. (Copyright (1999) Optical Society of America)

7.5 Coherent Anti-Stokes Raman Scattering Microscopy

Similar to THG, coherent anti-Stokes Raman scattering (CARS) is a third order four-wave mixing process but is resonantly enhanced at vibrational resonances, and is a powerful vibrational spectroscopic tool [72]. The process involves two input waves at frequencies ω_P and ω_S with $\omega_P - \omega_S$ tuned to a vibrational resonance of the medium. The two inputs overlap in the medium and induce a third-order nonlinear polarization in the medium at the anti-Stokes frequency $\omega_{AS} = 2\omega_P - \omega_S$:

$$\boldsymbol{P}^{(3)}_{\omega_{AS}} = \epsilon_0 \overset{\leftrightarrow}{\boldsymbol{\chi}}^{(3)} : \boldsymbol{E}(\omega_P)\boldsymbol{E}(\omega_P)\boldsymbol{E}^*(\omega_S), \tag{7.16}$$

which is the radiation source for CARS. Similar to SFG (7.11), the nonlinear susceptibility can be written as the sum of a non-resonant and a resonant term:

$$\overset{\leftrightarrow}{\boldsymbol{\chi}}^{(3)} = \overset{\leftrightarrow}{\boldsymbol{\chi}}^{(3)}_{NR} + \sum_q \frac{\overset{\leftrightarrow}{\boldsymbol{A}}_q}{(\omega_P - \omega_S) - \omega_q + i\Gamma_q}. \tag{7.17}$$

Under the plane-wave approximation, the CARS output intensity from a uniform sample of thickness d is given by

$$S(\omega_{AS}) \propto \left|\chi^{(3)}_{eff}\right|^2 I_P^2 I_S \sin^2\left(|\Delta \boldsymbol{k}| d/2\right) / |\Delta \boldsymbol{k}|^2, \tag{7.18}$$

where $\Delta \boldsymbol{k} = 2\boldsymbol{k}_P - \boldsymbol{k}_S - \boldsymbol{k}_{AS}$ and

$$\chi^{(3)}_{eff} = \left(\overset{\leftrightarrow}{\boldsymbol{L}}_{\omega_{AS}} \cdot \hat{\boldsymbol{e}}_{\omega_{AS}}\right) \cdot \overset{\leftrightarrow}{\boldsymbol{\chi}}^{(3)} : \left(\overset{\leftrightarrow}{\boldsymbol{L}}_{\omega_1} \cdot \hat{\boldsymbol{e}}_{\omega_1}\right)\left(\overset{\leftrightarrow}{\boldsymbol{L}}_{\omega_1} \cdot \hat{\boldsymbol{e}}_{\omega_1}\right)\left(\overset{\leftrightarrow}{\boldsymbol{L}}_{\omega_2} \cdot \hat{\boldsymbol{e}}_{\omega_2}\right). \tag{7.19}$$

In the case of a strongly focused geometry or a heterogeneous medium, (7.18) needs to be modified, but the characteristic dependence of the CARS output on the phase matching $\Delta \boldsymbol{k}$ is still approximately true. For CARS generated in the forward direction (F-CARS), $|\Delta \boldsymbol{k}|$ is generally much smaller than $1/\lambda_{AS}$. For CARS generated in the backward direction (E-CARS, where E stands for epi-detection), $|\Delta \boldsymbol{k}|$ is larger than $2\pi/\lambda_{AS}$. From a thin sample, F-CARS and E-CARS, both having $|\Delta \boldsymbol{k}|d \ll \pi$, are expected to have nearly the same signal intensity. From a thick sample with $|\Delta \boldsymbol{k}|d < \pi$ for F-CARS but $|\Delta \boldsymbol{k}|d \gg \pi$ for E-CARS, F-CARS should have a much stronger output than E-CARS.

Xie and coworkers [10,73] have studied theoretically the ratio of E-CARS and F-CARS outputs in strongly focusing geometry by varying the sample thickness d. They show that the E-CARS output reaches the relative maximum at $d=0.65\lambda_P$. Incorporation of CARS into optical microscopy provides microspectrometric imaging of materials. Duncan and coworkers first used CARS microscopy with psec laser pulses to image biological cells soaked in deuterated water [74–76]. Asides from being able to selectively map out distributions of molecular species via their vibrational resonances, it also has a number of other advantages compared to, in particular, Raman spectroscopy. (1) Since the CARS intensity (7.18) is proportional to $I_P^2 I_S$, both the lateral and the axial resolution are $\sim \sqrt{3}$ times better than that of linear optical microscopy. For $\omega_P - \omega_S = 3053\,\mathrm{cm}^{-1}$ (CH ring stretching mode), a lateral resolution around 300 nm and an axial resolution around 1.6 µm (even without a confocal pinhole) have been obtained [77]. (2) A moderate laser intensity and power can be used to avoid damaging of biological samples like living cells. (3) The anti-Stokes emission in CARS is spectrally separable from fluorescence as its wavelength is shorter than λ_P. (4) Transparency of the medium to the input waves allows 3-D microscopy and imaging of buried samples.

Recently, Zumbusch et al. [77] and Volkmer et al. [73] have developed F-CARS and E-CARS microscopy in the 2600–3300 cm^{-1} region using femtosecond pulses generated by a Ti:sapphire laser and an associated optical parametric oscillator/amplifier (Fig. 7.9a). The input beams are focused through a 60× oil-immersion objective with NA=1.4. A similar objective was used to collect the signal in both forward and backward directions. The broad spectral width of femtosecond pulses limits the spectral resolution to $\sim 60\,\mathrm{cm}^{-1}$, as seen in Fig. 7.9b. For better spectral resolution longer input pulses with a narrower spectral width are required.

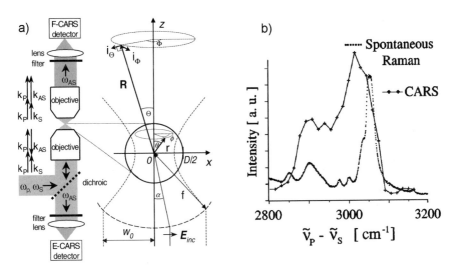

Fig. 7.9. (a) Optical Scheme of the collinear configurations for F- and E-CARS microscopy with collinear pump and Stokes beams and confocal detection in both forward and backward directions [73]. **(b)** CARS and spontaneous Raman spectra of a 910 nm polystyrene bead. For CARS, the pump wavelength (ν_P) was fixed at 854 nm and the Stokes wavelength (ν_S) was tuned from 1.12 to 1.17 µm. The spectral resolution is estimated to 60 cm^{-1} [77]

The 3-D imaging capability of CARS microscopy is shown in Fig. 7.10a where images of polystyrene beads are presented. The Raman transition probed is the CH ring stretch mode of polystyrene at 3038 cm^{-1}. The lateral spatial resolution obtained is around 300 nm.

If the sample is immersed in a liquid, the non-resonant background contribution from the solvent can significantly distort the CARS signal [78]. In such a case, since the overall thickness of the solution with the sample is much larger than the input wavelengths, the background contribution in F-CARS is much larger than in E-CARS. In fact, because of $|\Delta k| \gg 2\pi/\lambda_{AS}$, the background contribution in E-CARS from the solvent is largely suppressed. Figure 7.10b shows an E-CARS image taken at $\omega_P - \omega_S = 1570\,\text{cm}^{-1}$ of a human living cell ($75 \times 75\,\mu\text{m}^2$ in size) in an aqueous solvent. It clearly exhibits many detailed features. The total imaging time was about 8 min.

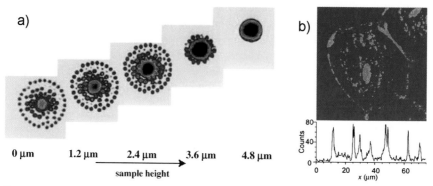

Fig. 7.10. (a) CARS section images ($20 \times 20\,\mu\text{m}^2$) of 910 nm polystyrene beads around a 4.3 μm bead taken at $\omega_P - \omega_S = 3038\,\text{cm}^{-1}$. Sectioning is in the z direction with 1.2 μm increments. The average powers of femtosecond pulses incident on the sample were 120 mW and 50 mW at 855 nm and 1.155 μm, respectively [77]. (**b**) E-CARS image with a Raman shift of 1570 cm^{-1} of an unstained human epithelial cell in an aqueous environment (size: 75×75 μm^2). The average powers of picosecond (f=100 kHz) Stokes (800 nm) and pump (741 nm) pulses were 1 an 2 mW, respectively. The lateral profile of the image displays features as small as 375 nm. Figure 7.10b is reprinted with permission from [10]. Copyright (2001) American Chemical Society

7.6   Multiphoton Excited Fluorescence Microscopy

Two-photon fluorescence microscopy was first developed by Webb and coworkers for applications to biological and medical sciences [9]. Its main asset comes from the 3-D imaging possibility to probe endogenous or exogenous fluorophores [79,80] with a good lateral and axial resolution using either the scanning or the confocal scheme. The technique has brought many new research opportunities to life science and is becoming a routine microscopy tool in many laboratories. Three-photon excited fluorescence microscopy has also been demonstrated. An excellent review on the subject can be found in [81].

Multiphoton-excited fluorescence microscopy offers several advantages over confocal one-photon fluorescence microscopy [82]. First, the pump beam is now in the transparent region that allows deeper probing into the sample [11,83]. The larger difference in wavelength between the pump and the fluorescence makes spectral filtering in signal detection much easier [84,85]. The much weaker absorption of pump radiation significantly reduces thermal distortion and photobleaching hazards on a sample. Second, because of the power dependence on the excitation intensity, the fluorescence originates mainly from the tight focal region. Therefore, sectioning of planes in 3-D imaging is limited only by the focal volume, and a confocal pinhole is often not required to suppress the out-of-focus fluorescence background. Probing volumes smaller than femto-liters have been reported .

Recent advances of multiphoton-excited fluorescence microscopy have been helped by the development of efficient fluorescent dyes with large multiphoton absorption cross-sections [86]. Work has been focused on improvement of the spatial resolution and reduction of the laser power needed. We review here a few most promising techniques and their combination with NSOM.

7.6.1 Two-Photon Excited Fluorescence (TPEF) Microscopy

We consider here both far-field and near-field versions of TPEF. Far-field TPEF is already widely used for characterization of biological samples because of its 3-D imaging capability. Near-field TPEF finds its interest in surface studies of microscope systems that allow simultaneous imaging of the surface topography and the different domains tagged by specific fluorophores.

Far-Field TPEF Microscopy Two-photon-excited fluorescence yield from a sample is proportional to the number of molecules excited per unit time by two-photon absorption and the fluorescence efficiency η. It can be written as [79]

$$F(t) = \eta\sigma_2 \int_V C(r,t)I^2(r,t)\mathrm{d}V, \tag{7.20}$$

where σ_2 is the two photon absorption cross-section, $C(r,t)$ is the density of molecules in the ground state at r and t, and I is the exciting laser intensity. For common chromophores with excitation wavelength ranging from 690 nm to 1050 nm, σ_2 is about 10^{-48} to 10^{-50} cm^{4}s/photon [79]. Owing to the quadratic dependence of TPEF on laser intensity, the fluorescence in TPEF comes from a much reduced volume as compared to one-photon excited fluorescence [87]. Consider a tightly focused gaussian laser beam in a medium. If the beam attenuation in the medium is neglected, the input intensity distribution is given by

$$I(\rho, z) = \frac{2P}{\pi w_0^2(z)} \exp\left(\frac{-2\rho^2}{w_0^2(z)}\right), \tag{7.21}$$

where z is in the propagation direction with z=0 at the center of the focus and

$$w_0(z) = \frac{\lambda}{\pi(\text{NA})} \sqrt{1 + \left(\frac{4\pi(\text{NA})^2 z}{\lambda}\right)^2} \tag{7.22}$$

is the beam radius with λ being the wavelength and NA the numerical aperture. Taking the values NA=0.45 and λ=794 nm, the calculation shows that one-photon fluorescence comes equally from mainly from a 1.9 µm section in depth around the focal region [87]. TPEF microscopy images are often obtained with a scanning microscope. The laser beam is focused to a diffraction-limited beam waist of about 1 µm and is raster-scanned across a specimen. To improve the axial resolution, a confocal pinhole placed in front of the detector can be used as a spatial filter to selectively detect fluorescence from a particular section plane in the focal region. However, owing to the intrinsic sectioning capability of TPEF mentioned earlier, the pinhole is not always required [88]. Because of the small two-photon absorption cross section of fluorophores, focused excitation intensity in the MW cm^{-2} to GW cm^{-2} range is often needed for detectable fluorescence. For fast scanning, CW mode-locked (femtosecond pulsed) lasers with moderate peak power but low mean power are better suited.

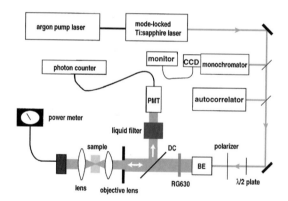

Fig. 7.11. Far-field TPEF microscopy setup. RG630, red-pass filter; BE, 5× beam expander; DC, dichroic mirror; PMT, photomultiplier tube [79]. (Copyright (1996) Optical Society of America)

Typical setup can be found in references [79,87,89]. An example is given in Fig. 7.11. Femtosecond pulses from a Ti:sapphire laser with an 80 MHz repetition rate is focused on a sample with a microscope objective of high NA. TPEF is epi-collected by the same objective and sent either directly to the detector [79,89] or after being spectrally resolved using a polychromator [87]. The 3-D image can be obtained by scanning of the laser focal spot on the sample in the $x - y$ plane with a galvanometer mirror and in the z direction by translation of the microscope objective [89] or the sample mounted on a piezoelectric ceramic. The spatial resolution of TPEF microscopy is similar to that of one-photon excited fluorescence microscopy despite the difference in excitation wavelengths, with typical values of 0.3 µm laterally and

0.9 μm axially [90]. The axial resolution can be improved by up to 50% with the addition of a confocal pinhole [91].

Near-Field TPEF Microscopy Coupling of NSOM microscopy with fluorescence measurement was initially developed for single-molecule spectroscopy imaging [92] or time-resolved studies [93,94]. It has been used to probe single molecules, quantum dots and macromolecules on surfaces with a spatial resolution better than 20 nm [95–98].

Near-field TPEF microscopy has also been developed in recent years. Jenei et al. [99] used picosecond pulses from a Nd:YVO$_4$ laser to induce TPEF from labelling dye in a sample and an uncoated fiber tip for both excitation and collection of the fluorescent signal from the sample. Figure 7.12 shows their setup and the observed surface topography and TPEF image of a labelled mitochondria cell. The spatial resolution obtained was better than 200 nm. Similar resolution was achieved by Lewis et al. using femtosecond pulses from a Ti:sapphire laser to probe individual Rhodamine 6G molecules [100].

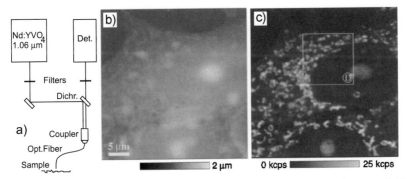

Fig. 7.12. (a) Optical configuration of TPEF combined with a scanning near-field optical microscope. The laser excitation of the sample is through an optical fiber tip which also collects the TPEF signal. The tip-sample distance is regulated by a shear force mechanism. (b) Topography of stained mitochondria cell. (c) TPEF image from the same area. Adapted from [99]. (Copyright (1999) Biophysical Society)

Hell et al. [15] and Kirsch et al. [101] showed that it is also possible to use a CW laser as the excitation source for TPEF microscopy and obtain good microscope images. An example is shown in Fig.7.13, where both far-field and near-field TPFE images of a chromosome stained with a dye are presented. The near-field image appears to have a better spatial resolution (150 nm).

Another variation of TPEF-NSOM has been developed by Sànchez et al. [102]. They used femtosecond laser pulses to illuminate a sample and a sharp gold tip near the surface to locally enhance the optical field at the sample (Fig. 7.14a). Such a local field enhancement can be very large and highly localized. In TPEF-NSOM, it

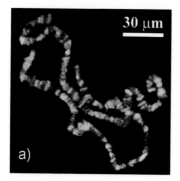 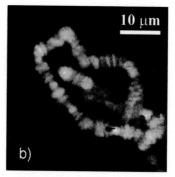

Fig. 7.13. (a) Far-field TPEF microscopy image of a stained chromosome using a CW ArKr laser excitation (λ=647 nm, P=100 mW, acquisition time 4 s). (b) Near-field TPEF microscopy image of the same stained chromosome [15]. (Copyright (1996) Optical Society of America)

strongly enhances the two-photon excitation and hence fluorescence from the sample area under the tip. (The same concept can be applied to other linear and nonlinear optical imaging techniques). Operating in both AFM and NSOM modes, the apparatus allows simultaneous imaging of surface topography and fluorescence microscopy. An example shown in Fig. 7.14b and c reveals features in fragments of photosynthetic membranes as well as J-aggregates with a resolution of 20 nm that roughly corresponds to the tip apex. As pointed out by the authors, the use of a metal tip a few nanometer away from a fluorophore may quench the fluorescence. Therefore, proper fluorophores with rapid energy transfer should be used in such a technique. Kawata et al. have studied theoretically the optical field enhancement from metallic and dielectric tips of various shapes. Laser heating of the tips does not appear to be important. The lateral resolution of such an apertureless near-field microscope is estimated to be in the nanometer range [103].

7.6.2   TPEF Far-Field Microscopy Using Multipoint Excitation

Introduced by Hell et al. [104,105] and Buist et al. [106], parallel microscopy imaging using a two-dimensional microlens array has been proved to be a powerful instrument for fast image acquisition without increasing excitation intensity. This is possible because TPEF is confined to the small focal volumes. The microlens array splits an incident laser beam into beamlets that separately focus into the sample; the number of focal points equals to the number of microlenses. Fluorescent signals from the focal regions are collected by a 2-D photodetection system. Shown in Fig. 7.15 is the apparatus developed by Bewersdorf et al. that permits real-time 3-D imaging with high efficiency and resolution [105]. The expanded beam from a femtosecond laser illuminates a section of microlenses etched on a disk each having a diameter of 460 μm and a focal length of 6 mm. The overall arrangement of the microlenses on the disk form a pattern of spirals with 10 rows. The multiple beamlets pass through intermediate optics and form an array of independent foci

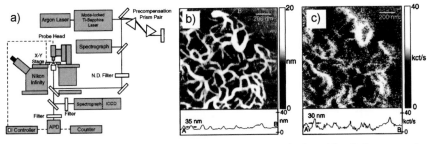

Fig. 7.14. (a) NSOM-TPEF set up with metallic tip excitation. The light source is a Ti:sapphire laser (λ=830 nm, τ=100 fs, f=76 MHz). The beam is sent into an inverted fluorescence microscope, reflected by a dichroic beam splitter and focused by a microscope objective (NA=1.4, 60×) on the sample surface. The metal tip is centered onto the focal spot. The TPEF is collected by the same objective lens and detected either by an avalanche photodiode or analyzed by a spectrometer in conjunction with a CCD camera. Simultaneous topography image (**b**) and near-field TPEF image (**c**) of J-aggregates of pseudoisocyanine dye embedded in polyvinyl sulfate were obtained using the apparatus [102]

on the sample with the help of an objective lens that is also used for collection and mapping of fluorescence from the multiple foci onto a CCD camera. Rotation of the microlens disk allows for scanning of foci in a sample plane. The image acquisition speed depends on the number of lenses arranged in rows and the rotation frequency of the disk. Bewersdorf et al. have used an acquisition speed 40 to 100 times faster than that of a single-beam scanning TPEF, permitting real-time TPEF imaging of living cells. The spatial resolution they obtained was similar to those of conventional TPEF microscope with an axial resolution of about 0.84 μm when an oil immersion objective lens was used.

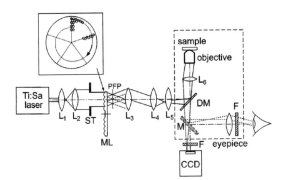

Fig. 7.15. Optical setup of a multifocal multipoint microscope for real-time direct-view TPEF microscopy. L: lenses, ML: microlens disk, M: mirror, DM: dichroic mirror. The inset shows the array of a spiral arrangement of microlenses on the disk [105]. (Copyright (1996) Optical Society of America)

7.6.3 4-Pi Confocal TPEF Microscopy

4-Pi confocal microscopy is 3-D TPEF microscopy technique developed to improve the axial resolution of far-field microscopy [107]. An axial resolution of about

100 nm has been demonstrated [108]. The scheme involves excitation of the sample by counter-propagating beams through two objective lenses with high numerical aperture (Fig. 7.16a). The overall excitation aperture of the system would approach 4π if the aperture extended by each lens were 2π. The acronym of the system 4-Pi was chosen as a reminder of this arrangement.

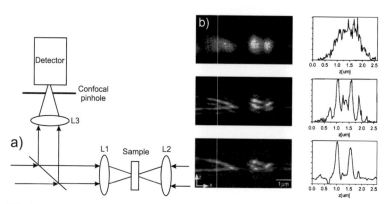

Fig. 7.16. (a) Setup of the TPEF 4-Pi confocal microscope. The numerical aperture of the 100× oil immersion objectives is 1.4 [109]. **(b)** Microscopy images obtained with various schemes: two photon confocal (top), two photon 4-Pi confocal (center), two photon 4-Pi confocal after side-lobe removal (bottom). The right column shows the intensity profiles of the fluorescence along the lines indicated in the respective images [108]. (Copyright (1999) Biophysical Society)

The two counter-propagating beams interfering in the focal region should yield an intensity distribution with a main peak several times narrower in the axial direction than that obtainable by a single focused beam. Furthermore, fluorescence from secondary interference peaks (axial lobes) along the axis in the focal region can be significantly reduced by a confocal pinhole (Fig. 7.16a) and by a deconvolution procedure using a restoration algorithm [109]. Figure 7.16b shows as example axial images of a fibroblast cell using various schemes (confocal, 4-Pi confocal, 4-Pi confocal with axial lobes removed). The excitation source was a mode-locked femtosecond Ti:sapphire. The 4-Pi confocal microscopy image with axial lobes suppressed obviously has the best axial resolution. It reveals details on a scale less than 200 nm.

7.6.4 Simultaneous SHG/TPEF Microscopy

SHG and TPEF can be simultaneously generated from a sample by a focused beam and detected in the same microscopy system. Both processes have an output proportional to the square of the input laser intensity, but one is coherent and the other incoherent. If they are pumped by the same laser input, then, as shown in Fig.

7.17a, both experience the same two-photon resonant excitation. However, SHG is surface-specific in media with inversion symmetry and TPEF is not. Thus the two processes can provide complementary information about the sample. Lewis and coworkers [37,38] and Moreaux et al. [36,89,110], have used combined SHG/TPEF microscopy to study biological membranes. In the setup of Moreaux et al., femtosecond pulses from a Ti:sapphire were used for excitation of a sample through a microscope objective. The SHG output in the forward direction was collected by a lens while TPEF was epi-collected with the help of a dichroic mirror. Three dimensional SHG and TPEF images were obtained by scanning the laser focal spot in the $x - y$ plane with galvanometer mirrors and in the axial direction by translation of the objective. Presented in Figs.7.17(b) and (c) are vesicles images acquired in 1.5 s with an excitation power less than 1 mW [89]. They provide an excellent example of complementarity of SHG and TPEF: the adhesion between the two vesicles appears to be centrosymmetric as it contributes to TPEF but not to SHG.

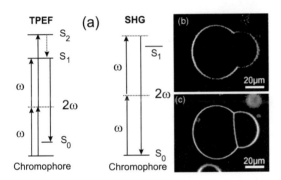

Fig. 7.17. (a) Level diagrams describing two-photon excited fluorescence and second harmonic generation. (b) SHG and (c) TPFE images of labelled vesicles excited by laser input at λ= 880 nm (τ=80 fs, f=80 MHz, P $\leq$1 mW) [89]. (Copyright (1996) Optical Society of America)

7.6.5   Three-Photon-Excited Fluorescence Microscopy

Three-photon-excited fluorescence microscopy has also been demonstrated. The cubic dependence of fluorescence on the local input laser intensity allows further improvement of the spatial resolution as compared to TPEF. For samples such as amino acids, proteins and neurotransmitters that are susceptible to photodamage by one-photon absorption in the ultra-violet and residual absorption in the visible-red range, the three-photon-excited fluorescence scheme with a near-infrared input beam could avoid the damage and seems to be most suitable for microscopy. Using femtosecond Ti:sapphire laser pulses for excitation, Gryczynski et al. successfully imaged the triptophan residues [111] and Maiti et al. imaged neurotransmitters in living cells [85].

A lateral and axial resolution of ~ 0.2 and ~ 0.6 μm has been achieved [80]. The main disadvantage of the technique is the narrow window of operation. Because the three-photon absorption cross-section of a sample is generally very small (10^{-75}– 10^{-84} cm^6 s^2 photon^{-2}), high peak laser intensity is required for excitation that likely causes damage of the sample.

7.6.6 Stimulated-Emission-Depletion (STED) Fluorescence Microscopy

Recently, Klar et al. proposed and demonstrated an interesting idea that can significantly improve the spatial resolution of fluorescence microscopy [112]. It involves fluorescence quenching of excited molecules at the rim of the focal spot through stimulated emission thus significantly reducing the focal volume that emits fluorescence. This is accomplished by passing the quenching laser beam through an optical phase plate, yielding a focused wave front that produces by destructive interference a central minimum at the focal point. In the experiment (Fig. 7.18a), two synchronized

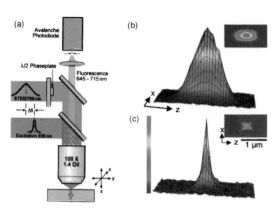

Fig. 7.18. (a) STED microscope setup. Excitation pulse is followed by fluorescence quenching pulse. Fluorescence is then detected after passing through dichroic filters and a confocal pinhole. (b) and (c), Intensity distributions of fluorescence along x and z (axial direction) in confocal and in confocal STED microscopy, respectively. The latter has an axial width of 97 nm which is 5 times narrower than in (b) [112] (Copyright (2000) National Academy of Sciences, U.S.A.)

pulses from a Ti:sapphire laser/optical parametric oscillator system with a 76 Mhz repetition rate were used: a visible pulse with a 0.2 ps pulsewidth for excitation followed by a near-infrared pulse with a 40 ps pulsewidth for fluorescence quenching. The visible pulse at λ=560 nm originated from the optical parametric oscillator with an intracavity frequency doubler and the near-infrared pulse at λ=765 nm came directly from the laser. The fluorescence was epi-collected by an avalanche photodiode after passing through a confocal pinhole. The lateral and axial resolutions obtained were ~100 nm, which is an improvement of a factor 5 in the axial direction and 2 in the lateral one, compared to ordinary confocal microscope (Fig. 7.18(b) and (c)).

Besides the improvement on spatial resolution which compete with most near-field optical microscopes, STED microscopy can also find other important applications in studies of ultrafast dynamics. The ultrafast control of fluorescence under microscopy conditions opens new opportunities for transient microscopy and spectroscopy of nano-objects. It should also have interesting perspectives in single molecule spectroscopy such as control of individual excited molecules with femtosecond time resolution [113].

7.7 Conclusion

In this contribution, we have briefly surveyed the field of nonlinear optical microscopy and presented a few examples of contemporary applications. The survey is by no means complete. The field is still at its infant stage. Further development of the techniques and their applications can be anticipated. As in the past, future advances would benefit from having better lasers, better optics and optical systems and better photodetectors. For example, an efficient high-speed detection and data acquisition system incorporated with an optimized pulse laser excitation scheme would allow *in-situ* probing of time-dependent surface reactions on nanostructures or *in-vivo* study of biological cells. Compared to linear-optics, nonlinear optics has the advantage that it probes a larger domain of material properties. SHG and SFG, for example, are unique in their abilities to probe selectively surfaces and interfaces. However, nonlinear optics also has the disadvantages that it generally requires stronger input laser intensity and the effects are more complex. The latter point is important for interpretation of the observed microscopy images. In particular, a clear understanding of the effects in near-field microscopy is essential for the future progress of nonlinear NSOM.

Acknowledgments
Work was supported by the Director, Office of Energy Research, Office of Basic Energy Sciences, Materials Science Division, of the US Departement of Energy under contract No.DE–AC03–765F00098. F.L.L. gratefully acknowledges the Centre National de la Recherche Scientifique (C.N.R.S., France) for his support during his stay at the University of California at Berkeley.

References

1. P.T.C. So: Opt. Express **3**, 312 (1998)
2. X. Zhuang, P.B. Miranda, D. Kim, Y.R. Shen: Phys. Rev. B **59**, 12632 (1999)
3. X. Wei, S.-C. Hong, X. Zhuang, T. Goto, Y.R. Shen: Phys. Rev. E **62**, 5160 (2000)
4. V. Vogel, Y.R. Shen: Annu. Rev. Mater. Sci. **21**, 515 (1991)
5. M. Flörsheimer: Phys. Status Solidi (a) **173**, 15 (1999)
6. G.J. Simpson: Appl. Spectrosc. **55**, 16 (2001)
7. J.A. Squier, M. Müller, G.J. Brakenhoff, K.R. Wilson: Opt. Express **3**, 315 (1998)
8. E.O. Potma, W.D. de Boej, D.A. Wiersma: J. Opt. Soc. Am. B **17**, 1678 (2000)
9. W. Denk, J.H. Strickler, W.W. Webb: Science **248**, 73 (1990)
10. J.-X. Cheng, A. Volkmer, L.D. Book, X.S. Xie: J. Phys. Chem. B **105**, 1277 (2001)
11. K. König: J. Microsc. – Oxford **200**, 83 (2000)
12. S.I. Bozhevolnyi, T. Geisler: J. Opt. Soc. Am. A **15**, 2156 (1998)
13. Y. Shen, J. Swiatkiewicz, J. Winiarz, P. Markowicz, P.N. Prasad: Appl. Phys. Lett. **77**, 2946 (2000)

14. R.D. Schaller, J.C. Johnson, R.J. Saykally: Anal. Chem. **72**, 5361 (2000)
15. S.W. Hell, M. Booth, S. Wilms, C.M. Schnetter, A.K. Kirsch, D.J. Arndt-Jovin, T.M. Jovin: Opt. Lett. **23**, 1238 (1998)
16. Y.R. Shen: *The Principles of Nonlinear Optics* (Wiley, New York 1984)
17. R. Boyd: *Nonlinear Optics* (Academic, New York 1992)
18. M. Oh-e, S.C. Hong, Y.R. Shen: J. Phys. Chem. B **104**, 7455 (2000)
19. S.-C. Hong, M. Oh-e, X. Zhuang, Y.R. Shen, J.J. Ge, F.W. Harris, S.Z.D. Cheng: Phys. Rev. E **63**, 517061 (2001)
20. R. Hellwarth, P. Christensen: Appl. Optics **14**, 247 (1975)
21. G.T. Boyd, Y.R. Shen, T.W. Hänsch: Opt. Lett. **11**, 97 (1986)
22. K.A. Schultz, E.G. Seebauer: J. Chem. Phys. **97**, 6958 (1992)
23. K.A. Schultz, I.I. Suni, E.G. Seebauer: J. Opt. Soc. Am. B **10**, 546 (1993)
24. M.S. Johal, A.N. Parikh, Y. Lee, J.L. Casson, L. Foster, B.I. Swanson, D.W. McBranch, D.Q. Li, J.M. Robinson: Langmuir **15**, 1275 (1999)
25. N. Kato, K. Saito, Y. Uesu: Thin solid films **335**, 5 (1999)
26. S.B. Bakiamoh, G.J. Blanchard: Langmuir **17**, 3438 (2001)
27. L. Smilowitz, Q.X. Jia, X. Yang, D.Q. Li, D. McBranch, S.J. Buelow, J.M. Robinson: J. Appl. Phys. **81**, 2051 (1997)
28. Y. Sonoda, G. Mizutani, H. Sano, S. Ushioda, T. Sekiya, S. Kurita: Jap. J. Appl. Phys. **39**, L253 (2000)
29. H. Sano, T. Shimizu, G. Mizutani, S. Ushioda: J. Appl. Phys. **87**, 1614 (2000)
30. K. Pedersen, S.I. Bozhevolnyi, J. Arentoft, M. Kristensen, C. Laurent-Lund: J. Appl. Phys. **88**, 3872 (2000)
31. I.I. Suni, E.G. Seebauer: J. Chem. Phys. **100**, 6772 (1994)
32. S. Kurimura, Y. Uesu: J. Appl. Phys. **81**, 369 (1996)
33. F. Rojo, F. Agulló-lópez, B. del Rey, T. Torres: J. Appl. Phys. **84**, 6507 (1998)
34. S.E. Kapphan: J. Lumin. **83**, 411 (1999)
35. I. Freund, M. Deutsch: Opt. Lett. **11**, 94 (1986)
36. L. Moreaux, O. Sandre, S. Charpak, M. Blanchard-Desce, J. Mertz: Biophys. J. **80**, 1568 (2001)
37. P.J. Campagnola, M.-D. Wei, A. Lewis, L.M. Loew: Biophys. J. **77**, 3341 (1999)
38. C. Peleg, A. Lewis, M. Linial, M. Loew: P. Natl. Acad. Sci. USA **96**, 6700 (1999)
39. M. Flörsheimer, D.H. Jundt, H. Looser, K. Sutter, M. Küpfer, P. Günter: Ber. Bunsenges. Phys. Chem. **9**, 521 (1994)
40. M. Flörsheimer, M. Bösch, Ch. Brillert, M. Wierschem, H. Fuchs: J. Vac. Sci. Technol. B **15**, 1564 (1997)
41. M. Flörsheimer, M. Bösch, Ch. Brillert, M. Wierschem, H. Fuchs: Supramol. Sci. **4**, 255 (1997)
42. M. Flörsheimer, C. Brillert, H. Fuchs: Langmuir **15**, 5437 (1999)
43. Y. Guo, P.P. Ho, H. Savage, D. Harris, P. Sacks, S. Schantz, F. Liu, N. Zhadin, R.R. Alfano: Opt. Lett. **22**, 1323 (1997)
44. R. Gauderon, P.B. Lukins, C.J.R. Sheppard: Opt. Lett. **23**, 1209 (1998)
45. M. Cernusca, M. Hofer, G.A. Reider: J. Opt. Soc. Am. B **15**, 2476 (1998)
46. G. Berkovic, Y.R. Shen, G. Marowsky, R. Steinhoff: J. Opt. Soc. Am. B **6**, 205 (1989)
47. G.A. Reider, M. Cernusca, M. Hofer: Appl. Phys. B. – Lasers O. **68**, 343 (1999)
48. P. Rechsteiner, J. Hulliger, M. Flörsheimer: Chem. Mater. **12**, 3296 (2000)
49. I.I. Smolyaninov, A.V. Zayats, C.C. Davis: Phys. Rev. B **56**, 9290 (1997)
50. I.I. Smolyaninov, A.V. Zayats, C.C. Davis: Opt. Lett. **22**, 1592 (1997)
51. I.I. Smolyaninov, C.H. Lee, C.C. Davis: J. Microsc. – Oxford **194**, 426 (1999)

52. S.I. Bozhevolnyi, B. Vohnsen, K. Pedersen: Opt. Commun. **150**, 49 (1998)
53. K. Karrai, R.D. Grobber: Appl. Phys. Lett. **66**, 1842 (1995)
54. B. Vohnsen, S.I. Bozhevolnyi: J. Microsc. – Oxford **202**, 244 (2001)
55. A. Liu, G.W. Bryant: Phys. Rev. B **59**, 2245 (1999)
56. S.I. Bozhevolnyi, V.Z. Lozovski: Phys. Rev. B **16**, 11139 (2000)
57. S.I. Bozhevolnyi, V.Z. Lozovski, K. Pedersen, J.M. Hvam: Phys. Status Solidi(a) **175**, 331 (1999)
58. A.V. Zayats, T. Kalkbrenner, V. Sandoghdar, J. Mlynek: Phys. Rev. B **61**, 4545 (2000)
59. Z.-Y. Li, B.-Y. Gu, G.Z. Yang: Phys. Rev. B **59**, 12622 (1999)
60. P.B. Miranda, Y.R. Shen: J. Phys. Chem. B **103**, 3292 (1999)
61. X. Wei, P.B. Miranda, Y.R. Shen: Phys. Rev. Lett. **86**, 1554 (2001)
62. R.D. Schaller, R.J. Saykally: Langmuir **17**, 2055 (2001)
63. B. Humbert, J. Grausem, A. Burneau, M. Spajer, A. Tadjeddine: Appl. Phys. Lett. **78**, 135 (2001)
64. Y. Barad, H. Eisenberg, M. Horowitz, Y. Silberberg: Appl. Phys. Lett. **70**, 922 (1997)
65. T.Y.F. Tsang: Phys. Rev. A **52**, 4116 (1995)
66. D. Yelin, Y. Silberberg, Y. Barad, J.S. Patel: Appl. Phys. Lett. **74**, 3107 (1999)
67. C.-K. Sun, S.-W. Chu, S.-P. Tai, S. Keller, U.K. Mishra, S.P. DenBaars: Appl. Phys. Lett. **77**, 2331 (2000)
68. M. Müller, J. Squier, C.A. De Lange, G.J. Brakenhoff: J. Microsc. – Oxford **197**, 150 (2000)
69. M. Müller, J. Squier, K.R. Wilson, G.J. Brakenhoff: J. Microsc. – Oxford **191**, 266 (1998)
70. L. Canioni, S. Rivet, L. Sarger, P. Barille, P. Vacher, P. Voisin: Opt. Lett. **26**, 515 (2001)
71. J.A. Squier, M. Müller: Appl. Optics **38**, 5789 (1999)
72. G.L. Eesley: *Coherent Raman Spectroscopy* (Pergamon, New York 1981)
73. A. Volkmer, J.-X. Cheng, X.S. Xie: Phys. Rev. Lett. **87**, 23901 (2001)
74. M.D. Duncan, J. Reintjes, T.J. Manuccia: Opt. Lett. **7**, 350 (1982)
75. M.D. Duncan, J. Reintjes, T.J. Manuccia: Opt. Commun. **50**, 307 (1984)
76. M.D. Duncan, J. Reintjes, T.J. Manuccia: Opt. Express **24**, 352 (1985)
77. A. Zumbusch, G.R. Holtom, X.S. Xie: Phys. Rev. Lett. **82**, 4142 (1999)
78. M. Hashimoto, T. Araki, S. Kawata: Opt. Lett. **25**, 1768 (2000)
79. C. Xu, W.W. Webb: J. Opt. Soc. Am. B **13**, 481 (1996)
80. C. Xu, W. Zipfel, J.B. Shear, R.M. Williams, W.W. Webb: P. Natl. Acad. Sci. USA **93**, 10763 (1996)
81. J.R. Lakowicz: *Topics in Fluorescence Spectroscopy: Nonlinear and Two-Photon Induced Fluorescence* (Plenum, New York 1997)
82. D.W. Piston, D.R. Sandison, W.W. Webb: SPIE **1640**, 379 (1992)
83. J. Ying, F. Liu, R.R. Alfano: Appl. Optics **39**, 509 (2000)
84. J.R. Lakowicz and I. Gryczynski, H. Malak, P. Schrader, P. Engelhardt, H. Kano, S.W. Hell: Biophys. J. **72**, 567 (1997)
85. S. Maiti, J.B. Shear, R.M. Williams, W.R. Zipfel, W.W. Webb: Science **275**, 530 (1997)
86. M.Barzoukas, M. Blanchard-Desce: J. Chem. Phys. **113**, 3951 (2000)
87. S. Andersson-Engels, I. Rokahr, J. Carlsson: J. Microsc. – Oxford **176**, 195 (1994)
88. R. Gauderon, P.B. Lukins, C.J.R. Sheppard: Microsc. Res. Techniq. **47**, 210 (1999)
89. L. Moreaux, O. Sandre, M. Blanchard-Desce, J. Mertz: Opt. Lett. **25**, 320 (2000)
90. W. Denk, D.W. Piston, W.W. Webb: 'Two-photon molecular excitation in laser-scanning microscopy', In: *Handbook of Biological Confocal Microscopy*, ed. by J.B. Pawley (Plenum, New York 1995) pp. 445–458

91. E. Stelzer, S. Hell, S. Lindek, R. Stricker, R. Pick, C. Storz, G. Ritter, N. Salmon: Opt. Commun. **104**, 223 (1994)
92. E. Betzig, R.J. Chichester: Science **262**, 1422 (1993)
93. X.S. Xie, R.C. Dunn: Science **265**, 361 (1994)
94. G. Parent, D. Van Labeke, F.I. Baida: J. Microsc. – Oxford **202**, 296 (2001)
95. R. Zenobi, V. Deckert: Angew. Chem. Int. Ed. **39**, 1746 (2000)
96. P.K. Yang, J.Y. Huang: Opt. Commun. **173**, 315 (2000)
97. H. Muramatsu, K. Homma, N. Yamamoto, J. Wang, K. Sakata-Sogawa, N. Shimamoto: Materials Science and Engineering C **12**, 29 (2000)
98. N. Hosaka, T. Saiki: J. Microsc. – Oxford **202**, 362 (2001)
99. A. Jenei, A.K. Kirsch, V. Subramaniam, D.J. Arndt-Jovin, T.J. Jovin: Biophys. J. **76**, 1092 (1999)
100. M.K. Lewis, P. Wolanin, A. Gafni, D.G. Steel: Opt. Lett. **23**, 1111 (1998)
101. A.K. Kirsch, V. Subramaniam, G. Striker, C. Schnetter, D.J. Arndt-Jovin, T.M. Jovin: Biophys. J. **75**, 1513 (1998)
102. E.J. Sánchez, L. Novotny, X.S. Xie: Phys. Rev. Lett. **82**, 4014 (1999)
103. Y. Kawata, C. Xu, W. Denk: J. Appl. Phys. **85**, 1294 (1999)
104. S.W. Hell, V. Andresen: J. Microsc. – Oxford **202**, 457 (2001)
105. J. Bewersdorf, R. Pick, S.W. Hell: Opt. Lett. **23**, 655 (1998)
106. A.H. Buist, M. Müller, J. Squier, G.J. Brakenhoff: J. Microsc. – Oxford **192**, 217 (1998)
107. P.E. Hänninen, S.W. Hell, J. Salo, E. Soini: J. Appl. Phys. **66**, 1698 (1995)
108. M. Schrader, K. Bahlmann, G. Giese, S.W. Hell: Biophys. J. **75**, 1659 (1998)
109. M. Schrader, S.W. Hell, H.T.M. van der Voort: J. Appl. Phys. **84**, 4033 (1998)
110. L. Moreaux, O. Sandre, J.W. Mertz: J. Opt. Soc. Am. B **17**, 1685 (2000)
111. I. Gryczynski, H. Malak, J.R. Lakowicz: Biospectroscopy **2**, 9 (1996)
112. T.A. Klar, S. Jakobs, M. Dyba, A. Egner, S.W. Hell: P. Natl. Acad. Sci. USA **97**, 8206 (2000)
113. M. Dyba, T.A. Klar, S. Jakobs, S.W. Hell: Appl. Phys. Lett. **77**, 597 (2000)

8 Parametric Nonlinear Optical Techniques in Microscopy

M. Müller and G.J. Brakenhoff

8.1 Introduction

A small revolution is taking place in microscopy. Nonlinear optical spectroscopic techniques – which in bulk applications have been around almost since the invention of the laser in the sixties – are rapidly being incorporated within high resolution microscopy. Starting with the successful introduction of two-photon absorption microscopy in 1990 [1], which has found important applications in biology, a whole series of other nonlinear optical techniques have been introduced: Three-photon absorption, Second Harmonic Generation (SHG), Third Harmonic Generation (THG), Coherent Anti-Stokes Raman Scattering (CARS), the optical Kerr effect, . . .

The startling progress of laser technology has enabled this new field in microscopy. Ultrashort pulses – which combine high peak power with moderate average powers – are required for efficient application of nonlinear optics at high numerical aperture (NA) conditions. Also, the lasers need to combine a low noise level of operation with high power output, "turn-key" operation and tunability over a large wavelength range. All these features are now available in current "state-of-the-art" laser technology.

A number of features make the introduction of nonlinear optical techniques into microscopy particularly interesting. The nonlinear dependence of the specimen's response to the incident laser power, provides inherent optical sectioning and thus three-dimensional imaging capability. A consequence of the fact that the interaction is limited to the focal region only, is the reduction of detrimental out-of-focus interactions, such as photobleaching and photo-induced damage. The use of nonlinear optics often enables the use of longer wavelengths, which reduces scattering and is – in some cases – less harmful to e.g. biological specimen. Another important reason for using nonlinear optics in microscopy is its possibility for spatially resolved spectroscopic measurements. Nonlinear optical spectroscopy, particularly in the form of four-wave mixing, can provide a wealth of information in both the time and frequency domain through the intricate interactions with the molecular energy levels.

Of particular interest here is the class of parametric nonlinear optical processes. Parametric processes are those in which during the interaction no net energy transfer occurs from the laser field to the sample, and *vice versa*. This absence of energy transfer makes these techniques potentially noninvasive imaging modes. In addition,

Török/Kao (Eds.): Optical Imaging and Microscopy, Springer Series in Optical Sciences
Vol. 87 – © Springer-Verlag, Berlin Heidelberg 2003

because of the absence of photo-induced damage to the sample, no fading of contrast takes place.

This chapter is organised as follows. We start out with a brief review of the theory of nonlinear optics in general and of parametric processes in particular. This highlights the similarities of the different techniques, while at the same time indicating the differences in response generation. All the different techniques probe different parts and orders of the nonlinear susceptibility of the material, thereby providing detailed information on molecular structure and – in some cases – intermolecular interactions. In the next chapters we consider two nonlinear optical parametric techniques in more detail: third harmonic generation and coherent anti-Stokes Raman scattering. These two techniques provide widely different information on the sample. The former can be applied to image – and analyse – changes in refractive index and/or nonlinear susceptibility in transparent materials. The latter can be used as a general "chemical microscope": the contrast is based on molecular structure and molecular interactions. We conclude with some general remarks with respect to the use nonlinear parametric processes in high numerical aperture (NA) microscopy.

8.2   Nonlinear Optics – Parametric Processes

The field of nonlinear optics is concerned with all light-matter interactions in which the response of the material depends in a nonlinear fashion on the input electromagnetic field strength. Excellent text books exist (e.g. [2–5]) that provide both a general overview of the field and a detailed theoretical analysis. Here some of the main features are reviewed which are relevant for the application of nonlinear optical techniques to high resolution microscopy. The discussion is limited to so-called parametric processes, which are processes for which the initial and final quantum-mechanical states of the system are equal. These processes include phenomena like harmonic generation and stimulated Raman scattering.

8.2.1   Introduction

The standard approach in nonlinear optics is to split the macroscopic polarisation – i.e. the result of an ensemble of oscillating dipole moments – in a linear and a nonlinear part and to expand it in successive orders of the electromagnetic field (E):

$$P(r,t) = P_{\mathrm{L}}(r,t) + P_{\mathrm{NL}}(r,t)$$
$$= \overleftrightarrow{\chi}^{(1)} : E + \overleftrightarrow{\chi}^{(2)} : EE + \overleftrightarrow{\chi}^{(3)} : EEE + \dots . \tag{8.1}$$

In (8.1), $\overleftrightarrow{\chi}^{(1)}$ and $\overleftrightarrow{\chi}^{(n)}$ denote the linear and nonlinear susceptibility respectively, which relate the complex amplitude of the electromagnetic fields with the polarisation. For example, for the third-order susceptibility:

$$P_i = \sum_{jkl} \chi^{(3)}_{ijkl} E_j E_k E_l, \tag{8.2}$$

where i, j, k and l run over all Cartesian coordinates. In general, for a material with dispersion and/or loss, the susceptibility is a complex tensor quantity. Parametric processes however can always be described by a real susceptibility. Note that from symmetry arguments it follows immediately that even orders of the nonlinear susceptibility – $\overleftrightarrow{\chi}^{(2n)}$ – vanish for materials that possess a centre of inversion symmetry. Various approaches have been developed to relate the macroscopic susceptibility to microscopic quantities (see e.g. [6,7]).

The nonlinear optical signal intensity follows from the macroscopic polarisation that is induced by a specific order of interaction in the expansion of (8.2). The macroscopic polarisation $P(r,t)$ acts as a source term in the wave equation, producing an electromagnetic signal wave. Accordingly, the signal intensity is proportional to the complex square of the field. As an example consider the process of third har-

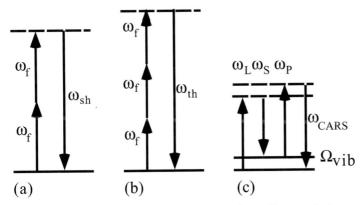

Fig. 8.1. Energy level diagrams for the parametric nonlinear optical processes of SHG **(a)**, THG **(b)** and CARS **(c)**. Common to all is that they are electronically nonresonant and – by definition – have no net energy transfer to the specimen. The various laser and signal frequencies represent: fundamental (ω_f), second-harmonic (ω_{sh}), third harmonic (ω_{th}), *Laser* (ω_L), *Stokes* (ω_S), and *Probe* (ω_P). $\Delta\Omega$ denotes the difference in energy between two vibrational energy levels

monic generation (THG). This process is depicted schematically in an energy level diagram in Fig. 8.1b. Three electromagnetic waves of fundamental frequency ω_f interact non-resonantly and simultaneously with the material to produce a signal wave at the third harmonic frequency $\omega_{th} = 3\omega_f$. Thus the THG process is related to the third-order susceptibility through[1]

$$P(\omega_{th}) \propto \overleftrightarrow{\chi}^{(3)}(\omega_{th} = \omega_f + \omega_f + \omega_f) : E(\omega_f)E(\omega_f)E(\omega_f), \tag{8.3}$$

where the explicit spatial and temporal dependence of the electromagnetic fields has been dropped. It follows that the THG signal intensity is proportional to the cube of

---

[1] The notation used here follows [2].

input power:

$$I_{th} \propto |P(\omega_{th})|^2$$
$$\propto \left| \overleftrightarrow{\chi}^{(3)}(\omega_{th} = \omega_f + \omega_f + \omega_f) : E(\omega_f)E(\omega_f)E(\omega_f) \right|^2 = \left| \chi^{(3)} \right|^2 (I_f)^3 . \tag{8.4}$$

In the second part of 8.4 it is assumed for simplicity that all fields have parallel polarisation, i.e. $\chi^{(3)} \equiv \chi^{(3)}_{1111}$. In the following the absence of explicit vector or tensor notation indicates the implicit assumption of a parallel polarisation condition for all fields involved. By definition parametric processes are coherent techniques which are electronically nonresonant with an, for all practical purposes, instantaneous interaction with the specimen. This results in the absence of a net energy transfer to the specimen, rendering these techniques potentially nondestructive.

8.2.2 Optical Sectioning Capability

Common to all nonlinear optical techniques is their capability of optical sectioning in microscopy applications. Due to the nonlinear dependence of the signal on the laser intensity – $I_{signal} \propto (I_{input})^n$ with $n > 1$ – the recorded signal is dominated by the in-focus contributions with negligible out-of-focus contributions.

This is in contrast to "conventional" linear microscopic techniques, where the signal response depends linearly on the input intensity – as is the case for e.g. single-photon absorption fluorescence. In this case a uniform sample layer contributes equally to the signal, whether it is in- or out-of-focus. Thus the axial position of such a layer cannot be determined with widefield fluorescence microscopy. The use of a confocal pinhole suppresses the contributions from out-of-focus planes permitting "optical sectioning" of the sample and thus a true axial resolution.

Nonlinear optics microscopy applications, on the other hand, have inherent optical sectioning capability. Even without the use of a confocal pinhole, these microscopic techniques provide three-dimensional imaging capability. This can provide significant advantages especially for imaging in turbid media.

8.2.3 Second Harmonic Generation (SHG)

Second harmonic generation is widely used to generate new frequencies, and has been used as a tool for imaging nonlinear susceptibilities in various materials (see for instance: [8–14]). SHG involves the interaction of light with the local nonlinear properties that depend on the molecular structure and hyperpolarizabilities. Therefore, the second harmonic intensity, and hence the contrast, is a function of the molecular properties of the specimen and its orientation with respect to both the direction and the propagation of the laser beam.

The induced macroscopic polarisation in second harmonic generation (SHG) is given by (see also the energy level diagram of Fig. 8.1a):

$$P(\omega_{sh}) \propto \overleftrightarrow{\chi}^{(2)}(\omega_{sh} = 2\omega_f) : E(\omega_f)E(\omega_f), \tag{8.5}$$

where ω_f and ω_{sh} denote the frequency of the fundamental and second harmonic field respectively. It follows that the intensity of the second harmonic signal has a square dependence on the input laser intensity:

$$I_{sh} \propto \left|\chi^{(2)}\right|^2 (I_f)^2 . \tag{8.6}$$

Since it is related to the second-order nonlinear susceptibility, SHG vanishes for a (locally) symmetric distribution of dipoles. The signal response is either confined to the surface of isotropic specimens or can probe local molecular asymmetries.

As characteristic for all coherent nonlinear phenomena, the SHG signal is strongly directional. The so-called phase anomaly of tightly focused laser beams – with a phase retardation of the focused wave relative to an unfocused plane wave – results in a specific, in some cases non-isotropic, signal emission patterns [12]. Resonant enhancement of the SHG signal occurs when the two-photon transition approaches an electronic state of the specimen. This effect can be used effectively through the addition of a specific "SHG label". Also, SHG contrast can be used to discriminate between various molecular conformations. For instance, it has been shown that chirality of the molecule can provide a two-fold enhancement of the signal in certain molecules, relative to an achiral form [11].

8.2.4 Third Harmonic Generation (THG)

In contrast to SHG, third harmonic generation (THG) is allowed in isotropic media. Nevertheless, interference – as a result of the Gouy phase shift – of radiation generated before and after focus results in efficient THG only in case of interfaces in either the third-order nonlinear susceptibility or the dispersion. Recently it was realised that this property makes THG a powerful three-dimensional imaging technique [15,16].

For THG the induced macroscopic polarisation is given by (see also the energy level diagram of Fig. 8.1b):

$$P(\omega_{th}) \propto \overleftrightarrow{\chi}^{(3)}(\omega_{th} = 3\omega_f) : E(\omega_f)E(\omega_f)E(\omega_f), \tag{8.7}$$

where ω_f and ω_{th} denote the frequency of the fundamental and second harmonic field respectively. It follows that the intensity of the third harmonic signal has a cube dependence on the input laser intensity:

$$I_{th} \propto \left|\chi^{(3)}\right|^2 (I_f)^3 . \tag{8.8}$$

The generation of the third harmonic under tight focusing conditions has been described in detail (see e.g.: [4,15,17]). THG is generally allowed in any material, since odd-powered nonlinear susceptibilities are nonvanishing in all materials. However, due to the Gouy phase shift of π radians that any beam experiences when passing through focus, THG is absent for $\Delta k = 3k_f - k_{th} \leq 0$, i.e. for exact phase matching or in case of a negative phase mismatch. The latter is the case for media with normal dispersion, i.e. where the refractive index decreases as a function of wavelength. Qualitatively, in a homogeneous medium the THG waves generated before and after the focal point destructively interfere, which results in the absence of a net THG production. However, in case of inhomogeneities near the focal point, efficient generation of the third harmonic is possible. This is especially the case for interfaces in the dispersion and/or third-order nonlinear susceptibility ($\overleftrightarrow{\chi}^{(3)}$). Note that THG is thus not restricted to the surface of the material only, but rather results from the bulk of the material contained within the focal volume and the presence of interfaces or inhomogeneities therein. This is confirmed also by the absence of a back-propagating THG signal [15].

These specific properties of THG are described well with Gaussian paraxial theory, even in the case of high NA focussing conditions [18]. With these approximations, an analytic solution can be derived which takes the form:

$$E_{th}(\boldsymbol{r}) = \eta A_{th}(z) \exp\left(\frac{-3\eta k_f r^2}{2z_R}\right), \tag{8.9a}$$

where

$$\eta(z) = \frac{1}{1 + iz/z_R},$$
$$A_{th}(z) = 2\pi i \omega_{th} \chi^{(3)} A_f^3 \frac{S(z)}{n_{th}c}, \tag{8.9b}$$
$$S(z) = \int^z \eta^2(\xi) \exp(i\Delta k\xi) d\xi.$$

In (8.9a), $E_{th}(\boldsymbol{r})$ is the third harmonic field envelope, A denotes the peak amplitude and z_R is the Rayleigh length, which is equal to half the confocal parameter ($b = 2z_R$).

8.2.5   Coherent Anti-Stokes Raman Scattering (CARS)

CARS is the nonlinear optical analogue of spontaneous Raman scattering. While amplifying the Raman signal by many orders of magnitude and enhancing the delectability of the signal against background luminescence, it retains the unique spectral specificity of Raman scattering. The Raman spectrum can provide a molecular "fingerprint" even at room temperature and in complex environments such as living cells. In addition it is sensitive to intermolecular interactions. Recent developments in laser technology have permitted introduction of this well known nonlinear spectroscopic technique into the field of high resolution microscopy [19,20].

For CARS the induced macroscopic polarisation is given by (see also the energy level diagram of Fig. 8.1c):

$$\boldsymbol{P}(\omega_{AS}) \propto \overleftrightarrow{\chi}^{(3)}(\omega_{AS} = \omega_L - \omega_S + \omega_P) : \boldsymbol{E}(\omega_L)\boldsymbol{E}(\omega_S)\boldsymbol{E}(\omega_P), \tag{8.10}$$

where ω_L, ω_S, ω_P and ω_{AS} denote the frequencies of the so-called *Laser*, *Stokes*, *Probe* and *Anti-Stokes* field respectively. It follows that the intensity of the CARS signal has a cube dependence on the combined input laser intensity:

$$I_{AS} \propto \left|\chi^{(3)}\right|^2 I_L I_S I_P. \tag{8.11}$$

CARS is a four-wave mixing technique in which two laser beams – *Laser* and *Stokes* – setup a grating, off which the third beam – *Probe* – undergoes a Bragg diffraction, generating an *anti-Stokes* signal beam. In the case of plane waves and a nonabsorbing medium the intensity of the CARS signal is given by (see e.g.: [21]):

$$I_{CARS} \propto \left(\frac{N}{V}\right)^2 \left(\frac{\partial\sigma}{\partial\Omega}\right)^2 A I_L I_S I_P d^2 \, \text{sinc}(\Delta k \cdot d/2), \tag{8.12}$$

where N/V is the particle number density, $\partial\sigma/\partial\Omega$ the differential Raman scattering cross section, A the area of the scattering volume perpendicular to the beam propagation, I_i - with $i = L, S, P$ – the intensity of *Laser*, *Stokes* and *Probe* beam respectively and d is the thickness of the scattering volume. Only (vibrationally) resonant excitation of a particular Raman active vibrational mode ($\Omega_v = \omega_L - \omega_S$) is considered here and energy conservation requires that $\omega_{AS} = \omega_L + \omega_P - \omega_S$. The phase mismatch Δk, which is defined as:

$$\Delta k = k_{AS} - k_L - k_P + k_S, \tag{8.13}$$

reflects the coherent nature of the process. In practice often $\Omega_v = \omega_L - \omega_S$ is used for convenience, while still permitting $\Delta k = k_{AS} - k_L - k_P + k_S$.

Various phase matching configurations can be conceived for which $\Delta k = 0$ in (8.12). The three most commonly used are depicted in Fig. 8.2. In collinear phase matching (Fig. 8.2a) all laser beams are parallel and the signal beam travels along the

(a)

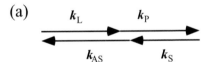

(b)

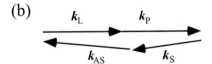

(c)

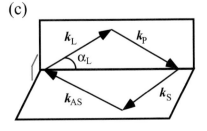

Fig. 8.2. Possible phase matching conditions in CARS. (a) collinear; (b) semi-collinear; and (c) folded BoxCARS

same path as the excitation beam. In this case usually multiple filters have to be used to separate the signal from the lasers. In addition, due to the general dispersion characteristics, a slight phase mismatch results in this case, which gets more pronounced when probing higher vibrational modes $\Delta\Omega = \omega_L - \omega_S$. The phase mismatch can be resolved by slightly displacing the *Stokes* laser with respect to the *Laser* and *Probe*, providing exact phase matching (Fig. 8.2b). The best separation of the CARS signal from the input laser beams – while retaining exact phase matching conditions – is achieved in the so-called folded BoxCARS phase matching scheme (Fig. 8.2c). The possibility to combine this type of phase matching with high numerical aperture (NA) microscopy has recently been demonstrated [20]. It should be noted that the effect of the phase mismatch that results from collinear phase matching becomes increasingly less severe at high NA focusing conditions due to the d^2 term in (8.12).

8.3   Third Harmonic Generation (THG) Microscopy

8.3.1   General Characteristics

The main properties of THG microscopy that follow from theory – most of which have now been verified in practice – are as follows:

- The contrast in THG microscopy is based on either a change in third-order susceptibility or in the dispersion properties of the material, within the focal volume of the fundamental radiation. In other words, an interface between two materials is needed for efficient THG. Denoting the two materials with 1 and 2 respectively, it is required that either $\chi_1^{(3)} \neq \chi_2^{(3)}$ or $\Delta n_1 \neq \Delta n_2$, where $\Delta n_i = n_i(\lambda_f) - n_i(\lambda_{th})$ describes the dispersion of material i. It follows that, in contrast to phase microscopy, THG imaging can discern the interface between two media on the basis of a difference in nonlinear susceptibility alone. An example of this is the fact that the boundary between immersion oil and a microscope coverglass – which have been matched in refractive index explicitly – is clearly imaged in THG microscopy. THG imaging is a transmission mode microscopy, similar to phase-contrast or DIC microscopy, but with inherent three-dimensional sectioning properties. Thus, whereas phase-contrast microscopy depends on accumulated phase differences along the optical path length, THG microscopy is sensitive to differences in specimen properties localised within the focal volume.
- The generation of third harmonic is restricted to the focal region. In particular, the full-width-at-half-maximum (FWHM) of the axial response of a THG microscope to an interface between two media with a difference in nonlinear susceptibility alone is equal to the confocal parameter at the fundamental wavelength [15].
- THG is a coherent phenomenon in which the third harmonic radiation is generated in the forward direction. For a linearly polarised input laser beam, the generated third harmonic is also linearly polarised in the same direction [22]. The third-order power dependence of THG on the input laser power, results in

an approximately inverse square dependence on the laser pulse width. Typical conversion efficiencies from fundamental to third harmonic are in the range of 10^{-7}–10^{-9} [22], and conversion efficiencies upto 10^{-5} have been reported for specific materials [23]. The efficiency of the THG process depends critically on the orientation of the interface relative to the optical axis [16].

- The noninvasive character of THG imaging has been demonstrated in various applications of microscopic imaging of biological specimens *in-vivo* [22,24,25]. In addition, fading of contrast – equivalent to the bleaching of fluorescence – is absent in THG imaging applications.

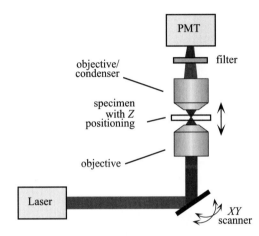

Fig. 8.3. Schematic of the experimental setup for THG microscopy. Laser light at the fundamental frequency ω_f is focussed onto the specimen by a high NA microscope objective. The third harmonic signal (at ω_{th}) is collected in the forward direction by a second microscope objective or condenser and detected on a photomultiplier tube (PMT). The specimen is raster scanned in the lateral plane by scanning of the laser beam and in the axial by movement of the specimen

The experimental setup for THG microscopy is shown schematically in Fig. 8.3. A near IR femtosecond laser beam – in our case the idler wave from an optical parametric amplifier – is focused by a high NA microscope objective onto the sample. Two scanning mirrors provide for *XY* beam scanning. The sample itself can be moved in the axial direction. The signal is emitted in the forward direction and collected by either a second microscope objective or a condenser lens. Note that the NA of the generated third harmonic signal is one-third of the input NA [18]. The signal passes a filter that blocks the residual laser light and is detected on a photomultiplier tube (PMT). It has also been shown [22] that the THG signal can readily be recorded directly on a video camera.

8.3.2   Selected Applications

In a first demonstration of the noninvasive character of THG microscopy, the technique was applied live cells [22]. In particular, the Rhizoids from the alga *Chara* were imaged Fig. 8.4, where the strong cytoplasmic streaming was used as an internal check for the survival of these tubular single cells. These cells have been studied widely, especially with respect to their response to gravity, which is thought to be

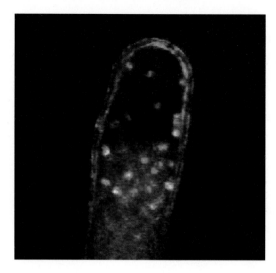

Fig. 8.4. THG image of the Rhizoid tip from the alga *Chara*. The so-called statoliths – which are vesicles containing BaSO$_4$ crystals – are clearly visible as well as the cellular membrane

related to the so-called statoliths – which are vesicles containing BaSO$_4$ crystals, contained in the tip of the root. In *in-vivo* imaging, the statoliths show dynamic motion while remaining anchored to the actin filament network. No disruption of the cytoplasmic streaming, nor any fading of contrast, has been observed for more than an hour of continuous exposure; a first indication that the cell remains functional.

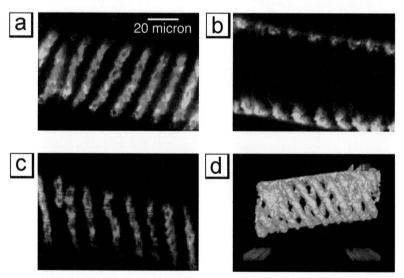

Fig. 8.5. Selection of THG images of *Spyrogira* out of a total of 36 optical sections at 1 μm axial intervals. (**a**)–(**c**) Images at the top, middle and bottom of the stack. (**d**) 3D reconstruction of the complete stack

Figure 8.5 shows the three-dimensional imaging capabilities of THG microscopy. A total of 36 optical sections – taken at 1 μm axial intervals – were taken of *Spyrogira*. Figure 8.5a–c shows different optical sections of top, middle and bottom part of the axial stack. Fig. 8.5d is a three-dimensional reconstruction. The diameter of the spiral is approximately 35 μm and the spacing between the individual spiral ribs ~10 μm. The shadow below the three-dimensional reconstruction is due to THG from the water-coverglass interface. We have checked – by measurement of the output wavelength and polarisation analysis –that the signal is truly from THG and not from some sort of autofluorescence from the sample. Also, we observed no fading of contrast over prolonged exposure times. Alternatively, THG microscopy can be used in the material sciences as demonstrated by the use of the technique to visualise the results of laser-induced breakdown in glass Fig. 8.6 [26]. In this data storage type application, high energy IR pulses (~0.7 μJ/pulse) were used to "write" specific patterns in different focal planes in glass. In this particular case, the letters "U", "C", "S" and "D" were written in planes with an axial separation of 19 μm. The "write" process is due to laser-induced breakdown, which is a multi-photon absorption processes that causes highly localised plasma formation. The rapid expansion of the plasma causes a microexplosion which results in observable damage. The created "damage structure" is considered to be due to either a small change in refractive index or a vacuum bubble [27]. The "written" pattern – consisting of these microstructures – can subsequently be "read" with THG microscopy using low energy IR pulses (~ 40 nJ/pulse). This energy level is below the threshold of laser-induced breakdown. THG microscopy can also be used for material characterisation. In principle, a quantitative measurement of the dispersion, $\Delta n_i = n_i(\lambda_\text{f}) - n_i(\lambda_\text{th})$, and

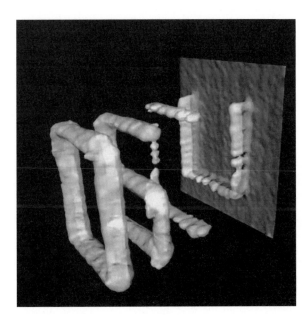

Fig. 8.6. 3D reconstruction from a series of axially sectioned THG images taken at 2 μm axial intervals. The letters are approximately 20 μm wide and were written in focal planes spaced 19 μm apart

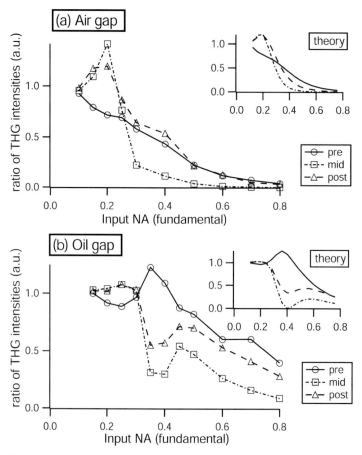

Fig. 8.7. Measured intensity ratios, R_{pre}, R_{mid}, R_{past}, for a double-interface geometry, the gap being filled with immersion air (**a**) and oil (**b**) respectively. The insets show the results from calculations based on Gaussian paraxial theory. For the calculations we used $\Delta n_{\text{oil}} = -0.028$ and $\Delta n_{\text{K5}} = -0.032$, with $n_{\text{oil}} = 1.50276$ and $n_{\text{K5}} = 1.51146$ at $\lambda_{\text{f}} = 1100$ nm and $\chi^{(3)}_{\text{oil}}/\chi^{(3)}_{\text{K5}} = 3$

the nonlinear susceptibility, $\chi^{(3)}$, is possible [18]. An indication of this potential is shown in Fig. 8.7. In these experiments the sample consisted of a thin layer of either immersion oil or air between two microscope coverglasses. The total THG intensity was recorded as a function of the axial position for various excitation numerical apertures. To characterise the functional shape of these THG axial profiles three intensity ratios were defined: $R_{\text{pre}} = I_{\text{THG}}(-L/4)/I_{\text{THG}}(0)$, $R_{\text{post}} = I_{\text{THG}}(+L/4)/I_{\text{THG}}(0)$ and $R_{\text{mid}} = I_{\text{THG}}(-L/2)/I_{\text{THG}}(0)$, where L is the separation between the two interfaces. The THG yield is represented by I_{THG}, with the z position of the interface as argument: the ratios represent the THG intensities resulting when the focus lies exactly halfway the interfaces $I_{\text{THG}}(+L/2)$, one quarter gap thickness before the first interface $I_{\text{THG}}(-L/4)$, and likewise behind the first interface $I_{\text{THG}}(+L/4)$; all normalised by the intensity at the first interface $I_{THG}(0)$. Theoretically these ratios are calculated with Gaussian paraxial theory. With "hat"-profile beams, the agreement between experiment and theory is only qualitative. Nevertheless, the calculations reproduce the basic features of the experimental data. For the calculations we used $\Delta n_{oil} = -0.028$ and $\Delta n_{K5} = -0.032$, with $n_{oil} = 1.50276$ and $n_{K5} = 1.51146$ at $\lambda_f = 1100\,nm$). The ratio of the nonlinear susceptibilities of K5 and immersion oil was chosen arbitrarily to be $\chi_{oil}^{(3)}/\chi_{K5}^{(3)} = 3$. The numerical calculations show that the functional form of the signal is dependent almost solely on the change in dispersion, whereas the magnitude of the THG signal depends primarily on the ratio of the nonlinear susceptibilities.

8.3.3 Summary

The examples of THG microscopy applications given above show the potential of this technique. Its unique ability to observe – with three-dimensional resolution – changes in the material properties of transparent specimen can be used effectively in both biological and material sciences applications. The lateral resolution of the technique is in the range of 400 nm, whereas the axial resolution scales with the confocal parameter of the fundamental beam, and is generally in the order of 1 μm. The fact that THG is an electronically nonresonant interaction with the specimen, renders it noninvasive, as has been shown in *in-vivo* cell biological studies. In addition this ensures that there is no fading of contrast, which is a general problem in fluorescence based techniques.

The contrast in THG microscopy is based on changes in dispersion and changes in the magnitude of the nonlinear susceptibility. The influences of both of these parameters can be largely decoupled. When measuring for instance THG axial profiles as a function of the numerical aperture, the functional shape is dependent primarily on changes in dispersion, whereas the total THG yield depends almost completely on the changes in nonlinear susceptibility alone. This provides the opportunity to access these parameters quantitatively. A prerequisite for these experiments is that the geometrical shape of the specimen is known in some detail. Being a coherent phenomenon – the magnitude of the THG signal is dependent on the specific structure of the specimen. Thus in principle, various structures may yield the same image. Note however, that due to the nonlinearity of the THG process, this sensitivity to the structure of the specimen extends only over a range of the order of the confocal parameter of the fundamental excitation beam.

8.4 Coherent Anti-Stokes Raman Scattering (CARS) Microscopy

8.4.1 General Characteristics

Raman spectroscopy is sensitive to molecular vibrations, which in turn reflect molecular structure, composition and inter-molecular interactions. As such, Raman spectroscopy is unique in that it provides detailed intra- and inter-molecular structural information and specificity at room temperature, even within extremely complex systems such as living cells. Raman spectroscopy suffers, however, particularly from a low scattering cross section as well as from interference from naturally occurring luminescence, limiting its applicability to high resolution microscopy. Stimulated Raman scattering – and in particular CARS – can overcome these drawbacks of spontaneous Raman scattering rendering the signal strength required for effective image acquisition in high resolution microscopy.

CARS microscopy was first introduced in the early eighties [28,29], but found limited application at that time, probably due to shortcomings in laser technology. The technique was recently reinvented [19,20], using novel approaches both in terms of the laser apparatus used and the application of non-collinear phase matching geometries at high numerical-aperture focusing conditions. Several approaches have now been developed for the application of CARS microscopy in high resolution microscopy. For instance, Xie and co-workers [30,31] showed that an epi-detection mode in CARS enhances the sensitivity of the technique to small features, by effectively removing the nonresonant background contribution. In a later publication this group showed that the nonresonant background can alternatively be suppressed by using polarised CARS [32]. Recently, we have shown [33] the potential of multiplex CARS microscopy for imaging the chemical and physical state of biological model system. In multiplex CARS a significant region of the vibrational spectrum is addressed simultaneously – rather than point-by-point as in "conventional" CARS. The general properties of CARS microscopy can be summarised as follows:

- The CARS spectrum generally consists of both a resonant and a nonresonant contribution The resonant part is proportional to the square of the Raman scattering cross section.
- CARS is a coherent process in which the signal is generated in a forward direction determined by the phase matching geometry ($k_{AS} = k_L - k_S + k_P$). The divergence of the generated emission scales directly with the lateral waist of the interaction volume.
- Because of the signal's cube dependence on the laser input intensity, CARS microscopy provides inherent optical sectioning, enabling three-dimensional microscopy. The resolution in CARS microscopy is determined by the interaction volume. While sacrificing some of the attainable spatial resolution, nonlinear phase matching configurations [20] can be implemented in high NA microscopy. This mode of operation is essential for the use of time-resolved CARS and may enhance the detection sensitivity in some cases.
- CARS microscopy is a form of "chemical imaging", which – through the Raman spectrum – is particularly sensitive to both the molecular vibrational signature and to intermolecular interactions.

8.4.2   Multiplex CARS

A typical CARS experimental setup is shown in Fig. 8.8. Two lasers (pico- and/or femtosecond) are synchronised in time and made collinear before being focused by a

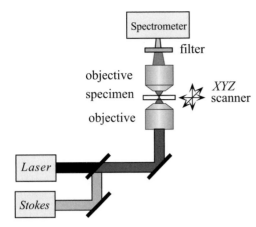

Fig. 8.8. Schematic of the experiment setup for CARS microscopy. The output of two lasers – denoted by *Laser* and *Stokes* – is made collinear and synchronised in time. The beams are focused by a high NA microscopy objective onto the sample, which is moved by piezo scanners in all three dimensions. The generated CARS signal is collected in the forward direction by a second microscope objective and measured on a spectrometer. A CARS spectrum is acquired at each specimen position

high NA microscope objective onto the sample. The CARS image is acquired point-by-point where the sample is scanned in all three spatial directions. The CARS signal is emitted in the forward direction and the, so-called, *Laser* and *Stokes* are blocked by a set of appropriate filters. In the multiplex CARS configuration a narrow bandwidth ($\sim$2 cm^{-1}) picosecond Ti:Sapphire *Laser* laser is used, in combination with a broad bandwidth ($\sim$150 cm^{-1}) femtosecond tunable Ti:Sapphire *Stokes* laser. For every image point, the CARS spectrum is measured with a spectrometer, with typical acquisition times in the order of 20-100 ms. As an example, Fig.

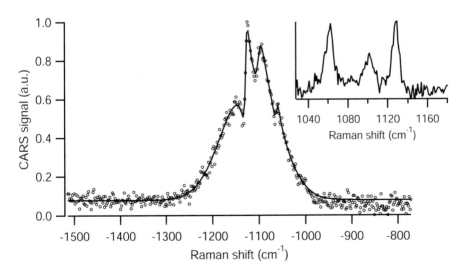

Fig. 8.9. Multiplex CARS spectrum of a DSPC multi-lamellar vesicle. The experimental data (*open circle*) are fitted (*solid line*) to a theoretical expression for the CARS signal which contains three vibrational resonances as deducted from the spontaneous Raman signal (*inset*)

8.9 shows the CARS spectrum of a di-stearoylphosphatidylcholine (DSPC) multi-lamellar vesicle. The solid line represents a fit of the theoretically expected signal to the data. For the theoretical expression use is made of parameters (linewidths and line positions) obtained from the spontaneous Raman scattering signal (shown as an inset). This effectively reduces the number of free fitting parameters for the CARS signal to the magnitude of both the resonant and nonresonant contribution.

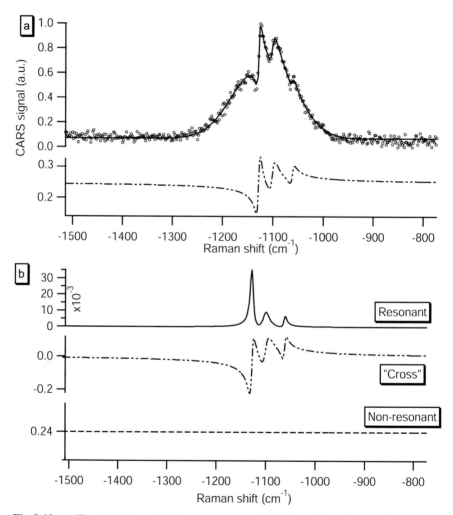

Fig. 8.10. (**a**) Top: Multiplex CARS spectrum of a DSOPC multi-lamellar vesicle. Bottom: CARS signal after dividing out the *Stokes* spectral intensity profile. (**b**) The three contributions that make up the signal of part a, bottom). Top: vibrationally resonant. Middle: "cross"-term. Bottom: vibrationally nonresonant

Figure 8.10 shows the various components that contribute to the measured multiplex CARS signal. In Fig. 8.10a both the original data are shown and the signal after dividing out the influence of the *Stokes* spectral intensity profile. The CARS signal is proportional to the absolute square of the nonlinear susceptibility – (8.11) – which in turn consists of a sum of a resonant and a nonresonant contribution [33]:

$$I_{AS} \propto \left| \chi_{NR}^{(3)} + \chi_{R}^{(3)} \right|^2 . \tag{8.14}$$

The total signal thus consists of three parts: a constant nonresonant background, a purely resonant contribution with a Lorentzian shape and amplitude proportional to the square of the Raman scattering cross sections and a cross term which yields a dispersive signal. The different contributions to the total CARS signal of Fig. 8.10a are shown in Fig. 8.10b as they are derived from the fit. Finally, Fig. 8.11 shows the

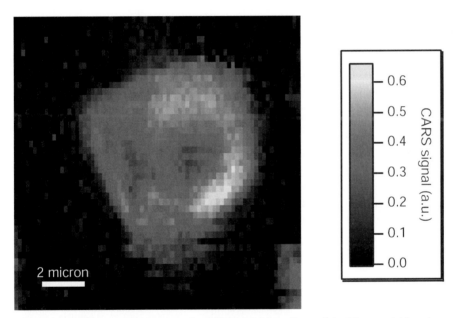

Fig. 8.11. Multiplex CARS images of a DSPC multi-lamellar vesicle. The acquisition time for the full CARS spectrum is 100 ms. Plotted is the amplitude of the $-1101\,\mathrm{cm}^{-1}$ mode as deduced from fitting the data to a theoretical expression for the CARS spectrum

image – one optical section – acquired from the DSPC multi-lamellar vesicle. The 50×50 pixel image was acquired in 250 seconds. This should be compared to a typical acquisition time – for the same excitation power and signal-to-noise ratio – of a single Raman spectrum of such vesicles in the order of 15 minutes. Note that the fitting procedure ensures that laser power fluctuations and time jitter between the lasers do not influence the signal-to-noise of the acquired image. The figure clearly shows that the density of lipids minimises in the centre of the vesicle.

As shown in these figures, multiplex CARS microscopy provides a CARS spectrum over a significant part of the vibrational spectrum in typically 20-100 ms. The signal-to-noise of these spectra is more than sufficient to deduce important information from the specimen. We have shown [33], for example, that within a single experimental run, the information obtained from the so-called skeletal optical mode region of the vibrational spectrum can be used to discriminate between lipid membranes which are in the liquid crystalline phase (i.e. above the phase transition temperature) or in the gel phase (below the phase transition temperature). Interestingly, this region of the spectrum is quite insensitive to the particular chemical structure of the acyl chains, providing a general probe to the physical structure of lipid membranes. Most importantly, multiplex CARS provides a more than four orders of magnitude increase in signal strength compared to spontaneous Raman spectroscopy, permitting high resolution microscopy with realistic (~minutes) image acquisition times.

8.4.3 Summary

CARS microscopy is a novel approach to obtain spatially resolved information about the chemical composition and physical structure of samples. While retaining the spectral specificity of spontaneous Raman spectroscopy, it provides a signal enhancement through the nonlinear interaction with the specimen that enables high resolution microscopy. The nonlinearity of the process also provides the technique with inherent optical sectioning capability. The contrast in the images is based on vibrational spectral features – i.e. to specimen inherent properties – and does not require the addition of specific labels. Also, since CARS is a parametric process, no energy transfer to the specimen occurs in the generation process of the signal providing generally mild exposure conditions for (biological) specimen and ensuring that no fading of contrast occurs.

In many application, CARS microscopy is accomplished using two picosecond lasers, in which case only a single point in the CARS spectrum can be addressed at a time. To obtain spectral information, one of the lasers needs to be tuned, making the CARS spectrum obtained in this way particularly sensitive to laser-induced fluctuations in the signal strength (e.g. due to power fluctuations, timing jitter, etc.). In the newly developed multiplex CARS microscopy technique, a significant part of the CARS spectrum is obtained simultaneously, through the use of a combination of a picosecond (small spectral bandwidth) laser and a femtosecond (broad bandwidth) laser. This eliminates the problems associated with laser-induced fluctuations, and provides CARS spectra of which the signal-to-noise is limited only by Poisson noise. This mode of operation enables – for the first time – the use of CARS spectral data to image highly specific features of the sample.

8.5   Conclusion

Advances in ultrashort pulsed laser technology has provided the opportunity to combine nonlinear optics with high NA microscopy. This opens the way to a rich field of molecular spectroscopic techniques for spatially resolved studies both in the material sciences and in biology. Through its nonlinearity of response, this class of optical techniques provides inherent optical sectioning, and thus three-dimensional imaging capability.

Here we have considered a sub-class of these techniques: parametric nonlinear optics. In parametric processes no energy transfer between the object and the laser field(s), and *vice versa*, takes place, providing the potential for noninvasive imaging modes, as well as the absence of fading of contrast in the images. Second harmonic generation microscopy exploits the fact that second-order nonlinear interactions are not observed in specimen that contain a centre of inversion symmetry. Thus this technique is particularly sensitive to surfaces or highly oriented sample. Third harmonic generation on the other hand is in principle a "bulk" phenomenon. However, under strong focusing conditions – as in high NA microscopy – the Gouy phase shift at focus causes destructive interference between the third harmonic radiation that is produced before focus with that produced after the focal plane. This implies that significant third harmonic generation is only observed near interfaces in dispersion or nonlinear susceptibility. This makes the technique highly attractive for material properties characterization studies.

Third harmonic generation is one nonlinear optical technique of the broad class of so-called four-wave mixing techniques that have been applied widely in spectroscopic studies. Recently other members of this class have been introduced to high resolution microscopy. Most prominent of these new additions is coherent anti-Stokes Raman scattering (CARS). CARS microscopy combines the unique vibrational spectroscopic specificity of spontaneous Raman scattering with a signal enhancement through the nonlinear interactions by several orders of magnitude. This signal enhancement enables – for the first time – to exploit the detailed Raman spectral information for imaging purposes. In the recently developed multiplex CARS microscopy configuration, the CARS signal is acquired simultaneously over a significant part of the vibrational spectrum. A consequence of this is that the accuracy with which the spectral information can be measured is limited only by Poisson noise, and not – as in conventional CARS – by laser induced signal strength fluctuations (power fluctuations, time jitter, etc.).

Another recent addition to nonlinear optical microscopy from the class of four-wave mixing techniques is the application of the nonlinear optical Kerr effect [34]. In this application, this novel technique has been applied to study the mobility of intracellular water. The technique permits the determination of rotational relaxation times, which for small molecules such as water are typically in the order of ~800 fs.

As a final remark, a word of caution with respect to the interpretation of images that result from parametric nonlinear optical techniques. All these techniques provide a coherent imaging mode, that is: the generated signals from the specimen interfere on the detector to give rise to a certain detected intensity. In general this may obscure the relation between object and image, since in principle different objects can give rise to similar images. However the situation in nonlinear optical microscopy is more favourable since the signals are generated only within the focal volume of the laser beams. Thus only sub-resolution structure within the specimen may complicate the image interpretation.

Acknowledgement

This research was financially supported in part by the Stichting voor Fundamenteel Onderzoek der Materie (FOM), The Netherlands, under grant no. 94RG02 and by the Stichting Technische Wetenschappen (STW), The Netherlands, under grant no. ABI.4929.

References

1. W. Denk, J.H. Strickler, W.W. Webb: Science **248**, 73 (1990)
2. Y.R. Shen: *The principles of nonlinear optics* (John Wiley & Sons, New York 1984)
3. M. Schubert, B. Wilhelmi: *Nonlinear optics and quantum electronics* (John Wiley & Sons, New York 1986)
4. R.W. Boyd: *Nonlinear optics* (Academic Press, Inc., Boston 1992)
5. S. Mukamel: *Principles of nonlinear optical spectroscopy* (Oxford University Press, Oxford 1995)
6. C. Flytzanis: 'Theory of nonlinear optical susceptibilities', In: *Quantum Electronics, vol. I: Nonlinear Optics*, ed. by H. Rabin, C. L. Tang (Academic Press, New York 1975) pp. 9–207
7. Y. Prior: IEEE J. Quantum Elect. **20**, 37 (1984)
8. R. Hellwarth, P. Christensen: Opt. Commun. **12**, 318 (1974)
9. J.N. Gannaway, C.J.R. Sheppard: Opt. Quant. Electron. **10**, 435 (1978)
10. R. Gauderon, P.B. Lukins, C.J.R. Sheppard: Opt. Lett. **23**, 1209 (1998)
11. P.J. Campagnola, M. Wei, A. Lewis, L.M. Loew: Biophys. J. **77**, 3341 (1999)
12. L. Moreaux, O. Sandre, M. Blanchard-Desce, J. Mertz: Opt. Lett. **25**, 320 (2000)
13. E.C.Y. Tan, K.B. Eisenthal: Biophys. J. **79**, 898 (2000)
14. L. Moreaux, O. Sandre, S. Charpak, M. Blanchard-Desce, J. Mertz: Biophys. J. **80**, 1568 (2001)
15. Y. Barad, H. Eisenberg, M. Horowitz, Y. Silberberg: Appl. Phys. Lett. **70**, 922 (1997)
16. M. Müller, J. Squier, K.R. Wilson, G.J. Brakenhoff: J. Microsc. – Oxford **191**, 266 (1998)
17. J.F. Ward, G.H.C. New: Phys. Rev. **185**, 57 (1969)
18. J.M. Schins, T. Schrama, G.J. Brakenhoff, M. Müller: Manuscript submitted for publication (2001)
19. A. Zumbusch, G.R. Holtom, X.S. Xie: Phys. Rev. Lett. **82**, 4142 (1999)
20. M. Müller, J. Squier, C.A. de Lange, G.J. Brakenhoff: J. Microsc. – Oxford **197**, 150 (2000)
21. G.L. Eesley: *Coherent Raman Spectroscopy* (Pergamon Press, New York 1981)
22. J. Squier, M. Müller, G.J. Brakenhoff, K.R. Wilson: Opt. Express **3**, 315 (1998)
23. D. Yelin, Y. Silberberg, Y. Barad, J.S. Patel: Phys. Rev. Lett. **82**, 3046 (1999)
24. D. Yelin, Y. Silberberg: Opt. Express **5**, 169 (1999)
25. A.C. Millard, P. Wiseman, D.N. Fittinghoff, J.A. Squier, M. Müller: Appl. Optics **38**, 7393 (1999)
26. J. Squier, M. Müller: Appl. Optics **38**, 5789 (1999)

27. C.B. Schaffer, E.N. Glezer, N. Nishimura, E. Mazur: 'Ultrafast laser induced microexplosions: explosive dynamics and sub-micrometer structures', In: *Proceedings of SPIE, San Jose Jan. 24-30, 1998*, ed. by M.K Reed (SPIE, Washington 1998) pp. 36–45
28. M.D. Duncan, J. Reijntjes, T.J. Manuccia: Opt. Lett. **7**, 350 (1982)
29. M.D. Duncan: Opt. Commun. **50**, 307 (1984)
30. J. Cheng, A. Volkmer, L.D. Book, X.S. Xie: J. Phys. Chem. B **105**, 1277 (2001)
31. A. Volkmer, J. Cheng, X.S. Xie: Phys. Rev. Lett. **87**, 023901 (2001)
32. J. Cheng, L.D. Book, X.S. Xie: Opt. Lett. **26**, 1341 (2001)
33. M. Müller, J.M. Schins: Manuscript submitted for publication (2001)
34. E.O. Potma, W.P. de Boeij, D.A. Wiersma: Biophys. J. **80**, 3019 (2001)

9 Second Harmonic Generation Microscopy Versus Third Harmonic Generation Microscopy in Biological Tissues

Chi-Kuang Sun

9.1 Introduction

Second-harmonic generation (SHG) and third-harmonic generation (THG) processes are both nonlinear processes, related to the interaction of intense light with matters. SHG process describes the generation of light wave that is twice the frequency (with half of the original wavelength) of the original one while THG process describes the generation of light wave that triples the frequency (with one third of the original wavelength) of the original one. The harmonic light wave generation is coupled from the excited nonlinear polarization P^{NL} under intense laser excitation. The interaction of nonlinear polarization P^{NL} and the excitation light is usually related through a nonlinear susceptibility χ, as previously described in Chaps. 7 and 8. SHG and THG can be visualized by considering the interaction in terms of the exchange of photons between various frequencies of the fields. According to this picture, which is previously illustrated in Figs. 8.1 (a), (b), two or three photons of angular frequency ω are destroyed and a photon of angular frequency 2ω (for SHG) or 3ω (for THG) is created in a single quantum-mechanical process. The solid lines in the figure represent the atomic ground states, and the dashed lines represent what are known as virtual levels. These virtual levels are not energy eigenlevels of the atoms, but rather represent the combined energy of one of the energy eigenstates of the atom and one or more photons of the radiation field. Due to its virtual level transition characteristics, harmonic generations are known to leave no energy deposition to the interacted matters, since no real transition involved and the emitted photon energy will be exactly the same as the total absorbed photon energy. This virtual transition characteristic provides the optical "noninvasive" nature desirable for microscopy applications, especially for live biological imaging.

Due to its nonlinearity nature, the generated SHG intensity depends on square of the incident light intensity, while the generated THG intensity depends on cubic of the incident light intensity. Similar to multi-photon induced fluorescence process, this nonlinear dependency allows localized excitation and is ideal for intrinsic optical sectioning in scanning laser microscopy. Usually the third-order nonlinear susceptibility $\chi^{(3)}(3\omega : \omega, \omega, \omega)$ needed for THG is much weaker than the second-order nonlinear susceptibility $\chi^{(2)}(2\omega : \omega, \omega)$ needed for SHG, thus THG is harder to observe. However, not all biological materials have second-order nonlinear susceptibility. For centro-symmetric media, the lowest order nonlinear susceptibility will be $\chi^{(3)}$ instead of $\chi^{(2)}$. This is why people have rarely heard of generation of

Török/Kao (Eds.): Optical Imaging and Microscopy, Springer Series in Optical Sciences
Vol. 87 – © Springer-Verlag, Berlin Heidelberg 2003

SHG from a bulk glass. The random distribution of oxides in glass creates optical centro-symmetry in optical wavelength scale, thus inhibiting the SHG generation. Only non-centro-symmetry media is allowed to generate SHG. On the other hand, all materials are allowed to create third-order susceptibility, of which magnitudes vary according to material property and wavelength.

According to photon momentum conservation, the generated harmonic photons will be emitted in the same direction as the incident photons, unless noncollinear phase matching process is involved. This forward direction emission property restricts the light collection geometry in microscopy applications. Nowadays most SHG and THG microscopes utilize transmission detection, which is different from the reflection detection in most laser scanning fluorescence microscopes.

9.2   SHG Microscopy

In the past four decades, SHG has been widely applied to the study of SHG materials and interfacial regions without a center of symmetry, and was later combined with a microscope for SHG scanning microscopy [1,2] in 1970s. Today, SHG scanning microscope has been widely used in material studies for surface monolayer detection [3], ferroelectric domain structures [4], and nonlinear crystal characterization with 3D resolution [5], as previously described in Chap. 6. With low frequency electric field breaking the centro-symmetry, SHG can also be generated due to low frequency electric field even in centro-symmetry media, providing a tool to image electric field. This process is usually described as a third-order nonlinear process with the nonlinear polarization described by [6]

$$P^{\mathrm{NL}}(2\omega) = \frac{3}{4}\varepsilon_o \chi^{(3)}(2\omega : \omega, \omega, 0)E(\omega)E(\omega)E(0) \tag{9.1}$$

where $E(0)$ is the low frequency electric field. This process is usually referred as Electric-Field-Induced-Second-Harmonic-Generation (EFISHG). Utilizing EFISHG in gallium nitride, we have successfully demonstrated electric field distribution imaging using SHG scanning microscopy [7,8]. Compared with traditional electro-optical sensor probe technique, EFISHG microscopy of electric field has all the advantage of scanning confocal microscope, with high optical resolution and great 3D sectioning power. It is also a background-free experiment if the EFISHG material possesses centro-symmetric properties. Recently we have successfully achieved 3D electric field distribution mapping in an IC circuit utilizing liquid crystal as EFISHG probe materials with sub-micron resolution [9,10].

SHG generation in biological specimen was first observed in 1971 by Fine and Hansen from collageneous tissues [11] by using a Q-switched ruby laser at 694 nm. In 1986, Freund, Deutsch, and Sprecher demonstrated SHG microscopy in connective tissues [12] based on a 1064 nm Q-switched Nd:YAG laser. They attributed the SHG generation to the polarity (that is one type of non-centro-symmetry) of rat-tail tendon, which was then correlated to polar collagen fibrils. In 1989, J. Y. Huang and his coworkers used SHG to probe the nonlinear optical properties of purple membrane–poly(vinyl alcohol) films [13]. The SHG is attributed to the naturally oriented dipole layers. Recently Alfano and coworkers have also reported SHG from animal tissues [14]. Even though they attributed the SHG from only the surface term that is due to the broken symmetry at the boundary, they also found some SHG intensity dependence on the tissue constitutes with asymmetric structures such as collagen and keratin. Combining this effect, they demonstrated SHG tomography for mapping the structure of animal tissues by use of 100-fs laser pulses at 625 nm [15].

Taking advantage of EFISHG effect, SHG was shown to have intrinsic sensitivity to the voltage (low frequency electric field) across a biological membrane [16] and was combined with microscopy for membrane potential imaging [17,18]. Membrane-staining dye was also used to demonstrate membrane imaging with SHG microscopy [17–19]. 1064 nm Nd:YAG lasers or 880 nm Ti:sapphire lasers were used in these studies.

We have also extensively studied SHG microscopy on biological tissues. Our recently study indicated that strong SHG images are corresponding to highly organized nano-structures. Numerous biological structures including stacked membranes and arranged protein structures are highly organized in optical scale and are found to exhibit strong optical activities through SHG interactions, behaving similar to man-made nonlinear nano-photonic crystals. Previous observations of SHG on collagen fibrils [11,12,14], purple membrane [13], and muscle fibrils [14,15] are just a few examples of it. This assumption was also supported by the recent SHG microscopy of tooth [20], where the SHG was attributed to the highly ordered structures in enamel that encapsulates the dentine.

9.3   Bio-Photonic Crystal Effect in Biological SHG Microscopy

Recent studies on man-made nano-periodic structures, e.g. super-lattices, indicate strong enhancement in SHG occurring only in noncentro-symmetric media [21]. Another recent study of SHG in a one-dimensional semiconductor Bragg mirror with alternating layers of ~ 100 nm thickness also supports the hypothesis of forward emission SHG enhancement [22] due to nonlinear photonic crystal effect [23]. It is interesting to notice that the periodicity of the observed nonlinear photonic crystal is not necessarily to be on the order of the wavelength or a fraction of the wavelength in the materials [24], but could be much smaller than the affected wavelength with a nanometer scale [21,25]. This strong SHG enhancement could be understood as the break down of optical isotropy within the materials. Therefore, it is reasonable to speculate that the highly organized biological nano-structures may also break optical isotropy and behave as SHG-active photonic crystals. Examples of highly organized biological nano-structures include stacked membranes, such as myelin sheath, endoplastic reticulum (ER), grana in the chloroplast, Golgi apparatus, microfibrils and collagen bundles. It is thus highly possible that these biological structures act as nonlinear "photonic" crystals and can be detected in SHG microscopy. Recently we have studied this bio-photonic crystal effect using a multimodal nonlinear microscopy, which reveals optical SHG activities in those naturally occurring biophotonic crystalline structures. Multimodal nonlinear microscopy [26], in conjunction with the use of a Cr:forsterite laser [27], is based on a combination of different imaging modalities including second-, third-harmonic generations, and multi-photon fluorescence. Due to optical isotropy created by randomly organized biological structures, we found that SHG microscopic images can reveal specially organized crystalline nano-structures inside biological samples where optical centro-symmetry is broken, similar to the nonlinear photonic crystal effect. Our conclusion is also supported by a recent paper, demonstrating that forward emission SHG in purple membrane is caused by nonlinear photonic crystal properties due orderly patched protein bacteriorhodopsin with nano-scaled periodicity [25]. Not only purple membrane, previous SHG observations on collagen and muscle fibrils [11–15] are all with nano-crystalline structures, thus all related to the nonlinear photonic crystal effect. Different from laser-induced fluorescence processes, only virtual states are involved in the harmonic generations (including both SHG and THG). The marked advantage of this virtual transition during wavelength conversion is the lack of energy deposition, thus no photo-damage or bleaching from the process is expected, can be considered as a truly "noninvasive" imaging modality. The SHG microscopy, together with THG microscopy compared in the next section, thus allows structural visualization with minimal or no additional preparation of the samples. Combining multi-photon fluorescence imaging modes, SHG microscopy is useful for investigating the dynamics of structure-function relationship at the molecular and sub-cellular levels.

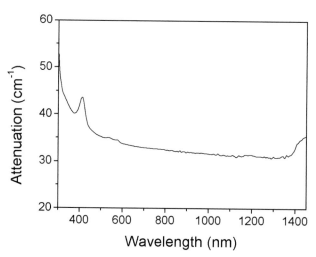

Fig. 9.1. Attenuation spectrum of porcine skin in the visible and near infrared regions. An optical window between 365 nm and 1400 nm is evident

In order to allow both SHG and THG (see next section) within the visible spectrum, but also provide advantages of low attenuation for illumination, we move the excitation wavelength to 1200–1350 nm regimes by using a 100-fs-pulsewidth Cr:forsterite laser. Fig. 9.1 shows that light attenuation (absorption and scattering) in porcine skin reaching a minimum around 1300 nm. Light attenuation in most biological specimen [28,29] reaches minimum in the region of 1300 nm due to the combination of diminishing scattering cross-section with longer wavelength and avoiding resonant molecular absorption of common tissue constituents such as water, protein, and carbohydrates. Due to the high absorption in the visible and near-infrared spectrum, this wavelength is particular useful for imaging plant material [29]. Due to the low attenuation coefficient around 1200–1350 nm, only 10-fold reduction in the signal (for SHG and two-photon fluorescence (2PF)) was observed with 360-μm depths into a maize stem segment fixed in 10% ethanol [26], in good agreement with light attenuation measurement of maize stems [29] that indicates low attenuation coefficient around our illumination with 1230 nm light. This superior depth performance agrees well with previous studies for optical coherent tomography [30] comparing penetration depth between 800 and 1300 nm light sources. In addition, the use of 1300 nm region allows fiber compatibility (due to zero dispersion at 1300 nm) and minimum background detection by Si-based detectors (due to the non-sensitivity of 1200–1350 nm light to Si-based detectors).

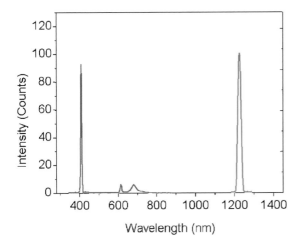

Fig. 9.2. Measured nonlinear emission spectrum from the cell wall of maize stem excited by a femtosecond Cr:forsterite laser with THG centered at 410 nm, SHG centered at 615 nm, and 2PF peaked at 680 nm. The laser spectrum centered at 1230 nm is also provided (not scaled)

Plant cell wall consists of cellulose microfibrils, which are orderly arranged individual to form macrofibrils with a dimension about 10 nm. Within the microfibrils, micelles represent another degree of highly ordered crystalline structure. These crystalline structures produce optical anisotropy and provide active SHG. Figure 9.2

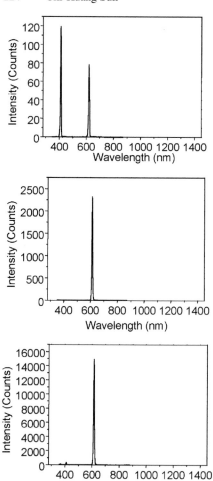

Fig. 9.3. Emission spectra from pear stone cell (*top*), mouse skeletal muscle (*middle*), and potato starch granule (*bottom*) are provided for comparison. All emission spectra (including Fig. 9.2) were taken with similar illumination intensity with normalized 0.1-second integration time

shows the nonlinear emission spectrum that was measured in the transmission direction from the cell wall of a parenchyma in maize (*Zea mays*) stem. Symmetric SHG and THG spectra centered at 615 and 410 nm can both be observed, with intensity similar to/or stronger than residual 2PF centered at 680 nm. SHG reflects the highly organized crystalline structures that break the three-dimensional (3D) optical centro-symmetry in the cell wall. The nature of SHG is further confirmed by the strong signal obtained from the stone cell of pear (*Pyrus serotina* R.) fruit (top of Fig. 9.3). The extensive secondary wall development of the sclerenchyma generates significant SHG signals.

Starch granule, a strong birefringent structure, consists of crystalline amylopectin lamellae organized into effectively spherical blocklets and large concentric growth rings. These structural features are presumably responsible for the strong SHG we observed (bottom of Fig. 9.3). For example, the SHG signal from potato (*Solanum tuberosum* L.) starch granule is so strong that is visible to the naked eyes [26]. Its alternating crystalline and semi-crystalline rings [27] with spatial modulated non-linear properties could behave as 3D nonlinear photonic "bandgap" crystals [31]. The unexpected strong SHG activity might not only be the result of its super-helical amylopectin nano-structures, but also suggest possible SHG non-collinear phase matching condition provided with its reciprocal lattice basis vectors of the 3D photonic bandgap crystals. Spatial frequency of high order structures between the order of 100 nm up to the order of 10 μm, depending on illumination wavelength and composition materials, could all provide the non-collinear phase matching base-vector for SHG process and can be considered as nonlinear biophotonic "bandgap" crystals [23], providing even stronger SHG activity. This significant enhancement of SHG interactions in a 3D structure could also be treated as a simultaneous availability of a high density of states and improvement of effective coherent length (similar to phase matching effect) due to these wavelength-scale photonic "bandgap" structures [22,24].

Figure 9.4 (a) shows the sectioned (x-y) SHG image taken from the ground tissue of maize stem at a depth of 420 μm from the sample surface. THG (see next

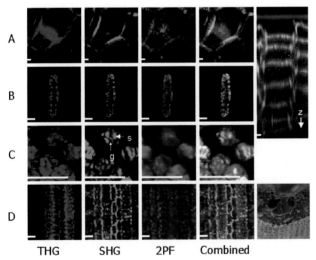

THG   SHG   2PF   Combined

Fig. 9.4. Scanning THG (shown in *blue*), SHG (shown in *green*), 2PF (shown in *red*), and the combined multimodal images. Images were taken from **(a)** (x-y images) ground tissue cells in a maize stem 420 μm from the sample surface **(b)** (x-y images) a live mesophyll cell from *Commelina communis* L. **(c)** (x-y images) enlarged view of (b) **(d)** (x-y images) adaxial surface of rice leaf. The dark region in the middle of THG image corresponds to the bulliform cell layers. Transverse sectional SEM image taken from a rice leaf is provided for correlation purpose (lower right). Multimodal x-z image (with image size of 200 × 540 μm) of ground tissue cells in a maize stem is also shown in the upper right corner. Scale bar: 15 μm. *g*: grana, *s*: starch granule

section) and 2PF of residue fluorescence images are also shown for comparison. SHG shows mainly the longitudinal cell walls while 2PF indicates the distribution of the auto-fluorescent molecules. The combined multimodal image is also provided with blue, green, and red colors representing THG, SHG, and 2PF respectively. A combined x-z-λ image (image size: $200 \times 540 \,\mu m$) of ground tissue cells is also presented in the upper right corner of Fig. 9.4. The intensity contrast of this specific image was processed with log scale so that bottom image corresponding to deeper part of the sample can be easily observed. The correspondence of SHG to longitudinal cell walls can be clearly observed.

A paradermal optical (x-y) section of the adaxial surface of rice (*Oryza sativa* L.) leaf (Fig. 9.4 (d)) also reveals that SHG picks up optically active crystalline structures, including the cuticular papillae and longitudinal cell walls, due to the orderly arrangement of cutin, waxes and cellulose microfibrils respectively. The THG (see next section) and 2PF (corresponding to chlorophyll emission wavelength) images are also shown. For correlation purpose, the lower right corner of Fig. 9.4 gives a cross-sectional SEM (black-and-white) image of a similar rice leaf.

There are numbers of structures in animal tissue that are good candidates for strong SHG. For instance, the orderly arranged sarcomeres in the skeletal muscle have structures fall into the spatial range of strong SHG activity. Fig. 9.5 (a) shows longitudinally sectioned x-y images obtained from the skeletal muscle of mouse. Strong SHG emission was recorded due to orderly packed actin/myosin complexes in sarcomeres (middle of Fig. 9.3). The strong SHG activity from actin/myosin complex can be a useful tool for the study of muscle cell dynamics. Other SHG sources include collagen fiber bundles in connective tissues, as previously reported. Collagen fibrils exhibit a sequence of closely spaced transverse bands those repeat every 68 nm along the length of the fiber, providing necessary condition for bio-photonic crystal effect of SHG activity. Figure 9.5 (b) shows cross-sectional x-z image taken from the mouse dermis. The collagen fibrils inside the connective tissue right underneath the epithelial cells can be clearly observed through SHG. Its corresponding x-y section taken at a constant depth of $30 \,\mu m$ was also shown in Fig. 9.5 (c) with wavy structures from collagen fiber bundles. Excellent contrast can be easily obtained through SHG microscopy without staining.

The strength of the SHG signals can vary according to the orientation of the structures which may have different $\chi^{(2)}$ matrix components. This mechanism provides opportunity for structural orientation studies using SHG with controlled illumination polarization. For instance, by varying the incident light polarization (Fig. 9.5 (d)–(f)), the concentric silica deposition in the dumb-bell-shaped silica cells of rice leaf produces orientated SHG images in respecting to the orientation of the illumination polarization (shown as arrows). For comparison, a paradermal SEM (black-and-white) image showing silica cells taken from a similar rice leaf is included in lower right corner of Fig. 9.5.

Apart from mineral deposition in plant cells, laminated membrane structures are also potential candidates for producing strong SHG activity. In chloroplasts, in addition to the 2PF signals generated from the highly auto-fluorescing photosynthetic

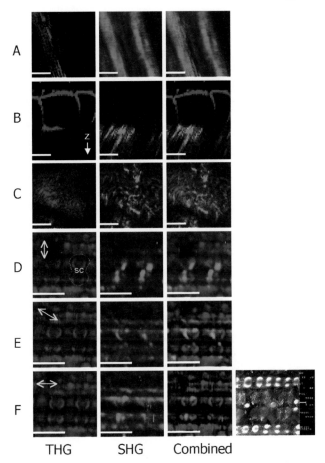

THG   SHG   Combined

Fig. 9.5. Scanning THG (shown in *blue*), SHG (shown in *green*), and the combined multi-modal images. Laser induced fluorescences are too weak to detect in these samples under our imaging condition. Images were taken from **(a)** (x-y images) longitudinal optical section of mouse skeletal muscle **(b)** (x-z images) optical cross-section of mouse dermis **(c)** (x-y images) corresponding to (b) at a constant depth of 30 μm with wavy structures showing collagen fiber bundles. **(c)**, **(e)**, **(f)** adaxial surface of rice leaf showing dumb-bell-shaped silica cells. Paradermal SEM image taken from a rice leaf is provided for correlation purpose (lower right). Scale bar: 15 μm. *Yellow arrows* in **(d)**, **(e)**, **(f)**: direction of illumination polarization. *sc*: silica cell

pigments, SHG appears in different sub-organelle compartments (Fig. 9.4 (c), (d)). Matching with TEM images of similar specimens, we concluded that the signals of SHG are the result of the orderly stacked thylakoid membranes in grana and the highly birefringent starch granules in the chloroplasts. The stacked thylakoid membranes of grana and the orderly deposited amylopectin in the starch granules provide the structural requirement for efficient SHG, resembling the behavior of photonic crystals.

Since SHG occurs from the nano-crystalline structures under study, it is highly desirable to use other noninvasive harmonic generation effect to image general cellular features so that proper correlation of structures can be obtained. Due to the optical dispersion property in biological tissues, THG was proven to be generated from regions with optical inhomogeneity [32–34] and was applied to image general cellular structures [26,35–37]. In the next section, we will discuss and compare THG microscopy with SHG microscopy in biological specimens.

9.4 THG Microscopy

THG is generally a weak process but is dipole allowed; therefore it occurs in all materials, including materials with inversion symmetry. In 1995, Tsang [32] reported that when using focused high intensity ultrashort laser pulses, this normally weak THG process becomes highly operative at a simple air-dielectric interface and is much stronger then the bulk of most dielectric materials. It was also reported [32] that the interface THG is a fundamental physical process occurring at all interfaces free from the constraint of a phase-matching condition and wavelength restriction. This could be explained by Boyd in his book "Nonlinear Optics" [33] by treating THG process with a focused Gaussian beam. According to Boyd, the efficiency of THG vanishes when focused inside a uniform sample for the case of positive phase mismatch ($\Delta k = k_3 - 3k_1 > 0$, true for normally dispersive materials) or even phase matching ($\Delta k = 0$). Efficient THG will only occur when the laser beam was focused in some inhomogeneous regions like interfaces, unless the materials possessing anomalous dispersion. Taking advantage of this characteristic, in 1997 Silberberg and his coworkers demonstrated THG microscopy for interface images in an optical glass fiber [34]. THG microscopy was then applied to image laser-induced breakdown in glass [38], liquid crystal structures [39], and defect distribution in GaN semiconductors [7].

For THG applications in biology, in 1996 Alfano and his coworkers found THG in chicken tissues using a 10 Hz 1064 nm Nd:YAG laser with 30 ps pulsewidth [14]. After the first demonstration of THG microscopy [34], it was then quickly applied to image biological specimen due to its interface sensitivity, by using 1200 nm femtosecond light from an optical parametric amplifier after a Ti:sapphire amplifier with 250 kHz repetition rate [34,35], 1500 nm femtosecond light from an optical parametric oscillator synchronously pumped by a Ti:sapphire laser with 80 MHz repetition rate [36], or with 1230 nm femtosecond light directly from a Cr:forsterite laser with 110 MHz repetition rate [26]. Detailed cellular and subcellular organelles can be clearly resolved with THG even with dynamic imaging. With THG imaging on Chara plant rhozoids, non-fading image can be achieved with continuous viewing, indicating prolonged viability under the viewing condition with 1200 nm light [35]. Recently THG microscopy even allows one to temporally visualize the release of Ca^{2+} [40]. With cubic dependence on the illumination intensity, THG provides even better optical sectioning resolution than SHG or 2PF using the same illumination wavelength, but is more sensitive to attenuation to the illumination light [26].

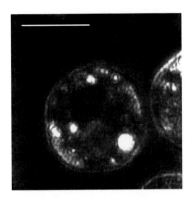

Fig. 9.6. Scanning (x-y) THG image of rat hepatocytes. General cellular and subcellular organelles can be clearly resolved. Scale bar: 20 μm

In order to allow THG within the visible spectrum, but also provide advantages of low attenuation for illumination, we move the THG excitation wavelength to 1230 nm regimes by using a 100-fs-pulsewidth Cr:forsterite laser. Like SHG microscopy, transmission detection in THG microscopy is preferred. Fig. 9.6 shows a THG image of rat hepatocytes taken with 1230 nm light from a Cr:forsterite laser. General cellular and subcellular organelles can be clearly resolved with THG due to its superior capability to image interfaces and inhomogeneity.

It is also interesting to compare THG with its corresponding SHG images. In general, SHG reflects the orderly arranged nano-structures while THG gives general interfacial profiles but requires inhomogeneity. For example, strong THG is induced due to optical inhomogeneity within and surrounding the cell walls (Fig. 9.4 (a)), providing images not just from the longitudinal cell walls but especially from the transverse cell walls due to their high spatial fluctuations in the light propagation direction (z-direction), with the ability to pick up the whole cell profile. In Fig. 9.5 (b), THG reveals the cell profile of epithelial cells. In Figs. 9.4 (d) and 9.5 (d)–(f), THG picks up structural interfaces, such as the papillae from the cuticular layer and the boundary of silica cells. THG can even pick up the bottom surface of cover glass as shown in the top of the upper-right corner image of Fig. 9.4. Not just cell profiles and structure interfaces, THG also provides information on the various sub-organellar interfaces (Fig. 9.5 (b), (c) and Fig. 9.6).

On the other hand, in highly organized skeleton muscles and connective tissues, SHG seems to provide better signal intensity (middle of Fig. 9.3) and contrast (Fig. 9.5 (a)–(c)) compared with THG. Due to optical homogeneity inside some animal tissues, very weak THG (middle of Fig. 9.3) can only be picked up in interfaces like the longitudinal interface of myofibril and surrounding cytoplasm due to induced optical inhomogeneity (Fig. 9.5 (a)). It is also interesting to notice that, since THG is allowed for isotropic materials and can be applied to image general interfaces, thus has weaker dependence on incident light polarization. No polarization dependency was found in THG images of Fig. 9.5 (d)–(f).

9.5   Conclusion

Light microscopy, fluorescence in particular, has been used extensively in the correlation between structures, molecules, and functions in modern biological sciences. Either intrinsic or extrinsic fluorescent probes are needed in fluorescence microscopy using single- or multi-photon excitation schemes. Hence, dye penetration, probe toxicity, and photo-bleaching/damage are the limiting factors frequently encountered. Different from fluorescence processes, SHG and THG processes are based on virtual transitions. The marked advantage of this virtual transition during wavelength conversion is the lack of energy deposition, thus no photo-damage or bleaching from the process is expected, can be considered as a truly "noninvasive" imaging modality. Due to its virtual nature, no saturation or bleaching in the harmonic generation signal is expected, thus ideal for live samples without preparation or with minimum staining under prolonged viewing condition. SHG microscopy can provide images on stacked membranes and arranged proteins with organized nanostructures due to the bio-photonic crystal effect or membrane potential imaging. On the other hand, THG microscopy can provide general cellular or subcellular interface imaging due to optical inhomogeneity.

In order to accomplish the "non-invasive" characteristic of harmonic generation imaging on live specimen, it is also essential to avoid other side effects induced by the excitation light. Our wavelength selection is based on previous studies of the attenuation spectra of living specimen indicating a light penetration window between 400–1300 nm. In order to have THG in the visible range, we choose the excitation light in the 1200–1300 nm range. Longer excitation wavelength will cause more water absorption and reduce its penetration capability. The quest for optimized wavelength and viewing condition will be a challenge for THG microscopy. Our recent studies comparing photo-damages induced by 1230 nm femtosecond light and 780 nm femtosecond light indicate suppressed multi-photon autofluorescence and suppressed multi-photon damages with longer 1230 nm light in plant materials [41], due to the fact that multi-photon absorption cross-section decreases with longer wavelength.

Sometimes it is also important to combine harmonic generation imaging modalities with multi-photon fluorescence modality due to the fact that the later is equipped with the capability to image different molecular distributions. For example, combining multi-photon fluorescence imaging modes, SHG microscopy is useful for investigating the dynamics of structure-function relationship at the molecular and sub-cellular levels. However, with an excitation wavelength longer than 1200 nm, the multiphoton absorption cross-section will also be decreased. For instance, blue-green fluorescence dyes and green-fluorescence protein will require 3-photon excitation, thus with much weaker fluorescence intensity [42]. It is therefore also important to develop efficient long excitation wavelength or long emission wavelength dyes for simultaneous multi-modal microscopy. Our recent study has indicated efficient multi-photon fluorescence for some common dyes under 1230 nm excitation condition [43].

Other restriction of harmonic generation microscopy includes its imaging geometry. Due to momentum conservation law, transmission detection is preferred for SHG and THG microscopies. Even though this might cause difficulties for some applications requiring reflection mode, it also possesses some advantages. The forward emission properties of harmonic generation allow efficient collection in transmission geometry, in contrast to inefficient collection of 4π fluorescence emission. The excitation light and the emission light will take different paths in transmission detection, thus allows thicker penetration depth with controlled sample thickness. On the other hand, in a highly scattering media, the reflection detection of the short-wavelength harmonic-generation light should also be possible due to the fact that light scattering increases with shorter wavelength.

References

1. R. Hellwarth and P. Christensen: Opt. Commun. **12**, 318 (1974)
2. C. J. R. Sheppard, R. Kompfner, J. Gannaway, and D. Walsh: IEEE J. Quantum Electron. **13**, 100 (1977)
3. G. T. Boyd, Y. R. Shen, and T. W. Hänsch: Opt. Lett. **11**, 97 (1986)
4. Y. Uesu, S. Kurimura, and Y. Yamamoto: Appl. Phys. Lett. **66**, 2165 (1995)
5. R. Gauderon, P. B. Lukins, and C. J. R. Sheppard: Opt. Lett. **23**, 1209 (1998)
6. We follow the notation of H. A. Haus: *Waves and Fields in Optoelectronics* (Prentice-Hall, Inc., New Jersey 1984)
7. C.-K. Sun, S.-W. Chu, S.-P. Tai, S. Keller, U. K. Mishra, and S. P. DenBaars: Appl. Phys. Lett. **77**, 2331 (2000)
8. C.-K. Sun, S.-W. Chu, S. P. Tai, S. Keller, A. Abare, U. K. Mishra, and S. P. DenBaars: J. Scanning Microscopies **23**, 182 (2001)
9. F. Bresson, F. E. Hernandez, J.-W. Shi, and C.-K. Sun: 'Electric-field induced second harmonic generation in nematic liquid crystals as a probe for electric field', In: *Proceeding of Optics and Photonics Taiwan '01*, ed. by (Kaohsiung, Taiwan ROC 2001) pp. 713–715
10. F. Bresson, C.C. Chen, F.E. Hernandez, J.-W. Shi, and C.-K. Sun: submitted to Rev. Sci. Instrum.
11. S. Fine and W. P. Hansen: Appl. Optics **10**, 2350 (1971)
12. I. Freund, M. Deutsch, and A. Sprecher: Biophys. J. **50**, 693 (1986)
13. J. Y. Huang, Z. Chen, and A. Lewis: J. Phys. Chem. **93**, 3314 (1989)
14. Y. Guo, P. P. Ho, A. Tirksliunas, F. Liu, and R. R. Alfano: Appl. Optics **35**, 6810 (1996)
15. Y. Guo, P. P. Ho, H. Savage, D. Harris, P. Sacks, S. Schantz, F. Liu, N. Zhadin, and R. R. Alfano: Opt. Lett. **22**, 1323 (1997)
16. O. Bouevitch, A. Lewis, I. Pinevsky, J. P. Wuskell, and L. M. Loew: Biophys. J. **65**, 672 (1993)
17. P. J. Campagnola, M. D. Wei, A. Lewis, and L. M. Loew: Biophys. J. **77**, 3341 (1999)
18. G. Peleg, A. Lewis, M. Linial, and L. M. Loew: P. Natl. Acad. Sci. USA **96**, 6700 (1999)
19. L. Moreax, O. Sandre, and J. Mertz: J. Opt. Soc. Am. B **17**, 1685 (2000)
20. F.-J. Kao, Y.-S. Wang, M.-K. Huang, S. L. Huang, P.-C. Cheng: P. Soc. Photo–Opt. Inst. **4082**, 119 (2000)
21. T. Zhao, Z.-H. Chen, F. Chen, W.-S. Shi, H.-B. Lu, and G.-Z. Yang: Phys. Rev. B **60**, 1697 (1999)

22. Y. Dumeige, P. Vidakovic, S. Sauvage, I. Sagnes, J. A. Levenson, C. Sibilia, M. Centini, G. D'Aguanno, and M. Scalora: Appl. Phys. Lett. **78**, 3021 (2001)
23. V. Berger: Phys. Rev. Lett. **81**, 4136 (1998)
24. N. G. R. Broderick, G. W. Ross, H. L. Offerhaus, D. J. Richardson, and D. C. Hanna: Phys. Rev. Lett. **84**, 4345 (2000)
25. K. Clays. S. Van Elshocht, M. Chi, E. Lepoudre, and A. Persoons: J. Opt. Soc. Am. B **18**, 1474 (2001)
26. S.-W. Chu, I-H. Chen, T.-M. Liu, B.-L. Lin, P. C. Cheng, and C.-K. Sun: Opt. Lett. **26**, 1909 (2001)
27. T.-M. Liu, S.-W. Chu, C.-K. Sun, B.-L. Lin, P. C. Cheng, and I. Johnson: J. Scanning Microscopies **23**, 249 (2001)
28. R. R. Anderson and J. A. Parish: J. Invest. Dermat. **77**, 13 (1981)
29. P. C. Cheng, S. J. Pan, A. Shih, K.-S. Kim, W. S. Liou, and M. S. Park: J. Microscopy **189**, 199 (1998)
30. B. E. Bouma, G. J. Tearney, I. P. Bilinsky, B. Golubovic, and J. G. Fujimoto: Opt. Lett. **21**, 1839 (1996)
31. D. J. Gallant, B. Bouchet, and P. M. Baldwin: Carbohydrate Polymers **32**, 177 (1997)
32. T. Y. F. Tsang: Phys. Rev. A **52**, 4116 (1995)
33. R. W. Boyd: *Nonlinear Optics* (Academic Press, San Diego, CA 1992)
34. Y. Barad, H. Eisenberg, M. Horowitz, and Y. Silberberg: Appl. Phys. Lett. **70**, 922 (1997)
35. M. Müller, J. Squier, K. R. Wilson, and G. J. Brakenhoff: J. Microscopy **191**, 266 (1998)
36. J. A. Squier, M. Müller, G. J. Brakenhoff, and K. R. Wilson: Opt. Express **3**, 315 (1998)
37. D. Yelin and Y. Silberberg: Opt. Express **5**, 169 (1999)
38. J. A. Squier and M. Müller: Appl. Optics **38**, 5789 (1999)
39. D. Yelin, Y. Silberberg, Y. Barad, and J. S. Patel: Appl. Phys. Lett. **74**, 3107 (1999)
40. L. Canioni, S. Rivet, L. Sarger, R. Barille, P. Vacher, and P. Voisin: Opt. Lett. **26**, 515 (2001)
41. I-H. Chen, S.-W. Chu, C.-K. Sun, P. C. Cheng, and B.-L. Lin: 'Wavelength dependent cell damages in multi-photon confocal microscopy'. In: *Optical and Quantum Electronics* (in press 2002)
42. T.-M. Liu, S.-W. Chiu, I-H. Chen, C.-K. Sun, B.-L. Lin, W.-J. Yang, P.-C. Cheng, and I. Johnson: 'Multi-photon fluorescence of green fluorescence protein (GFP) and commonly used bio-probes excited by femtosecond Cr:forsterite lasers', In: *Proceedings of 2nd International Photonics Conference*, ed. by Hsinchu (TAIWAN, 2000 119) pp. 121–
43. T.-M. Liu, S.-W. Chu, C.-K. Sun, B.-L. Lin, P. C. Cheng, and I. Johnson: J. Scanning Microscopies **23**, 249 (2001)

Part III

Miscellaneous Methods in Optical Imaging

10 Adaptive Optics

Chris Dainty

10.1 Introduction

The principle of adaptive optics is straightforward and is depicted in Fig. 10.1 in the context of astronomical imaging. Because of atmospheric turbulence, the wavefront

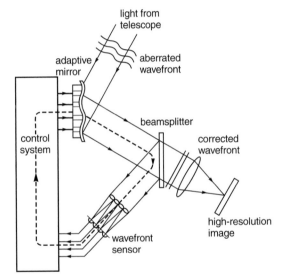

Fig. 10.1. Schematic arrangement of an adaptive optics system

from a pointlike object is continuously distorted. The purpose of the adaptive optics (AO) system is to remove this distortion and provide diffraction-limited imaging if possible. This is achieved by sensing the distortion using a *wavefront sensor* and compensating for it using a *deformable mirror*. The whole system operates in closed loop with the aid of a suitable *control system*. These are the three key elements of an adaptive optics system. The principle of operation is simple: the control system drives the deformable mirror so that the error in the wavefront is minimised. The "devil lies in the detail" of every individual part of the system and of the way they interact with each other and with the incoming dynamically distorted wave.

In this chapter, I review the basic elements of adaptive optics with special reference to the fundamental limitations rather than to the technology. This is preceeded by a brief (and selective) discussion of the history of adaptive optics. In order to

Török/Kao (Eds.): Optical Imaging and Microscopy, Springer Series in Optical Sciences
Vol. 87 – © Springer-Verlag, Berlin Heidelberg 2003

move the subject into the mainstream of optics – so that it becomes another tool for the construction of optical system – it is important to understand why some people felt (and others still believe) that it was necessary to spend millions of dollars to build adaptive optics systems. This chapter concludes with a short description of two of the key current issues in astronomical adaptive optics (laser guide stars and multiconjugate AO) and of the potential applications of adaptive optics in the human eye.

10.2   Historical Background

The origins of adaptive optics in the "modern" era can be traced back nearly fifty years to a paper by an astronomer, Horace Babcock [1]. The angular resolution of groundbased telescopes is limited by the optical effects of atmospheric turbulence. The loss in resolution is very large as the following discussion illustrates.

The statistics of the phase perturbation due to atmospheric turbulence are characterised by a scale length r_0, the Fried parameter [2] . A telescope with a diameter equal to the Fried parameter r_0 has a mean square (piston-removed) phase fluctuation of approximately 1 rad^2, and the image quality, provided that the two tilt components are removed, is high (a Strehl ratio [3] of ≈ 0.87, as shown by Noll [4]). The quantity r_0 is therefore also approximately equal to the diameter of a diffraction-limited telescope that gives the same angular width of the point spread function as that produced when imaging through turbulence whose phase statistics are characterised by r_0. If the atmospheric turbulence is characterised by a small value of r_0, then the image produced in a large telescope in the presence of turbulence will only have an angular image size given by that of a small telescope (of diameter r_0).

As is so often the case in adaptive optics, the key issues depend critically on the numerical values of certain parameters ("the devil lies in the detail"). What is a typical value for the Fried parameter r_0? Using the Kolmogorov model [5] for the refractive index fluctuations in the atmosphere and the assumption that the turbulence is "weak", it can be shown [6] that the Fried parameter is given by the expression:

$$r_0 = \left(0.42 k^2 \, (\cos \gamma)^{-1} \int_0^\infty C_N^2(z) dz \right)^{-3/5} \tag{10.1}$$

In this equation, k is the wavenumber ($= 2\pi/\lambda$), γ is the angle of observation measured from the zenith and $C_N^2(z)$ in the refractive index structure function "constant" as a function of altitude z. One of the consequences of (10.1) is that r_0 is proportional to the six-fifths power of the wavelength:

$$r_0 \propto \lambda^{6/5}. \tag{10.2}$$

At an excellent observing site, the numerical value of r_0 in the visible region of the spectrum is between 10 and 20 cm, and this corresponds to values in the range 50-100 cm in the K band ($\approx 2.2\,\mu m$). So, returning now to Babcock's era, the 5 m diameter telescope at Mt. Palomar (where r_0 in the visible would typically be ≈ 10 cm) would have an angular resolution roughly equal to that of a 10 cm diameter amateur telescope, that is, about 1 arc-second, compared to its theoretical limit some fifty times smaller, 0.02 arc-sec. The effect of atmospheric turbulence in large telescopes totally dominates their angular resolution, and the degradation in resolution caused by turbulence is very severe indeed.

In order to overcome the deleterious effects of atmospheric turbulence, Babcock suggested the closed loop arrangement schematically shown in Fig. 10.1 with a Schlieren optical system as the wavefront sensor and an Eidophor (a system using electric charge to deform an oil film) as the corrective device. A working system was not built. At least two problems existed at the time: firstly, the technology, particularly that of the corrective device was simply not up to the job, and secondly there was the suspicion that this principle would only work on bright stars. We return to this latter issue in the following Section.

Although Babcock did not make an adaptive optics system, he had suggested how the effects of atmospheric turbulence might be overcome in real-time in astronomy and indeed in any application which involved imaging through turbulence. As illustrated above, the potential gain in angular resolution is very large, and this benefit helped to justify the large investment made by the US Government from the early 1970s onwards in adaptive optics. Some details of this are given by J.W. Hardy [7] of Itek Corporation, who pioneered the construction first of a 21-actuator "real time atmospheric compensator" and later of a 168-actuator "Compensated Imaging System" on the 1.6 m diameter satellite tracking telescope at Mt. Haleakala in Maui, Hawaii, commissioned in 1982. Even by modern standards, this is an impressive system, although its initial limiting magnitude – an issue of crucial importance for astronomy – was only about $m_v \approx 7^{th}$.

The challenge of implementing adaptive optics in astronomy was first taken up by a group of French observatories working together with the European Southern Observatory. The first system was a 19-actuator system called COME-ON [8] installed at the 3.6 m diameter telescope at La Silla, Chile, later upgraded to 52-actuators and with user-friendly software (ADONIS) with a limiting magnitude on the order of $m_v \approx 13^{th}$. These systems, in common with all other astronomical AO systems built to date, operate primarily in the infrared. As a result of the wavelength dependence of r_0 (10.2), and also because the required bandwidth is smaller and the isoplanatic angle is larger, operation at longer wavelengths is easier, and, in particular, brighter guide stars can be used. Figure 10.2 shows one image obtained from this system [9], illustrating the remarkable increase in information that the increased resolution – even in the infrared – that adaptive optics brings to astronomy.

Not surprisingly, all large telescopes are now equipped with adaptive optics systems operating in the near infrared. The case for implementing adaptive optics is overwhelming, and justifies the relatively large cost of $1–5M. Astronomical AO systems do not yet operate routinely in the visible region of the spectrum and in a later section we explore some of the current research issues in astronomical AO which are working towards wide field, full sky coverage in the visible.

Come On + OFF   Come On + ON

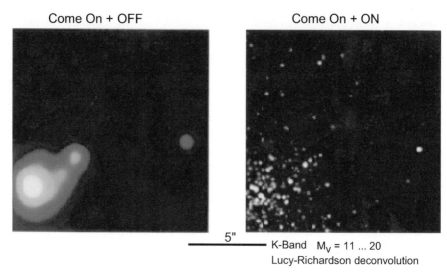

5"
——————————— K-Band M_v = 11 ... 20
Lucy-Richardson deconvolution

Fig. 10.2. Starburst region R136 in Doradus 30 [9]. Left: Adaptive optics off, Right: Adaptive optics on, with additional Richardson–Lucy deconvolution. Courtesy of Bernhard Brandl and Andreas Quirrenbach

The adaptive optics systems built for astronomy and for military systems are individually designed and constructed "one-off" instruments, operating to the highest possible performance given current technology. Particularly in astronomy, the building of an AO system is a high-profile project that cannot be allowed to fail. In addition, AO systems in astronomy have to be "user-instruments", i.e. be able to be operated by visiting astronomers not familiar with the detailed workings of adaptive optics: this means that the user-interface software has to be of a very high standard. All these facts contribute to the very large cost of astronomical AO systems, but these high costs are justifiable in terms of the scientific productivity of the finished instruments.

However, there are many subjects other than astronomy that would benefit enormously from adaptive optics, but in order to justify the use of AO in these cases it is essential that the cost be reduced dramatically, certainly to thousands if not hundreds of dollars. In order to reduce the costs by several orders of magnitude needs both a proper understanding of the essentials of adaptive optics as well as a different approach to that used in astronomy and in military applications. In 2000, two groups demonstrated that low cost systems could be built successfully [10,11] and now it is generally accepted that high quality, high bandwidth adaptive optics is achievable at much lower cost than previously imagined. This is due in part to developments in technology and in part to a more risk-oriented approach to research not found in the big-science area of astronomy.

10.3   Strehl Ratio and Wavefront Variance

The effect of a *small* aberration in an imaging system is to lower the central intensity of the point spread function and redistribute the energy, in particular, filling in the zeros, as shown schematically in Fig. 10.3. The **Strehl ratio** S [3] is the ratio of

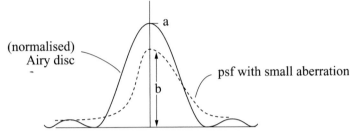

Fig. 10.3. Effect of a small aberration: Strehl ratio

the central intensity with the aberration to the central intensity in the absence of the aberration and a value of 0.8 was proposed by Strehl as providing a suitable criterion for the quality of an imaging system: if $S > 0.8$, the system is *effectively* diffraction-limited whereas if $S < 0.8$ the aberration is significant. For *small* aberrations, the Strehl ratio S and the wavefront phase variance σ_ϕ^2 are related by

$$S \approx 1 - \sigma_\phi^2 \qquad (10.3)$$

and thus a value $S = 0.8$ corresponds to a wavefront phase variance $\sigma_\phi^2 = 0.2\,\text{rad}^2$, or an actual wavefront variance $\sigma_W^2 = \lambda^2/200$.

In adaptive optical systems, like other high-quality imaging systems, it is customary to use the Strehl ratio and/or the wavefront variance as a measure of the image quality. The sources of error – and there are many of them – in an AO system are frequently uncorrelated, so that the total variance of the wavefront error is the sum of the variances from each source of error, and because $1 - \sigma^2 \approx \exp{-\sigma^2}$, the Strehl factors for each source of error are multiplicative. (It should be stressed that this latter result is true only for small errors.)

It is very difficult to make an AO system for astronomical use with a Strehl ratio as high as 0.8, and more typical values for these systems are in the range $0.2 - 0.6$.

In working out how to design an AO system, it is useful to know how different components affect the wavefront error. One might start with an understanding of the wavefront to be corrected. In his classic paper, Noll [4] evaluated the magnitudes of the mean square Zernike components for Kolmogorov turbulence, and then calculated the residual mean square phase error that would result if the Zernike components in the wavefront were successively corrected. Table 10.1 summarises his results. For example, if the telescope diameter $D = 5r_0$, i.e. $(D/r_0)^{5/3} \approx 15$, then exact correction of the first 10 Zernikes would yield a residual wavefront error of $\approx 15 \times 0.04 \approx 0.60\,\text{rad}^2$.

Table 10.1. The residual phase variance (in rad^2) when the first J Zernike coefficients of Kolmogorov turbulence are corrected exactly (after Noll [4]). The values of the phase variance given are for $D = r_0$ and scale as $(D/r_0)^{5/3}$

J	Squared error	J	Squared Error
1	1.030	12	0.0352
2	0.582	13	0.0328
3	0.134	14	0.0304
4	0.111	15	0.0279
5	0.0880	16	0.0267
6	0.0648	17	0.0255
7	0.0587	18	0.0243
8	0.0525	19	0.0232
9	0.0463	20	0.0220
10	0.0401	21	0.0208
11	0.0377		

Whilst Noll's paper sheds light on the idea of correcting successive components of the aberration, and thus improving the final image quality in an adaptive optics system, it has also been the source of some misunderstanding, since it focuses on the Zernike expansion of the wavefront. It is well-known that the Zernike expansion is particularly suited to the description of the aberrations of a rotationally symmetrical optical system, and for this reason it has a special place in traditional optics. However, as we shall see in Sec. 10.6, the Zernike expansion has no special role in adaptive optics, and other expansions, such as those involving sensor or mirror modes have a much more important role. In particular, when operating a closed loop AO control system, there is no benefit or need to consider Zernike modes.

10.4   Wavefront Sensing

The purpose of the wavefront sensor in an adaptive optics system is to measure the wavefront with the required spatial and temporal sampling. A generic wavefront sensor consists of some kind of optical system followed by a detector (or set of detectors) and gives rise to (at least) three sources of error, each specified as a wavefront variance: a spatial wavefront sampling error σ_s^2, a temporal bandwidth error σ_t^2 and a detector noise error σ_d^2. The temporal error is usually lumped together with other temporal errors (for example, those in the control system) as a bandwidth error σ_{bw}^2, and the detector error frequently divided into several terms, such as the errors due to CCD read noise, photon noise, etc. The important point here is to realise that these errors are of crucial importance in determining the effectiveness of the adaptive optics system, and in general the variance of each error has to be *very* much smaller than one rad^2, so that the combined error from all sources adds up to less than approximately one radian squared.

The *ideal* wavefront sensor would have no detector noise, would not introduce any time delay into the measurement process, and would be capable of sensing only those wavefront deformations that the mirror can produce. The first two requirements are clearly unattainable. The third requirement can be met in some cases by closely matching the wavefront sensor to the deformable mirror, for example using a wavefront curvature sensor for a bimorph or membrane mirror. The sensor must be capable of sensing all the mirror modes but there is little point in it being capable of sensing modes that the mirror cannot correct (see Sec. 10.6). Thus a wavefront sensor optimised for a particular closed-loop AO system is usually quite different from a wavefront sensor designed simply for measuring wavefronts in an open loop system (e.g. a wavefront sensor designed for metrology). In the latter case, one typically would use a much higher degree of spatial sampling.

Shack-Hartmann Wavefront Sensor

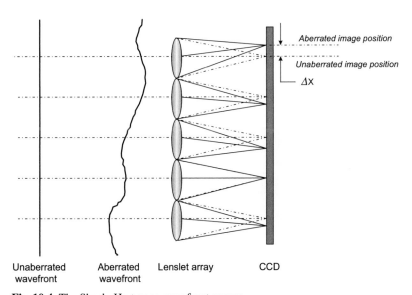

Fig. 10.4. The Shack–Hartmann wavefront sensor

There are a number of wavefront sensors that have been used in AO systems and several others have been suggested. One way of classifying them is by the quantity that is directly measured, and on this basis one can make the following classification:

- **Wavefront**: Interferometry, Common Path Interferometry [12]
- **Wavefront Slope**: Shearing Interferometer [13], Shack–Hartmann Sensor [14], Pyramid Sensor [15]
- **Wavefront Curvature**: Curvature Sensor [16], Phase Diversity [17,18]
- **Other**: Modal Sensing [19,20]

This is only a partial list, and special devices might be more suitable for specialist AO systems. Furthermore, devices can be used in different ways: for example, the

Shack–Hartmann sensor can be modified to yield the wavefront curvature as well as the wavefront slope [21].

From an operational point of view, a better classification might be whether the wavefront sensor is adaptable in its sensitivity as an AO loop is closed. On this basis, the pyramid and curvature sensors would be in a class of their own, and this variable sensitivity allows them to reach limiting magnitudes up to ten times fainter in astronomical applications. The most common wavefront sensor in adaptive optics is the Shack–Hartmann device, largely because of its simplicity to implement and to understand. The principle of operation is shown in Fig. 10.4. It consists of a lenslet array in front of a pixellated detector such as a CCD. A plane wave incident upon the device produces an array of spots on a precise matrix determined by the geometry of the lenslet array, and the centroids of these spots can be found by computation. A distorted wave also gives an array of spots but the centroid of each spot in now determined by the average slope of the wavefront over each lenslet, and hence by finding the centroids of every spot, the wavefront slope at each point can be found. The wavefront can be reconstructed from the slope information by a least squares or other method [22]. Figure 10.5 gives an example of a spot pattern and the reconstructed wavefront for the case of measuring the wavefront aberration of the eye. The detector error in wavefront sensing is one that dominates the error budget

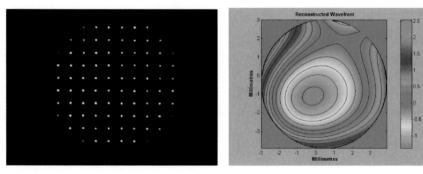

Fig. 10.5. Sample spot pattern in a Shack–Hartmann sensor (*left*) and the reconstructed wavefront (*right*). Courtesy of D. P. Catlin

of astronomical adaptive optics systems. In this case, the read noise of the CCD is significant, and special high-speed, low noise multi-port CCDs are specially developed for this purpose. The effect of detector noise on wavefront sensing is discussed in [23]. As a rule-of-thumb for astronomy, at least 100 detected photons per spot are required to close the loop with a low wavefront variance due to photon noise: however, one can still, in principle, close the loop with as few as 10 detected photons per spot provided one accepts a higher wavefront variance due to photon noise. In other applications, noise sources other than detector noise may be more important, such as the speckle noise in the case of wavefront sensing in the eye at high speed.

The spatial sampling error of the Shack–Hartmann sensor is governed by the lenslet dimensions. As a guideline, in astronomical applications, the lenslet side is on the order of the Fried parameter r_0, and the uniform mesh of the lenslet array is well-suited to the isotropic nature of the atmophere-induced aberration to be corrected. In contrast, for aberrations that are non-isotropic, e.g. increase in magnitude towards the edge of the pupil, as in the eye, the uniform sampling is not ideal. In non-astronomical applications, there are probably significant new methods of wavefront sensing to be invented for adaptive optics.

10.5   Deformable Mirrors and Other Corrective Devices

There are many possible devices that can be used to modulate the wavefront and hence provide correction for dynamic aberrations. These include:

- Zonal Continuous Facesheet Mirrors [24]
- Bimorph Mirrors [25,26]
- Membrane Mirrors [27]
- MEMS Mirrors [28,29]
- Liquid Crystal Devices [30–32]

In a zonal mirror, the influence function $r(x, y)$ of the mirror is localised around the location of the actuator, whereas in a modal device (such as a membrane or bimorph mirror) each actuator has an influence function that extends over the whole mirror. For low order AO, there is general agreement that modal corrective devices offer a number advantages; for example, focus correction can be very simply implemented. Until recently, the cost of a deformable and its power supply was formidable, in

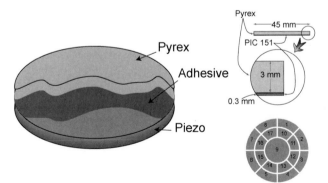

Fig. 10.6. The construction of a bimorph mirror

excess of $1000 per actuator but now low cost membrane mirrors [27] are available at approximately $100 per actuator for a 37-actuator device. These mirrors are rather fragile, and another modal technology is the bimorph mirror which has the potential for very low cost. In one form of bimorph mirror, shown in Fig. 10.6, a slice of piezo-electric material with an electrode pattern etched onto one of its conducting surfaces, is bonded to a glass substrate. This device tends to be temperature sensitive due to the differing coefficients of expansion of the glass and piezo material, and in an alternative structure two piezo slices are bonded to each other and an optical mirror replicated onto one of the surfaces. There is also great interest in the development of MEMS mirrors, although these are still in development and are not yet low cost devices [29,33].

How many actuators are required to correct a given type of aberration? This is not a simple question to answer, the result clearly depending on the statistics of the aberration and the influence functions of the mirror (or liquid crystal) actuators. For atmospheric turbulence, there is a rule-of-thumb that the number N of actuators (and Shack–Hartmann lenslets in the wavefront sensor) is given by $N \approx (D/r_0)^2$, but this is a crude approximation that ignores the effectiveness of the influence functions to correct Kolmogorov wavefronts.

In order to be more specific, consider a corrective device with N_A actuators whose influence functions are $r_j(x, y)$, one for each actuator, labelled j. For a set of control signals, w_j applied to these actuators, the resulting phase[1] given to the surface of the mirror can be assumed to be the linear superposition of the influence functions

$$\Phi_M(x, y) = \sum_{j=1}^{N_A} r_j(x, y) w_j. \tag{10.4}$$

The phase $\Phi_M(x, y)$ can be expanded in an orthonormal series and for convenience we use the Zernike basis as the pupils (and the mirrors) are usually circular and this basis is orthonormal over the unit circle. The Zernike coefficents ϕ_i are given by

$$\phi_i = \int\int W(x, y) \Phi_M(x, y) Z_i(x, y) \mathrm{d}x \mathrm{d}y$$
$$= \sum_{j=1}^{N_A} \left[\int\int W(x, y) Z_i(x, y) r_j(x, y) \mathrm{d}x \mathrm{d}y \right] w_j$$

or in matrix form

$$\phi = M w \tag{10.5}$$

where $W(x, y)$ is the circular aperture function. The matrix M is called the influence matrix, and in this case each column of M is the Zernike expansion coefficients of a single influence function of the mirror. It is easily measured, by applying a fixed signal to each electrode sequentially, recording the wavefront phase, and computing the Zernike expansion of each influence function. Note that this particular version of the influence matrix M uses the Zernike expansion, but this is not esential.

Given a particular wavefront aberration whose Zernike expansion is specified by the vector ϕ_0, the control signals w required to give the best least squares fit are given by

$$w = M^{-1*} \phi_0,$$

---

[1] We consider the continuous phase in the interval $-\infty$ to $+\infty$.

where $\mathbf{M}^{-1*}$ is the least squares inverse of $\mathbf{M}$. The residual wavefront error after correction is

$$\boldsymbol{\phi}_e = \mathbf{M}\mathbf{M}^{-1*}\boldsymbol{\phi}_0 - \boldsymbol{\phi}_0 = \left[\mathbf{M}\mathbf{M}^{-1*} - \mathbf{I}\right]\boldsymbol{\phi}_0. \tag{10.6}$$

Equation 10.6 forms the basis of the assessment of a given mirror to correct given wavefronts or given statistics of wavefronts and the mirror fitting error σ_{fit}^2 can be evaluated. It turns out that the "best" set of influence functions for any mirror is that which spans the subspace covered by the first N_A Karhunen–Loève functions of the aberration space. For atmospheric turbulence with Kolmogorov statistics, curvature mirrors (membrane, bimorph) have sets of influence functions which are much closer to the Karhunen–Loève functions than segmented or other zonal mirrors.

10.6   The Control System

The control system involves both spatial and temporal aspects, and in general these are coupled and should be considered together. Here, for simplicity, we separate the two aspects.

Before discussing the mathematics of the control system, some remarks on the hardware required are relevant. The first "controllers" were custom built circuits housed in several 19-inch racks. A second generation, found in a number of current astronomical AO systems, used many (20+) Texas C-40 or similar digital signal processing (DSP) chips. Such an approach may still be needed for systems with a very high bandwidth and a large number (several hundred) of actuators, but nowadays a single processor can handle the control for modest bandwidths and numbers of actuators. For example, in [11], a 500MHz Pentium chip was used for a 5×5 Shack–Hartmann wavefront sensor and a 37-actuator mirror at a frame rate of 780 Hz: only a fraction of the processor time was actually used and calculations indicate that this could have handled a 10×10 Shack–Hartmann array (with an appropriate increase in the number of mirror actuators) at the same frame rate. In fact the major practical problem is the input rate (and, more importantly, any latency) from the wavefront sensor detector, rather than the matrix operations of the control system.

The operation of a wavefront sensor on an aberration described by the vector $\boldsymbol{\phi}_0$ (of, for example, Zernike coefficients) can be expressed by the matrix equation

$$s = \mathbf{S}\boldsymbol{\phi}_0, \tag{10.7}$$

where s is the vector of sensor signals and $\mathbf{S}$ is the wavefront sensor response matrix. For example, for a Shack–Hartmann sensor, the components of s are the $x-$ and $y-$slopes of the wavefront averaged over each lenslet (for N lenslets the vector has length $2N$), and the elements S_{ij} of $\mathbf{S}$ are the average $x-$ and $y-$differentials of the Zernike polynomials for the ith lenslet and the jth polynomial averaged over each lenslet.

The actuator control signals w_j form a vector w that are assumed to be a linear combination of the sensor signals, i.e.

$$w = Cs \qquad (10.8)$$

where C is the control matrix.

There are many approaches to finding C. One approach, the least squares method, is outlined in the following. In a closed loop AO system, the wavefront entering the sensor is the sum of the aberrated wave ϕ_0 and the mirror-induced phase $\phi = Mw$. In the absence of noise, the wavefront sensor signal that results is given by

$$s = S[\phi_0 + Mw] = s_0 + Bw \qquad (10.9)$$

where $B = SM$ is the response matrix for the mirror-sensor system. The matrix B plays an important rôle in least squares control. Note that although both the sensor response matrix S and the actuator response matrix M were defined, in the above, in terms of a Zernike expansion, the overall response matrix is independent of this choice and does not involve Zernike polynomials. The matrix B is found by applying a fixed signal to each electrode of the mirror in turn and recording the sensor signals (for example, the $x-$ and $y-$slopes or simply the spot positions in a Shack–Hartmann sensor).

Since the wavefront sensor measures the net wavefront aberration after correction, we shall use this as a measure of the correction error. Using a standard least squares approach, it can be shown that the control matrix C that minimises the wavefront sensor error is

$$C = -\left[B^T B\right]^{-1} B^T \qquad (10.10)$$

where the quantity $\left[B^T B\right]^{-1} B^T$ is known as the least squares inverse, or psuedo-inverse, of B.

From 10.9 it can be seen that the matrix B defines a mapping between the actuator (w, length N_A) and sensor (s, length N_S) vectors. Any matrix of dimensions $N_S \times N_A$ can be written as a product of three matrices, the so-called *singular value decomposition*:

$$B = U\Lambda V^T \qquad (10.11)$$

where (1) U is an $N_S \times N_S$ orthogonal matrix, (2) V is an $N_A \times N_A$ orthogonal matrix and (3) Λ is an $N_S \times N_A$ diagonal matrix.

Equation 10.11 can be expanded as

$$B = \left(u_1 \ u_2 \ \ldots\right) \begin{pmatrix} \lambda_1 & & & \\ & \lambda_2 & & \\ & & \ldots & \\ & & & \lambda_N \end{pmatrix} \begin{pmatrix} v_1^T \\ v_2^T \\ \ldots \end{pmatrix} \qquad (10.12)$$

where the vectors u_i and v_i form complete sets of modes for the sensor signal and mirror control spaces respectively. The diagonal elements of Λ, λ_i are the singular values of the matrix **B**. Each non-zero value of λ_i relates the orthogonal basis component v_i in w, the control signal space, to an orthogonal basis component u_i in s, the sensor signal space.

We can now distinguish three possible situations:

- **Correctable Modes** For these modes, $\lambda_i \neq 0$, the actuator control signal $w = v_i$ results in the sensor signal $s = \lambda_i u_i$. This mode can be corrected by applying the actuator model $w = \lambda_i^{-1} v_i$, the singular value λ_i being the sensitivity of the mode (clearly we not want λ_i to be too small).
- **Unsensed Mirror Modes** These are modes v_i for which there is no non-zero λ_i. Unsensed mirror modes would cause a big problem in an AO system and are to be avoided if at all possible by proper design of the wavefront sensor: some of the early zonal AO systems suffered from this defect, producing so-called "waffle" modes in the mirror.
- **Uncorrectable Sensor Modes** These are modes u_i for which there is no non-zero λ_i. Nothing the mirror does affects these modes, and arguably there is no point in measuring them.

The control matrix **C** is now given by

$$\mathbf{C} = -\mathbf{B}^{-1*} = -\mathbf{V}\Lambda^{-1*}\mathbf{U}^T \tag{10.13}$$

where Λ^{-1*} is the least-squares pseudo-inverse of Λ formed by transposing Λ and replacing all non-zero diagonal elements by their reciprocals λ_i^{-1}, which now can be interpreted as the *gains* of the system modes. From a practical point of view, one discards modes which have a small value for λ_i as clearly they are susceptible to noise. An example of the effect of discarding modes is presented in Sec. 10.7.

The least squares approach is simple and does not require much prior information about either the AO system or the statistics of the aberration to be corrected. It also has the advantage that the key matrix **B** can easily be measured. But surely prior knowledge, for example about the aberration to be corrected or the noise properties of the sensor, could be an advantage in a well-designed control system? There are several other approaches to "optimal reconstructors", originating with the work of Wallner [34], and described in [35]. These are not so easy to implement but may offer improved performance. The approach to understand the temporal control aspects is through the transfer functions of each component [36]. Figure 10.7 shows four of the key elements which can play an important rôle in the time behaviour.

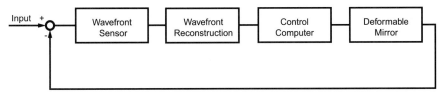

Fig. 10.7. Schematic of a closed loop adaptive optics system

The wavefront sensor detector integrates for a time T to record the data, and therefore has a transfer function (Laplace transform),

$$G_{WS}(s) = \frac{1 - \exp(-Ts)}{Ts}$$

whereas the wavefront reconstructor introduces a lag τ:

$$G_{WR}(s) = \exp(-\tau s).$$

We assume that the deformable mirror has a unit transfer function (i.e. it responds immediately)

$$G_{DM}(s) = 1.$$

Finally, the control computer, if it implements a simple integrator, has a transfer function

$$G_C(s) = \frac{K}{s},$$

where K is the gain of the integrator. The open loop transfer function G_{ol} is the product of the transfer functions:

$$\begin{aligned} G_{ol}(s) &= G_{WS}(s) \times G_{WR}(s) \times G_{DM}(s) \times G_C(s) \\ &= \frac{K[1 - \exp(-Ts)]\exp(-\tau s)}{Ts^2} \end{aligned} \tag{10.14}$$

and the closed loop transfer function $G_{cl}(s)$ is given by:

$$G_{cl}(s) = \frac{G_{ol}(s)}{1 + G_{ol}(s)H(s)} \tag{10.15}$$

where $H(s)$ is the feedback parameter.

The closed loop bandwidth can be defined in a number of ways. One way is to define it as the frequency at which the closed loop transfer function falls to the -3 dB level. However a more informative measure is found by plotting the ratio of the input and output signals as a function of frequency, and defining the closed loop bandwidth as the frequency at which this ratio is unity (i.e. above this frequency the system makes no correction). In general, this frequency is lower than the one given by the -3 dB definition.

10.7 Low Cost AO Systems

It is possible to build a low cost adaptive optics system from commercially available components, as demonstrated in [10,11]. Depending upon the requirements, a reasonable system based around a 37-actuator membrane mirror [27] costs in the region of $10K – $20K in parts. Given this choice of deformable mirror, and assuming (following an analysis based on 10.6) that this mirror has sufficient degrees-of-freedom to provide useful improvements (i.e. that its fitting error variance is less than approximately 1 rad^2), there are (at least) three further decisions to be made:

- **Type of Wavefront Sensor** A curvature sensor is best matched to the membrane mirror. However, both the systems referred to above used a Shack–Hartmann sensor, probably resulting in sub-optimal performance (see Fig. 10.10).
- **Speed of Operation** A system based on a video-rate detector in the wavefront sensor will be much less expensive to build than one based on a high speed CCD such as the Dalsa camera used in [11], which had a frame rate of 780 fps. The choice of frame grabber is important, as this could introduce additional delay into the control loop.
- **Choice of Control Computer** A standard PC (Pentium or Power PC chip) will easily handle the control of a 37-actuator membrane mirror and a well-matched wavefront sensor for frame rates up to 1 kHz. A choice has to be on the operating system (e.g. Linux or Windows). Alternatively, a compact design might use an embedded processor using a DSP chip. We have recently taken delivery of a custom-designed CMOS detector and control computer package based on SHARC DSPs that is very compact.

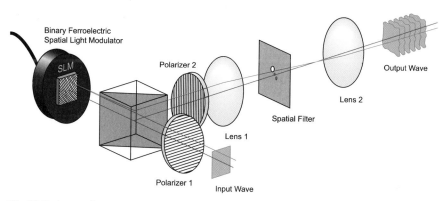

Fig. 10.8. A wavefront generator for testing AO systems [37]

For testing the performance of an AO system we have used a wavefront generator that uses a ferro-electric liquid crystal device to create dynamic binary holograms of any desired phase aberration [37]. This is used as shown in Fig. 10.8. The virtue of using a wavefront generator is that the spatial and temporal performance of the AO system can be studied quantitatively. Figures 10.9 and 10.10 show typical results from the first system built in our laboratory.

In Fig. 10.9 the open loop frame rate was 270 Hz, $D/r_0 \approx 7.5$ and $v/r_0 \approx 5$ Hz. The maximum Strehl ratio was approximately 0.4 in this case, compared to a value for the reference arm of approximately 0.8: we have not carried out detailed modeling of our system, but this Strehl is of the same order-of-magnitude that we would expect for this value of D/r_0 and a 37-actuator membrane mirror.

Figure 10.10 provides a more quantitative description of the performance of the system. The left hand side shows the effect of discarding higher order modes, for a

AO OFF   AO ON

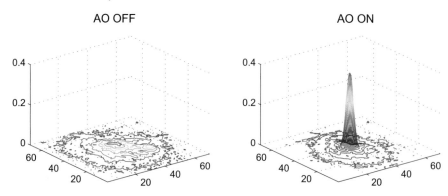

Fig. 10.9. The time-averaged point spread function. Left: AO system off. Right: AO system on (see text for parameters)

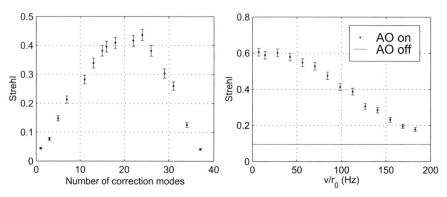

Fig. 10.10. Left: Strehl ratio as a function of the number of modes used in the control loop. Right: Strehl ratio as a function of the "wind speed" (see text for parameters)

value of D/r_0 of 7.5. We typically find that it is optimum to retain 20-25 modes in this system. The right hand side shows the temporal behaviour for an 780Hz frame rate and $D/r_0 \approx 5$ (note the higher Strehl value at the origin). There is still a good Strehl ratio at value of $v/r_0 \approx 100$ Hz, corresponding to a wind speed of 50ms^{-1} for a value of $r_0 \approx 0.5$ m (typical of a good observing site in the near IR).

10.8   Current Research Issues in Astronomical Adaptive Optics

Adaptive optics systems that operate in the near infrared are now installed on all the World's major groundbased optical/IR telescopes. The reason that these first systems all operate in the near IR is threefold:

- As shown in (10.2), the Fried parameter r_0 is proportional to $\lambda^{6/5}$. This is means that the lenslet area in a Shack–Hartmann sensor, and hence the flux collected by a lenslet, is proportional to $\lambda^{2.4}$: this in turn means that fainter guide stars can be used in the IR.

- The correlation time of the atmospherically induced phase pertubation is proportional to $\lambda^{6/5}$: thus longer sample times, and fainter guide stars, can be used in the IR.
- The isoplanatic angle is also proportional to $\lambda^{6/5}$, giving an increased area (proportional to $\lambda^{2.4}$) over which a guide star can be found. The isoplanatic angle (see the left hand side of Fig. 10.11) is the angle over which the variance of the wavefront aberration due to differing angles of incidence varies by approximately 1 rad^2. The isoplanatic angle is strongly influenced by the $C_N^2(z)$ profile of the refractive index structure function "constant".

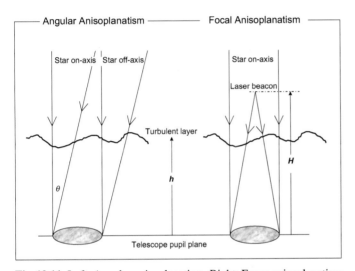

Fig. 10.11. Left: Angular anisoplanatism. Right: Focus anisoplanatism

The first two factors combine to make the required brightness of the guide star proportional to $\lambda^{3.6}$, so that, for a AO system to operate at 550 nm, the guide star has to be approximately $4^{3.6} = 150$ times brighter than at 2.2 µm. Futhermore, because of the third factor above, the region of the sky over which this much brighter guide star has to be found is approximately 28 times smaller in area.

This question of the required guide star brightness at different wavelengths translates to one of sky coverage – the fraction of the sky over which it is possible to find a bright enough natural guide star within the isoplanatic patch. For observation in the 1–2 µm region, the sky coverage, with an extremely high efficiency, low noise detector in the wavefront sensor, is on the order of 1% – 10%, an acceptable value. However, in the visible, the sky coverage is so low that only "targets of opportunity" can be observed using an AO system.

A related problem is the actual value of the isoplanatic angle: this is only a few arc-seconds in the visible (and is proportional to $\lambda^{6/5}$), and thus the field of view of an AO system is so small as to be of little interest to astronomers as a general purpose enhancement of image quality.

The solution is to make an artificial point source – a laser guide star – either by Rayleigh scattering or by resonantly exciting atomic sodium in the mesosphere. In each case, it is technically feasible to make a bright enough source but a new problem is now introduced, that of focus anisoplanatism. This is illustrated on the right hand side of Fig. 10.11: the wavefront sensed by the laser guide star is not the same as that experienced by the light from the astronomical object under observation. Since the mesospheric sodium guide star is at a much higher altitude (≈ 85 km) than the Rayleigh guide star (up to ≈ 20 km), the former minimises this effect. Unfortunately, a high power (10–20 W) laser precisely tuned to the sodium D_2 line is rather complex and expensive.

In the visible, focus anisoplanatism severely reduces the Strehl ratio for a 8 m diameter telescope and must be overcome. By using several laser guide stars, one can reduce the error due to focus anisoplanatism considerably, and, as a bonus, slightly increase the isoplanatic angle. Several laser guide stars means several wavefront sensors, and thus the system complexity is greatly increased. By further adding additional deformable mirrors, each conjugate to a different height in the atmosphere, i.e. doing so-called *multi-conjugate adaptive optics*, MCAO, one can increase the field of view significantly.

At present, MCAO (using either laser or natural guide stars) is the key research area in astronomical adaptive optics. Two approaches are being explored, one in which there is one wavefront sensor for each guide star ("conventional" MCAO) and the other in which all the guide stars contribute to each sensor, which is conjugated to a certain height above the telescope (the "layer-oriented" approach). In both cases, the AO system will be very complex.

Looking beyond the time when MCAO is implemented in astronomy, there will be the need to make MCAO work in low cost applications, for example, AO-assisted binoculars. Clearly, the complexity of the astronomical systems will have to be eliminated: there is still much to invent in this area before it becomes a reality.

10.9 Adaptive Optics and the Eye

The concepts of wavefront control, wavefront sensing and adaptive optics have many applications apart from astronomy, although in order to justify the use of AO the costs have to be several orders of magnitude smaller. Some of these potential applications are:

- *Line of sight optical communication* Building to building line of sight optical communication is limited in part by atmospherically induced intensity fluctuations (scintillation). AO may offer a means of reducing this and hence improving the reliability and bandwidth of the link.
- *Confocal microscopy* In confocal microscopy, the scanning probe is degraded as it propagates into the depth of an object, and also spherical aberration is induced. AO may offer a means of reducing this effect and thus improving the

transverse resolution, the depth discrimination and signal-to-noise ratio of the image. Since optical data storage may use confocal imaging, there are potential applications in his field as well.

- *High power lasers* Thermally induced effects in high power lasers reduce the beam quality and range of operating conditions. Intracavity AO may offer improved beam quality and a greater range of operating conditions.
- *Beam shaping* In lasers of all powers, AO may allow minor beam imperfections to be cleaned up and permit the dynamic shaping of beams, for example in laser materials processing.
- *Scanned imaging systems* AO may be usede to maintain the integity of a scanning probe, for example, in a laser printing system.

Another particularly promising area of application of adaptive optics and its technologies is in the eye. At the simplest level, wavefront sensing can be used to determine the exact refractive state of the eye, and several commercial wavefront sensors are becoming available for use in connection with laser refractive surgery (Lasik, PRK). Closed loop adaptive optics has two main areas of application in the eye. The first is for retinal imaging, and the second is to enhance vision.

When focus and astigmatism errors are well-corrected, the eye is more-or-less diffraction limited for a pupil diameter of approximately 2 mm. With this pupil size, the point spread function is ≈5 µm is diameter, compared to the cone photoreceptor spacing at the fovea on the order of 2 µm: this spacing is consistent with the Shannon or Nyquist sampling theorem. In order to image the cones at the fovea, one would need to dilate the pupil and then ensure that the aberrations over the pupil, which are typically severe for a dilated pupil, are corrected using adaptive optics. When the aberrations are corrected, we should have diffraction-limited retinal imaging, and also we may have enhanced visual acuity: indeed there are a number of interesting psychophysical experiments that can be carried out on AO-corrected eyes.

Initial studies in AO-assisted retinal imaging used open-loop adaptive optics, and combining AO with the technique of retinal densitometry, Roorda and Williams [38] were able to obtain remarkable images showing the spatial arrangement of the short, medium and long wavelength sensitive cones about 1 deg from the fovea, reproduced in Fig. 10.12. The temporal variation of the wavefront aberration is much slower in the eye than for atmospheric turbulence [39] and so that aspect of building a closed loop system is relatively straightforward. However, alignment and other issues due to the fact that a human subject is involved make adaptive optics in the eye quite difficult to achieve. Recently, three groups have published results showing closed-loop AO systems operating in the eye [40–42].

Finally, two groups [43,44] have implemented "poor man's AO", using wavefront data combined with direct retinal imagery in a deconvolution scheme [45].

Adaptive optics for retinal imaging and vision enhancement is on the threshold of a number of breakthroughs and improvements that can be expected to have a significant impact in clinical vision science.

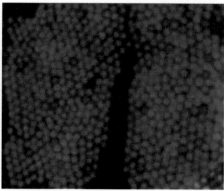

Fig. 10.12. Spatial arrangement of short (S), medium (M) and long (L) wavelength sensitive cones using retinal densitometry combined with open loop adaptive optics [38]. Left: 1 deg temporal. Right: 1 deg nasal. Courtesy of Austin Roorda and David Williams

Acknowledgements

I am grateful to my present and former students and post-docs for increasing my understanding of adaptive optics, particularly to Dr Carl Paterson who gave me a basic understanding of some aspects of control systems and whose lecture notes form the basis of part of Sec. 10.5 and all of Sec. 10.6, and to Dr Luis Diaz and Ian Munro who built our first closed loop AO system for the eye.

Our research on adaptive optics is funded by the European Union, the UK Engineering and Science Research Council, the UK Particle Physics and Astronomy Research Council, The Wellcome Trust, and several industrial partners, and I am grateful for their support.

References

1. H.W. Babcock: Pub. Astron. Soc. Pac. **65**, 229 (1953)
2. D.L. Fried: J. Opt. Soc. Am. **56**, 1372 (1966)
3. M. Born, E. Wolf: *Principles of Optics* (Cambridge University Press, Cambridge 1999)
4. R.J. Noll: J. Opt. Soc. Am. **66**, 207 (1976)
5. V.I. Tatarski: *Wave Propagation in a Turbulent Medium* (Dover Press, New York 1967)
6. F. Roddier: 'Effects of Atmospheric Turbulence in Astronomy', In: *Progress in Optics*, ed. by E. Wolf (Elsevier, Amsterdam 1981) pp. 281–376
7. J.W. Hardy: *Adaptive Optics for Astonomical Telescopes* (Oxford University Press, New York 1998)
8. G. Rousset et al.: Astron. Astrophys. **230**, L29 (1990)
9. B. Brandl et al.: Astrophys. J. **466**, 254 (1996)
10. D. Dayton et al.: Opt. Commun. **176**, 339 (2000)
11. C. Paterson, I. Munro, J.C. Dainty: Opt. Express **6**, 175 (2000)
12. J.R.P. Angel: Nature **368**, 203 (1)994
13. C.L. Koliopoulis: Appl. Optics **19**, 1523 (1980)

14. R.V. Shack, B.C. Platt: J. Opt. Soc. Am. **46**, 656 (1971) (abstract only)
15. R. Ragazzoni: J. Mod. Optics **43**, 289 (1996)
16. F. Roddier: Appl. Optics **29**, 1402 (1990)
17. R.A. Gonsalves: Opt. Eng. **21**, 829 (1982)
18. P.M. Blanchard et al.: Appl. Optics **39**, 6649 (2000)
19. M.A.A. Neil, M.J. Booth, T. Wilson: J. Opt. Soc. Am. A **17**, 1098 (2000)
20. E.N. Ribak, S.M. Ebstein: Opt. Express **9**, 152 (2001)
21. C. Paterson, J.C. Dainty: Opt. Lett. **25**, 1687 (2000)
22. R. Cubalchini: J. Opt. Soc. Am. **69**, 972 (1979)
23. G. Rousset: 'Wavefront Sensing', In: *Adaptive Optics for Astronomy*, ed. by D.M. Alloin, J.-M. Mariotti (Kluwer Academic Publishers, Dordrecht 1994) pp. 115–137
24. See http://www.xinetics.com/ and http://www.cilas.com/englais3/index
25. C. Schwartz, E. Ribak, S.G. Lipson: J. Opt. Soc. Am. **11**, 895 (1994)
26. A.V. Kudryashov, V.I. Shmalhausen: Opt. Eng. **35**, 3064 (1996)
27. G. Vdovin, P.M. Sarro, S. Middelhoek: J. Micromech. & Microeng. **9**, R8 (1999) (See also http://www.okotech.com/)
28. M. Horenstein et al.: J. Electrostatics **46**, 91 (1999)
29. T. Weyrauch et al.: Appl. Optics **40**, 4243 (2001)
30. See http://www.meadowlark.com/index.htm
31. T.L. Kelly, G.D. Love: Appl. Optics **38**, 1986 (1999)
32. S.R. Restaino et al.: Opt. Express **6**, 2 (2000)
33. M.A. Helmbrecht et al.: 'Micro-mirrors for adaptive optics arrays', In: *Transducers*, (Proc. 11th International conference on Solid State Sensors and Actuators, Munich, Germany 2001) pp. 1290–1293
34. E.P. Wallner: J. Opt. Soc. Am. **73**, 1771 (1983)
35. M.C. Roggemann, B. Welsh: *Imaging through Turbulence* (CRC Press, Boca Raton, Florida 1996)
36. M. Demerlé, P.Y. Madec, G. Rousset: 'Servo-loop analysis for adaptive optics', In: *Adaptive Optics for Astronomy*, ed. by D.M. Alloin, J.-M. Mariotti (Kluwer Academic Publishers, Dordrecht 1994) pp. 73–88
37. M.A.A. Neil, M.J. Booth, T. Wilson: Opt. Lett. **23**, 1849 (1998)
38. A. Roorda, D.R. Williams: Nature **397**, 520 (1999)
39. H. Hofer et al.: J. Opt. Soc. Am. A **18**, 497 (2001)
40. E.J. Fernández, I Iglesias, P Artal: Optics Lett. **26**, 746 (2001)
41. H. Hofer et al.: Opt. Express **18**, 631 (2001)
42. J.F. Le Gargasson, M. Glanc, P. Lena: C. R. Acad. Sci. Paris **t. 2 Série IV**, 1131 (2001)
43. I. Iglesias, P. Artal: Optics Lett. **25**, 1804 (2000)
44. D.P. Catlin: *High Resolution Imaging of the Human Retina* (PhD Thesis, Imperial College, London 2001)
45. J. Primot, G. Rousset, J.C. Fontanella: J. Opt. Soc. Am. A **7**, 1598 (1990)

11 Low-Coherence Interference Microscopy

C.J.R. Sheppard and M. Roy

11.1 Introduction

Confocal microscopy is a powerful technique that permits three-dimensional (3D) imaging of thick objects. Recently, however, two new techniques have been introduced, which are rivals for 3D imaging in reflection mode. Coherence probe microscopy (CPM) is used for surface inspection and profiling, particularly in the semiconductor device industry [1–10]. Optical coherence tomography (OCT) is used for medical diagnostics, particularly in ophthalmology and dermatology [11,12]. These are both types of low coherence interferometry (LCI), in which coherence gating is used for optical sectioning. LCI has many advantages over the conventional (narrow-band source) interferometric techniques, including the ability to reject strongly light that has undergone scattering outside of a small sample volume, thus allowing precise non-invasive optical probing of dense tissue and other turbid media. LCI can be used to investigate deep, narrow structures. This feature is particularly useful for in-vivo measurement of deep tissue, for example, in transpupilliary imaging of the posterior eye and endoscoping imaging. Another application area of LCI is in optical fibre sensors [13–15].

Although CPM and OCT are both forms of LCI, their implementations differ. In CPM, an interferometer is constructed using a tungsten-halogen or arc lamp as source, Fig. 11.1. Interference takes place only from the section of the sample located within the coherence length relative to the reference beam. This technique is usually used for profilometry of surfaces. A series of 2D images is recorded using a CCD camera as the sample is scanned axially through focus. For each pixel an interferogram is recorded, which can be processed to extract the location of the surface either from the peak of the fringe visibility, or from the phase of the fringes. Use of a broad-band source avoids phase-wrapping ambiguities present in laser interferometry with samples exhibiting height changes of greater than half a wavelength. If a microscope objective of high numerical aperture is employed in CPM, interference occurs only for regions of the sample which gives rise to a wave front whose curvature matches that of the reference beam [16,17]. For this reason CPM has been termed phase correlation microscopy[4]. An optical sectioning effect results which is analogous to that in confocal microscopy. This phenomenon been used by Fujii [18] to construct a lensless microscope, and by Sawatari [19] to construct a laser interference microscope. Overall optical sectioning and signal collection performance [20] is slightly superior to that of a confocal microscope employing a pinhole. In order to extract the height information from the interferogram, a complete axial scan

Török/Kao (Eds.): Optical Imaging and Microscopy, Springer Series in Optical Sciences
Vol. 87 – © Springer-Verlag, Berlin Heidelberg 2003

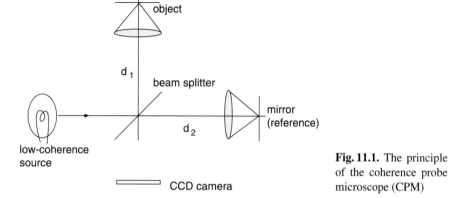

Fig. 11.1. The principle of the coherence probe microscope (CPM)

through the sample must be recorded. An alternative method is based on phase shifting, in which the visibility at a particular axial position can be determined by three or more measurements for different values of reference beam phase. This avoids the necessity to scan through the complete depth of the image, and is particularly advantageous with thick structures. Phase shifting is traditionally performed with laser illumination, and with a broad-band source a problem is encountered in that the phase shift of the different spectral components, when reflected from a mirror, varies. This problem has been overcome by performing the phase shifting by use of geometric phase shifting (GPS) [21,22], in which polarisation components are used to effect an achromatic phase shift [23].

Although the principle of OCT is identical to that of CPM, the embodiment is usually quite different [11]. In OCT the signal, scattered from bulk tissue, is very weak. Hence the source is often replaced by a super-luminescent laser diode, which is very bright but exhibits a smaller spectral bandwidth than a white-light source. The source is used in a scanning geometry, in which the sample is illuminated point-by-point, Fig. 11.2, so that considerable time is necessary to build up a full 3-D image. Therefore, usually x-z cross-sections are recorded. In OCT the numerical aperture (NA) of the objective lens is small and its depth of focus large, so z scanning can be achieved by translation of the reference beam mirror. Instead of phase-shifting, the alternative techniques of heterodyning is used, which can result in shot-noise limited detection performance. In OCT optical sectioning results from the limited coherence length of the incident radiation, but if the NA of the objective lens is large, an additional optical sectioning effect similar to that in confocal imaging results [24]. The resulting instrument is called the optical coherence microscope (OCM). In OCM, because the depth of focus of the objective is small, the sample has to be scanned relative to the objective lens to image different z positions, so scanning is usually performed in x-y sections.

Both OCT and OCM are usually implemented using single mode optical fibres for illumination and collection. A scanning microscope using single mode fibres behaves as a confocal microscope [25–27], and the geometry of the fibre spot influ-

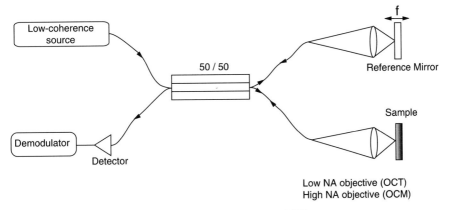

Fig. 11.2. The principle of optical coherence tomography (OCT)

ences the imaging performance of the system [28–32]. Confocal interference microscopes using monochromatic light and a pinhole in front of the detector were constructed long before the proposal of OCT [33–35]. Similarly single-mode optical fibre implementations of the confocal interference microscope with a monochromatic source have also been reported [36–38]. In these systems, a confocal optical sectioning effect is present. This optical sectioning behaviour is similar to, but exhibits some differences from, the correlation effect in the CPM. Either low-coherence light or ultra short pulses [39] can be used to measure internal structure in biological specimens. An optical signal that is transmitted through or reflected from a biological tissue contains time-of flight information, which in turn yields spatial information about tissue microstructure. In contrast to time domain techniques, LCI can be performed with continuous-wave light, avoiding an expensive laser and cumbersome systems. Summarizing, we distinguish between:

(a) CPM uses a medium/high NA, giving sectioning from a combination of a coherence and a correlation effect. Complete x-y images are recorded using a CCD detector.
(b) OCT uses a low NA, giving sectioning from coherence only. The low NA is used in OCT to provide a long depth of focus that enables $x - z$ cross-sectional imaging using coherence gate scanning.
(c) OCM uses a high NA, giving sectioning from a combination of a coherence and a confocal effect. The high NA is used in OCM to create a small focal volume in the sample that combines with the coherence gate, resulting in high resolution with rejection of out-of-focus or multiply scattered light.

11.2 Geometry of the Interference Microscope

There are several interference microscope available. Fig. 11.3 shows the Linnik, Mirau, Michelson and confocal interferometers, each with various advantages and

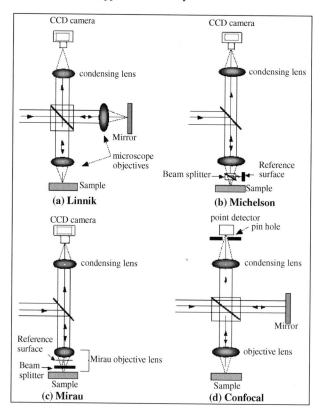

Fig. 11.3. Different types of interference microscope

disadvantages. For high magnification it is often advantageous to use a Linnik interference microscope [1], in which two matched microscope objectives with high NA are used. Using matched objectives, spherical and chromatic aberration can be removed. The main disadvantage of this type of microscope is that it is very sensitive to vibration and air currents, because it involves a long beam path. However, these vibrations are avoided in OCT by employment of heterodyning techniques that reject low frequency variations in signal.

Mirau [4] and Michelson interference microscopes require only one microscope objective. The difference between these two types is the positioning of the beam splitter and the reference surface. In the Mirau system, the beam splitter and reference mirror are positioned between the objective and test surface, whereas in the Michelson type, a long working distance microscope objective is used to accommodate the beam splitter between the objective and the test surface. These two methods, particularly the Michelson geometry, are aperture limited compared to the Linnik and confocal systems and therefore they are used for low magnifications.

On the other hand the concept of the confocal interference microscope [33] is quite different from the other methods, as shown in Fig. 11.3. This type is based on a point-by-point detection technique which requires a pinhole in front of a detector and because of this the wave-front distortion of the reference beam can be ignored. In practice of course the pinhole must have some finite size, and the wave front of the reference beam over the pinhole area does affect the overall performance. The pinhole can also be replaced by a single-mode fibre [36,38]. In this case the fibre acts as a coherent detector, so that only the component of the object and reference beams that match the mode profile are detected. In this case, aberrations on the reference beam only affect the strength of the reference beam.

11.3   Principle of Low-Coherence Interferometry

A schematic diagram of CPM is shown in Fig. 11.1, in which white, incoherent light is used as illumination. The sample is placed in one interferometer arm, and other arm provides a reference beam. The reflections from the sample are combined with the reflection from the mirror. The electric field of the light arriving at the camera, $U(t)$, is the superposition of the light traversing the two arms of the interferometer represented by

$$U(t) = U_s(t) + U_r(t + \tau) , \tag{11.1}$$

where U_s and U_r are the optical amplitude of signal and reference beam. The quantity τ is the time delay due to the difference in the length of the optical paths traversed by the beams. Normally any detector system for optical frequencies actually measures the time averaged intensity, which is the square of the electric field amplitude U, and expressed by

$$I_d(\tau) = \langle |U|^2 \rangle = \langle [U_s^*(t) + U_r(t + \tau)][U_s^*(t) + U_r^*(t + \tau)] \rangle , \tag{11.2}$$

where $\langle \rangle$ denotes the time average. Thus

$$I_d(\tau) = I_s + I_r + 2(I_s I_r)^{1/2} \mathcal{R}\{\gamma(\tau)\} , \tag{11.3}$$

where $\gamma(\tau)$ is the normalized form of the mutual coherence function, the complex degree of coherence, that can be represented by

$$\gamma(\tau) = \frac{\langle U_s(\tau)U_r(t + \tau) \rangle}{(I_s + I_r)^{1/2}} . \tag{11.4}$$

Equation (11.3) is the generalized interference law for partially coherent light.

In general, the complex degree of coherence includes both temporal and spatial coherence effects. For a Michelson interferometer which is an amplitude splitting interferometer, for a single point on the spatially incoherent source, spatial coherence (e.g. as in the Young interferometer) can be neglected, so that the mutual coherence reduces to self coherence, or temporal coherence, which has the same behaviour for

CPM and OCT. When the complete incoherent source is considered, partial spatial coherence must also be considered, and then CPM and OCT behave differently. Confining our attention to temporal coherence for the present, (11.3) can be written

$$I_d(\tau) = I_s + I_r + 2(I_s I_r)^{1/2} \Re\{\gamma_{tc}(\tau)\} , \qquad (11.5)$$

where $\Re\{\gamma_{tc}(\tau)\}$ is the real part of the complex degree of temporal coherence of the light source which is the normalized form of the self-coherence function $\gamma_{11}(\tau)$,

$$\gamma_{11}(\tau) = \frac{\Gamma_{11}(\tau)}{\Gamma_{11}(0)} , \qquad (11.6)$$

where

$$\Gamma_{11}(\tau) = \langle U(t+\tau)U^*(\tau)\rangle . \qquad (11.7)$$

Here the subscript 11 corresponds to a single point on the incoherent source. The normalized complex degree of coherence of the light source is given by the Fourier transform of the power spectrum of the light source.

Now for a polychromatic light source with a Gaussian power spectral density

$$G(f) = G_0 \exp\left[-\left(\frac{f-\bar{f}}{\Delta f}\right)^2\right] , \qquad (11.8)$$

where G_0 is constant, Δf represents the spectral width of the source, and $\bar{f}$ is the mean frequency. Thus

$$\gamma_{11}(\tau) = G_0 \int_{-\infty}^{\infty} \exp\left[-\left(\frac{f-\bar{f}}{\Delta f}\right)^2\right] \exp(-i2\pi f\tau)df , \qquad (11.9)$$

or

$$\gamma_{11}(\tau) = G_0\pi\Delta f \exp\left[-(\pi\tau\Delta f)^2\right] \exp(-i2\pi\bar{f}\tau) . \qquad (11.10)$$

But we know that $\gamma_{11}(0) = 1$, so we have

$$\gamma_{tc}(\tau) = \exp\left[-(\pi\tau\Delta f)^2\right] \exp(-i2\pi\bar{f}\tau) . \qquad (11.11)$$

Equation (11.3) can be therefore be written as

$$I(\tau) = I_s + I_r + 2(I_s I_r)^{1/2} \exp\left[-(\pi\tau\Delta f)^2\right] \cos(2\pi\bar{f}\tau) , \qquad (11.12)$$

or

$$I_d(\tau) = I_0\left[1 + V(\tau)\cos(2\pi\bar{f}\tau)\right] , \qquad (11.13)$$

where I_0 is the background intensity, V is the fringe contrast function or envelope of the observed fringe pattern, given by

$$V(\tau) = \frac{2(I_s I_r)^{1/2}}{I_s + I_r} \exp\left(-(\pi\tau\Delta f)^2\right) \ . \tag{11.14}$$

If we consider white light with wavelength range from 400 nm to 700 nm, the coherence time is then roughly 6.6 fs and hence the spectral bandwidth is about 1.5×10^{14} radians s^{-1}, giving a coherence length of about 1–2 μm. In OCT, the coherence time of the source is roughly 50 fs, and the coherence length typically 10–15 μm.

11.4 Analysis of White-Light Interference Fringes

Figure 11.4 shows the variations in intensity at a given point in the image as an object consisting of a surface is scanned along the depth axis (z axis). We assume that the origin of coordinates is taken at the point on the z axis at which the two optical paths are equal, and that the test surface is moved along the z axis in a series of steps of size Δz. With a broad-band spatially incoherent source, the intensity at any point (x, y) in the image plane corresponding to a point on the object whose height is h can be written as

$$I_d(\tau) = I_s + I_r + 2(I_s I_r)^{1/2}\gamma\left(\frac{p}{c}\right)\cos\left[\left(\frac{2\pi}{\bar\lambda}\right)p + \phi_0\right] , \tag{11.15}$$

where I_s and I_r are the intensities of the two beams acting independently, $\gamma(p/c)$ is the complex degree of coherence (corresponding to visibility of the envelope of the interference fringes), and $\cos\left[(2\pi/\bar\lambda)p + \phi_0\right]$ is a cosinusoidal modulation. In (11.15), $\bar\lambda$ corresponds to the mean wavelength of the source, $p = 2(z - h)$ is the difference in the lengths of the optical paths traversed by the two beams, and ϕ_0 is the difference in the phase shifts due to reflection at the beam-splitter and the mirrors.

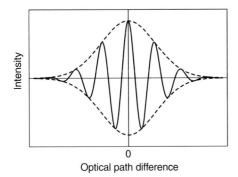

Intensity

0

Optical path difference

Fig. 11.4. The output from a white-light interferometer as a function of the position of one of the end mirrors along the z axis. The *dashed lines* represent the fringe envelope

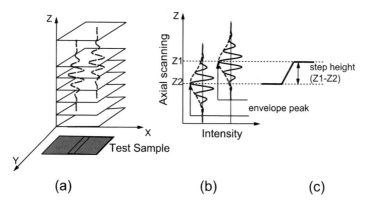

Fig. 11.5. A representation of different stages in white-light surface profiling: (**a**) A series of interferogram images is obtained by moving the sample along the z axis, (**b**) Processing of two sample pixels in the interferogram uses peak detection. (**c**) The location of the peak along the z axis corresponds to the height of the surface, allowing reconstruction of the surface profile

Each of these interference patterns in z for each pixel (see Fig. 11.4) can be processed to obtain the envelope of the intensity variations (the visibility function). We can therefore determine the peak amplitude of the intensity variations and the location of this peak along the scanning axis (Fig. 11.5). The values of the peak amplitude give an autofocus image of the test object (analogous to the similar technique in confocal microscopy [40], while the values of the location of this peak along the scanning axis yield the height of the surface at the corresponding points [40].

A major objective of the LCI is to find the height of a surface from the location of the peak fringe visibility by using variety of algorithms that can be categorized in two main groups:

11.4.1   Digital Filtering Algorithms

The method most widely used to recover the fringe visibility function from the sampled data has been digital filtering in the frequency domain [4]. This procedure involves two discrete Fourier transforms (forward and inverse) along the z direction for each pixel in the sample. In order to recover the interference fringes the step size along the z axis, Δz often corresponds to a change in the optical path difference, p of a quarter of the shortest wavelength or less. Typically, a step Δz around 50 nm is used. Consequently this procedure requires a large amount of memory and processing time.

These requirements can be reduced to some extent by modified sampling and processing techniques. For example, according to the sampling theorem, the signal need be sampled only at a rate of twice the bandwidth, rather than for the highest frequency in the fringe pattern. This has been called sub-Nyquist sampling [7].

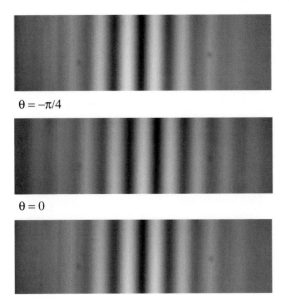

$\theta = -\pi/4$

$\theta = 0$

$\theta = +\pi/4$

Fig. 11.6. White-light fringes for different angular settings of the geometric phase shifter, showing that the fringes move relative a stationary envelope

Other approaches include using a Hilbert transform [5], communication theory [41], or nonlinear filter theory [42].

11.4.2   Phase Shift Algorithms

A more direct approach, which is computationally much less intensive, involves shifting the phase of the reference wave by three or more known amounts for each position along the z axis and recording the corresponding values of the intensity. These intensity values can then be used to evaluate the fringe contrast directly at that position [6].

However, the usual method for introducing these phase shifts has been by moving the reference mirror in the interferometer along the axis by means of a piezo-electric translator (PZT) so as to change the optical path difference between the beams. This procedure has the drawback that the dynamic phase introduced by a change in the optical path p varies inversely with the wavelength. In addition, since the value of p is changing, the maximum value of the envelope function (fringe visibility) $\gamma(p/c)$ is not the same at a given position of the object. Both these factors can lead to systematic errors in the values of fringe visibility obtained. These problems can be avoided by using a phase shifter that generates the required phase shifts without any change in the optical path difference. The only phase shifter that can satisfy this condition is a geometric phase shifter (GPS) [22], an achromatic phase-shifter operating on the principle of the geometric phase [23], which can be achieved by using a polarising beam splitter, a quarter wave plate and a rotating polariser [43]. Figure 11.6 shows how the white light fringes move relative to a stationary envelope as the geometric phase is altered. For a given point on the envelope, the fringe pattern changes without bound as the geometric phase is changed. Thus for this variation the bandwidth is effectively zero, and only three discrete measurements are needed to recover the amplitude and phase of the fringe. This is then repeated at different points on the envelope, but to recover the envelope the distance between these points is fixed by the bandwidth of the envelope rather than the fringes.

A variety of different phase shifting algorithms can be used to extract the envelope, based on measurements for three or more values of phase step. Five step algorithms appear to be good for contrast or visibility measurement [42,44]. Five measurements are made of the intensity as follows:

$$I_1 = I_0[1 + V \cos(\phi - \pi)] = I_0[1 - V \cos \phi] , \tag{11.16}$$
$$I_2 = I_0[1 + V \cos(\phi - \pi/2)] = I_0[1 + V \sin \phi] ,$$
$$I_3 = I_0[1 + V \cos \phi] ,$$
$$I_4 = I_0[1 + V \cos(\phi + \pi/2)] = I_0[1 - V \sin \phi] ,$$
$$I_5 = I_0[1 + V \cos(\phi - \pi)] = I_0[1 - V \cos \phi] .$$

Since the additional phase differences introduced by the GPS are the same for all wavelengths, the visibility of the interference fringes at any given point in the field can be expressed as [42]

$$V^2 = \frac{1}{4 \cos^4 \delta} \left[(I_2 - I_4)^2 - (I_1 - I_3)(I_3 - I_5) \right] , \tag{11.17}$$

where δ is the error in the phase step. This is a very efficient algorithm, which is also error correcting, so that if the phase step is not $\pi/2$ it still gives the visibility correctly, but scaled by a factor that depends on the phase step. This is satisfactory for determination of the peak position, but as the phase step can change if the spectral distribution is altered on reflection it can result in errors in the autofocus image. However, the scaling factor can be determined from the relationship

$$\sin \delta = \frac{I_5 - I_1}{2(I_4 - I_2)} , \tag{11.18}$$

and the intensity of the autofocus image corrected accordingly.

11.5   Spatial Coherence Effects

In a conventional interference microscope, nearly coherent illumination, produced by stopping down the condenser aperture, is used in order to give strong interference fringes [45]. However, in an interference microscope with digital storage, the interference term of the image can be extracted from the non-interference background terms by digital processing. In this case the relative strength of the fringes is not so important. If the aperture of the condenser is opened up until it is equal to that of the objective, as in a CPM, the relative strength of the fringes decreases. At the same time, a correlation effect is also introduced: only light which has a phase front which matches that of the reference beam results in an appreciable interference signal [16,17]. The result is an optical sectioning effect additional to that from the low-coherence source. The visibility of the fringes $V(p/c)$ is given by the product of the envelopes resulting from the low-coherence and the correlation effects [1]. The strength of the optical sectioning that results from the correlation effect increases rapidly as the numerical aperture (NA) of the system is increased. Typically, the envelope from the low-coherence effect has a width of about 5 μm for a broad-band source, while that from the correlation effect can be about 1 μm for high NA lenses.

In OCT, the system behaves as a confocal system, which can be regarded as a special case of a scanning interference microscope [46]. Using an optical fibre implementation, the mode profile of the fibre selects, for both the object beam and the reference beam, the overlap integral of the beam profile.

11.6   Experimental Setup

A schematic diagram of our LCI microscope [47] is depicted in Fig. 11.7. The optical system can be split into two parts: the illumination system, and the interferometer.

11.6.1   The Illumination System

A tungsten halogen lamp (12 V, 100 W) is used as a source. To illuminate the object uniformly, a Köhler illumination system is used, consisting of lenses L1-L4 (Fig. 11.7) together with a microscope objective. The system incorporates both aperture and field stops. The aperture stop is placed in a plane where the light source is brought to a focus, as shown by thick lines in Fig. 11.8. So by adjusting the aperture stop, the angle of illumination, and therefore the effective NA of the condenser, can be controlled. The field stop is placed in a plane where a back-projected image of the sample is formed, shown by thin lines in Fig. 11.8. Adjusting the field stop limits the area of illumination at the object, thereby reducing the strength of scattered light.

This system allows separate control of both the illumination aperture stop and the field stop. Stopping down the illumination aperture allows the system to be operated as a conventional interference microscope, with high spatial coherence. A 3 mW He-Ne laser is also provided for alignment. Since the coherence length of the laser is much longer than that of the white-light source, it can be used for finding the interference fringes.

11.6.2   The Interferometer

The interferometer used in our system is based on the Linnik configuration (Fig. 11.3). The linearly polarised beam transmitted by the polariser is divided at the polarising

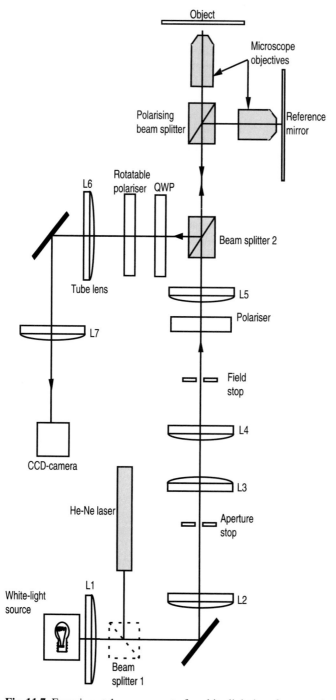

Fig. 11.7. Experimental arrangement of a white-light interference microscope using geometric phase shifter

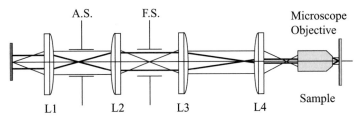

Fig. 11.8. The Köhler illumination system

beam splitter into two orthogonally polarised beams which are focused onto a reference mirror and a test surface by two identical infinity tube-length $40 \times$ microscope objectives with numerical aperture 0.75. After reflection at reference mirror and test surface these beams return along their original paths to a second beam-splitter which sends them through a second polariser to the CCD array camera.

The phase difference between the beams is varied by a geometric phase-shifter (GPS) consisting of a quarter-wave plate (QWP) with its axis fixed at an azimuth of $45°$, and a polariser which can be rotated by known amounts [21,22]. The rotation of the polariser is controlled by a computer via a stepper-motor. In this case, if the polariser is set with its axis at an angle θ to the axis of the QWP, the linearly polarised beam reflected from the reference mirror and the orthogonally polarised beam from the test surface acquire a geometric phase shift equal to 2θ. This phase difference is very nearly independent of the wavelength. Achromatic phase-shifting based on a rotating polariser has better performance than a rotating half-wave plate, placed between two quarter-wave plates, used in white-light interferometry to measure the fringe contrast function [21,43].

Measurements and data analysis were automated using a graphic code LabView program. The hardware was composed of a Macintosh Quadra 900 with a Neotech Frame Grabber and a digital-to-analog card. All the interference images obtained by the CCD video camera were transferred to the computer via a Neotech Frame Grabber which was used to convert the analogue video signal to digital images. The CCD camera has a pixel resolution of 768×512.

11.7   Experimental Results

A test object of an integrated circuit chip was scanned along the z axis by means of the PZT in steps of a magnitude $\Delta z = 0.66\,\mu m$ over a range of $5\,\mu m$, centred approximately on the zero-order white light fringe. In practice a single grabbed image taken at TV rates (1/25 seconds) exhibits significant noise, which can be reduced by averaging. Each grabbed image is thus in fact the average of N grabbed images, which reduces the random electronic noise of the capture and improving significantly the resolution of the system. In practice ten averages is a good compromise between speed and noise. Fig. 11.9 shows the effect of averaging frames on the random electronic noise.

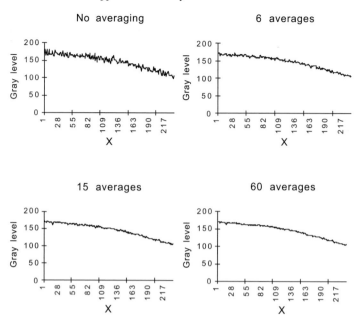

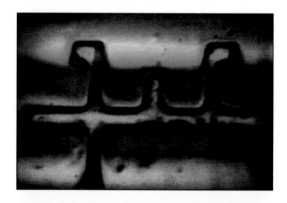

Fig. 11.9. Effect of averaging N images on the random noise

Fig. 11.10. A measured two-dimensional interference image

Fig. 11.11. A surface profile of an integrated circuit measured with our system using NA microscope objectives. The dimensions of the images are $170 \times 100\,\mu m^2$ with $1\,\mu m$ height

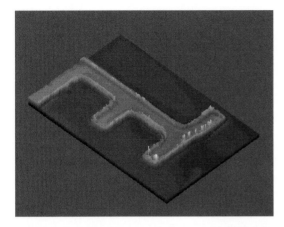

Fig. 11.12. A surface profile of an integrated circuit measured with our system using NA microscope objectives. The dimensions of the images are $25 \times 43\,\mu m^2$ with 1 μm height

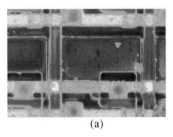

(a)

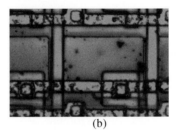

(b)

Fig. 11.13. (a) Surface profile, visualized as grey levels, of an integrated circuit measured with the system using ×10, 0.25 NA microscope objectives. (b) The corresponding autofocus (peak projection) image. The dimensions of the image are $100g \times 70\,\mu m^2$ with 1 μm height

A typical image of the white-light fringes is shown in Fig. 11.10. The surface height can be extracted by finding, for each pixel, the peak of the visibility variation. This can be done by various algorithms, such as evaluation of the first moment or a parabolic fit. We obtained good results by fitting a parabola through three points. Figures 11.11 and 11.12 show examples of the surface profile of a CCD chip obtained by our white-light interference microscope using a pair of 10×0.25 NA and 40×0.75 NA microscope objectives, respectively. Apart from the surface profile, we can simultaneously obtain an autofocus image, similar to a confocal autofocus image (maximum projection). In this, for every pixel, an image is formed from the maximum value of the visibility curve. This has the effect of bringing to focus all the different planes of the image, showing the reflectivity of the sample at any pixel. A surface profile of an integrated circuit, and the corresponding autofocus image, using a pair of 10×0.25 NA microscope objectives is shown in Fig. 11.13.

11.8 Discussion and Conclusion

We have demonstrated the successful use of a GPS that is close to achromatic in nature, for white-light interferometric surface profiling on a microscopic scale. Our system incorporate Köhler illumination, with control over illumination aperture and field diaphragms, and high numerical aperture microscope objectives. With high aperture optics an additional optical sectioning mechanism is provided by the spatial correlation effect. This property is not present in conventional laser interferometers because they use a low aperture condenser system. Thus separate control of the illumination (condenser) and imaging apertures allows us to investigate the properties of these different sectioning mechanisms. 3-D images of a variety of specimens including integrated circuit chips and thin film structures have been obtained by extracting the visibility envelope of the interferograms. The range of surface heights that can be profiled with this technique is limited only by the characteristics of the PZT used to translate the test specimen along the z axis and the available computer memory. However, since the steps between height settings at which data have to be taken can correspond to changes in the optical path difference of the order of a wavelength or more, a much smaller number of steps are required to cover a large range of depths.

If both fringe and envelope data is available, further information about the sample can be extracted, for example on the material properties of the sample. By knowing the fringe information, the phase of the signal can be recovered, the imaging system behaving as a coherent imaging system. Alteration of the condenser aperture alters the properties of this coherent system, the use of matched condenser and imaging apertures resulting in three-dimensional imaging performance similar to that in a confocal reflection microscope. However, the low temporal coherence of the illumination results in improved rejection from out-of-focus regions and improved three-dimensional imaging of thick objects. This could result in important applications in the biomedical area, similar to the current uses of optical coherence tomography, with higher resolution albeit for reduced imaging depth.

References

1. M. Davidson, K. Kaufman, I. Mazor and F. Cohen: P. Soc. Photo–Opt. Inst. **775**, 233 (1987)
2. B. S. Lee and T. C. Strand: Appl. Optics **29**, 3784 (1990)
3. B. L. Danielson and C. Y. Boisrobert: Appl. Optics **30**, 2975 (1991)
4. S. S. C. Chim and G. S. Kino: Opt. Lett. **15**, 579 (1990)
5. S. S. C. Chim and G. S. Kino: Appl. Optics **31**, 2550 (1992)
6. T. Dresel, G. Häusler, and H. Venzke: Appl. Optics **31**, 919 (1992)
7. P. de Groot and L. Deck: Opt. Lett. **18**, 1462 (1993)
8. L. Deck and P. de Groot: Appl. Optics **33**, 7334 (1994)
9. P. Sandoz and G. Tribillon: J. Mod. Optics **40**, 1691 (1993)
10. P. Sandoz: J. Mod. Optics **43**, 1545 (1996)
11. D. Huang, E. A. Swanson, C. P. Lin, J. S. Schuman, W. G. Stinson, W. Chang, M.R. Lee, T. Flotte, K. Gregory, C.A. Puliafito and Fujimoto, J. G.: Science **254**, 1178 (1991)
12. J. Schmitt, A. Knüttel and R. F. Bonner: Appl. Optics **32**, 6032 (1993)
13. S. Chen, A. W. Palmer, K. T. V. Grattan and B. T. Meggitt: Appl. Optics **31**, 6003 (1992)
14. Y. J. Rao, Y. N. Ning and D. A. Jackson: Opt. Lett. **18**, 462 (1993)

15. M. V. Plissi, A. L. Rogers, D. J. Brassington and M. G. F. Wilson: Appl. Optics **34**, 4735 (1995)
16. V. J. Corcoran: J. Appl. Phys. **36**, 1819 (1965)
17. A. E. Siegman: Appl. Optics **5**, 1588 (1966)
18. Y. Fujii, and H. Takimoto: Opt. Commun. **18**, 45 (1976)
19. T. Sawatari: Appl. Optics **12**, 2768 (1973)
20. M. Kempe and W. Rudolf: J. Opt. Soc. Am. A **13**, 46 (1996)
21. P. Hariharan: J. Mod. Optics **4**, 2061 (1993)
22. M. Roy and P. Hariharan: P. Soc. Photo–Opt. Inst. **2544**, 64 (1995)
23. M. V. Berry: J. Mod. Optics **34**, 1401 (1987)
24. J. A. Izatt, M. R. Hee, G. M. Owen, E. A. Swanson, J.G. Fujimoto: Opt. Lett. **19**, 590 (1994)
25. Y. Fujii and Y. Yamazaki: J. Microsc. **158**, 145 (1990)
26. L. Giniunas, R. Juskaitis and S. V. Shatalin: Electron. Lett. **27**, 724 (1991)
27. T. Dabbs and M. Glass: Appl. Optics **31**, 705 (1992)
28. M. Gu, C. J. R. Sheppard and X. Gan: J. Opt. Soc. Am. A **8**, 1755 (1991)
29. M. Gu and C. J. R. Sheppard: J. Mod. Optics **38**, 1621 (1991)
30. R. Gauderon and C. J. R. Sheppard: J. Mod. Optics **45**, 529 (1998)
31. M. D. Sharma and C. J. R. Sheppard: Bioimaging **6**, 98 (1998)
32. M. D. Sharma and C. J. R. Sheppard: J. Mod. Optics **46**, 605 (1999)
33. D. K. Hamilton and C. J. R. Sheppard: Opt. Acta **29**, 1573 (1982)
34. D. K. Hamilton and H. J. Matthews: Optik **71**, 31 (1985)
35. H. J. Matthews, D. K. Hamilton and C. J. R. Sheppard: Appl. Optics **25**, 2372 (1986)
36. M. Gu and C. J. R. Sheppard: Opt. Commun. **100**, 79 (1993)
37. M. Gu and C. J. R. Sheppard: Micron **24**, 557 (1993)
38. H. Zhou, M. Gu and C. J. R. Sheppard: Optik **103**, 45 (1996)
39. J. G. Fujimoto, S. De Silvestri, E. P. Ippen, C. A. Puliafito, et al.: Opt. Lett. **11**, 150 (1986)
40. I. J. Cox and C. J. R. Sheppard: Image and Vision Computing **1**, 52 (1983)
41. P. J. Caber: Appl. Optics **32**, 3438 (1993)
42. K. Larkin: J. Opt. Soc. Am. A **13**, 832 (1996)
43. H. S. Helen, M. P. Kothiyal and R. S. Sirohi: Opt. Commun. **154**, 249 (1998)
44. P. Hariharan, B. F. Oreb and T. Eiju: Appl. Optics **26**, 2504 (1987)
45. D. Gale, M. I. Pether and J. C. Dainty: Appl. Optics **35**, 131 (1996)
46. C. J. R. Sheppard and T. Wilson: P. Roy. Soc. Lond. A **295**, 513 (1980)
47. M. Roy and C. J. R. Sheppard: Optics and Lasers in Engineering **37**, 631 (2002)

12 Surface Plasmon and Surface Wave Microscopy

M.G. Somekh

12.1 Introduction

Surface plasmons (SPs) – or surface plasmon polaritons – are guided electromagnetic surface waves [1]. In their simplest form these are bound in a metallic or conducting layer sandwiched between two dielectrics. The confinement of these waves makes them particularly attractive as chemical and biological sensors being especially sensitive probes of interfacial properties [2,3]. On the other hand, the fact that SPs have characteristics distinct from bulk waves means that the optimum way to perform high resolution microscopy differs significantly from other microscopy techniques.

The purpose of this chapter is not to provide an exhaustive review of SPs or indeed even SP microscopy, but rather to examine approaches to obtaining high lateral resolution and sensitivity with SPs. The emphasis will be how the particular characteristics of SPs affect the way microscope systems perform. Most of the observations and conclusions in this chapter will also be applicable not only to SP microscopy, but more generally to microscopies exploiting surface waves. We will not discuss near field methods for SP microscopy [4–6], which can, of course, give very high lateral resolution since these are not readily compatible with biological microscopes and also involve inconvenient probing geometries as we will discuss later.

SP microscopy in the far field with lateral resolution approaching that of good quality optical microscopes is a relatively young field; for this reason the reader will not find an exhaustive range of applications where the techniques have *yet* proved themselves. The article should be read in the spirit of a preview and an encouragement for techniques that promise to have enormous impact.

This chapter is organised along the following lines. In Sect. 12.2 we will describe the underlying physics necessary to understand the key issues that make SP microscopy unique. In particular, we discuss excitation and detection of SPs in the context of the variation of reflectance functions with angle of incidence. We also give a brief discussion of the field distributions associated with these waves. Section 12.3 will give a brief overview of the prism–based approaches used to harness spatial resolution with the sensitivity associated with surface plasmons. We will conclude that although these methods are often valuable the lateral resolution that may be obtained is far short of that of a good quality optical microscope, so that the techniques while interesting have not yet become mainstream. Section 12.4 will discuss how high lateral resolution may be obtained using an high numerical aperture fluid

Török/Kao (Eds.): Optical Imaging and Microscopy, Springer Series in Optical Sciences Vol. 87 – © Springer-Verlag, Berlin Heidelberg 2003

coupled objective lens, restricting our discussion to 'amplitude' based methods. Section 12.5 will extend the idea of high resolution SP imaging to consider so-called phase dependent methods where the phase of the exciting beam and in some cases also the detected beam play a crucial role. The methods discussed in Sects. 12.4 and 12.5 are capable, at least in principle, of giving lateral resolution comparable to good quality optical microscopy. Section 12.6 considers the measures necessary to cope with aqueous samples, where we discuss the use of ultra high numerical aperture objectives as well as using generalised surface waves where an extra dielectric layer is deposited. Section 12.7 will briefly discuss the prospects for SP microscopy including the feasibility of high resolution imaging using a wide field microscope.

12.2   Overview of SP and Surface Wave Properties

The simplest system on which SPs can exist is between the interface of a relatively thin (in terms of the free space wavelength) conducting layer located between two dielectric layers. This situation is depicted schematically in Fig. 12.1 (a). A condition for the existence of SPs is that the real part of the dielectric permittivity, ε, of the conducting layer is negative. Physically SPs [1] may be thought of as collective oscillations of charge carriers in the conducting layer, with the result that energy is transferred parallel to the interface. In practice noble metals such as gold and silver are excellent metals to support the propagation of SPs and indeed the great majority of SP studies in the literature use these metals. Aluminium also supports SP propagation although the attenuation with this metal is very severe, since the positive imaginary part of the dielectric permittivity is relatively large. A key property of

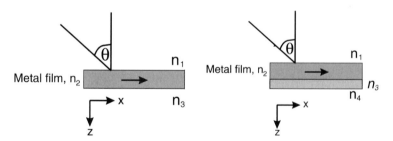

Fig. 12.1. (a) Simple three-layer system allowing the excitation of surface plasmons, consisting of a thin metal layer sandwiched between two dielectrics. Excitation occurs from the layer of higher refractive index. (b) A four layer system with an additional dielectric layer

SPs is that the associated k-vector is greater than the free space k-vector. This means that SPs cannot be excited from free space unless evanescent waves or a structured surface such as a grating is used. In order to overcome this problem light from a region of high refractive index can be used to illuminate the sample.

The variation of reflection coefficient with incident angle from a thin metal film gives considerable insight into the behaviour of SPs [7]. Figure 12.2 shows the modulus of the amplitude reflection coefficient for 633 nm incident wavelength when $n_1 = 1.52$, $n_2 = \varepsilon^{1/2} = (-12.33 + 1.21\mathrm{i})^{1/2} = 0.17 + 3.52\mathrm{i}$, and $n_3 = 1$. These values correspond to the refractive index of glass or coupling oil, gold and air. The thickness of the gold layer in the simulations is 43.5 nm. We see that for s-incident polarization there is not a great deal of structure in the modulus of the reflectivity. On the other hand, for p-incident polarization there is a sharp dip at an incident angle, θ_p close to 43.6 degrees; this corresponds to excitation of SPs, which can only be excited for p-incident polarization. The reduction of reflection coefficient relates to the fact that when SPs are excited some of the energy is dissipated in ohmic losses in the film. If the refractive index of the metal film is entirely imaginary the dip disappears, although SPs are still excited. Excitation at an angle of approximately 44 degrees is indicative of the observation made earlier that the k-vector of the SPs is greater than the free space k-vector. From the position of dip we can see that the k-vector of the SP is $n_1 \sin \theta_p = 1.05$ times greater than the free space propagation constant.

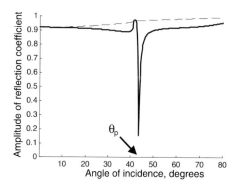

Fig. 12.2. Modulus of the reflection coefficient for a three material system consisting of a gold layer sandwiched between two semi-infinite layers of refractive indices. $n_1 = 1.52$ and $n_3 = 1$. For gold layer $n_2 = \varepsilon^{1/2} = (-12.33 + 1.21\mathrm{i})^{1/2}$ layer thickness= 43.5 nm. optical wavelength 633 nm. *Solid line p*-incident polarization, *fine dashed line s*-incident polarization

The modulus of the reflection coefficient for p-polarized shown in Fig. 12.2 is indicative of the excitation of SPs, but as we mentioned above the dip in the modulus is not an essential manifestation of the SP excitation. The phase of the reflection coefficient, in fact, gives real evidence of the excitation of surface waves, we will show, moreover, that it is the phase of the reflection (and transmission) coefficient, which allows one to exploit new imaging modes. Figure 12.3 shows the phase of the reflection coefficient for both s-polarized and p-polarized light. The phase behaviour of the s-polarized light, like the amplitude, shows little structure, but the phase variation of the p-polarized light shows a strong feature close to the angle θ_p, around this angle the phase changes through close to 2π radians. This is a more fundamental feature of SP excitation than the dip in the modulus of the amplitude reflection coefficient, since it will occur even when the refractive index is purely imaginary, when the dip disappears.

It is useful to consider in more detail the physical meaning of the phase shift, we do this using a simple angular spectrum approach. Consider a weakly confined cylindrical beam illuminating the sample from the medium of refractive index n_1

278   M.G. Somekh

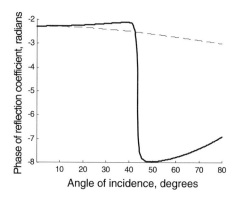

Fig. 12.3. Phase of reflection coefficient corresponding to case described in Fig. 12.2, note for the reflection coefficients the phase of the p-polarization is shifted by π radians, so that they match at normal incidence. *Solid line* p-incident polarization, *fine dashed* line s-incident polarization

shown in Fig. 12.1. The confinement of the beam is considered to be sufficiently weak that the whole angular spectrum is contained in the narrow region (of the order of a degree or less) where the phase of the reflection coefficient varies approximately linearly with incident angle. Let the angular spectrum of the incident wave be $A(k_x)$, where k_x is the spatial frequency along the x-direction, the spectrum is assumed to be centered on the spatial frequency, k_{x0}, which is the position of the minimum of the modulus of the reflection coefficient. The profile of the incident distribution is given by:

$$E_i(x) = \int_{-\infty}^{\infty} A(k_x) \exp(ik_x x)\, dk_x \,, \tag{12.1}$$

where $E_i(x)$ represents the variation of electric field with spatial position x. In this argument we assume that the E-field can be represented as a scalar. In general this is not the case when the incident angles are large, and we will need to consider the vectorial nature of the field in Sect. 12.3. In this case this approach is justified simply by the assumption that the range of spatial frequencies is small so that any obliquity factors will be virtually the same for all components. The reflected beam, $E_r(x)$ will be given by:

$$E_r(x) = \int_{-\infty}^{\infty} A(k_x) R(k_x) \exp(ik_x x)\, dk_x \,. \tag{12.2}$$

We approximate $R(k_x)$ as follows:

$$R(k_x) \approx \left[R_0 + a\Delta k_x^2\right] \exp\left[i(\phi_0 + b\Delta k_x)\right] \,. \tag{12.3}$$

Where R_0 is the modulus of the minimum reflection coefficient and a fits the amplitude distribution to a parabola, centered at k_{x0}. Δk_x is equal to the difference in the spatial frequency from k_{x0} that is $k_x - k_{x0}$. For the curve corresponding to gold in Fig. 12.3 we obtain $a \approx 1.54 \times 10^4$ and $b \approx -92$. The fit to a parabola is highly approximate as the modulus of the reflectivity is not symmetrical about the minimum.

Substituting for $R(k_x)$ in (12.3) into (12.2) gives:

$$E_r(x) \approx \exp\left[i(\phi_0 - bk_{x0})\right]$$

$$\times \left\{ R_0 \int_{-\infty}^{\infty} A(k_x) \exp\left[ik_x(x+b)\right] dk_x + a \int_{-\infty}^{\infty} A(k_x)\Delta k_x^2 \exp\left[ik_x(x+b)\right] dk_x \right\} \quad (12.4)$$

It can be readily seen that the first term in {} is a scaled and displaced version of the original distribution. This displacement is simply $-b$, that is proportional to the gradient of the phase distribution. For the situation where there is no loss present a broad beam propagates along the surface and reradiates with little distortion. The second term in the {} represents a displaced and distorted beam, which after interference with the first term also accounts for the attenuation of the wave. The effect we have just described is effectively the Goos Hanschen shift [8] associated with light or the Schoch displacement [9] that occurs with surface acoustic wave excitation in ultrasonics. We can therefore see that the phase variation of the reflection coefficient around the angle for SP excitation contains much of the information concerning the propagation of the SPs. We will show later that the phase structure of both the reflection and transmission coefficients play a crucial role in developing high resolution SP microscopy methods.

The propagation of SPs is nicely demonstrated in the simple experiment depicted in Fig. 12.4 (a). This shows SPs excited in the metal by a weakly focused beam from a 633 nm HeNe laser source. This arrangement for excitation of SPs is the so-called Kretschmann configuration [1], this is described in more detail at the end of this section. The SPs propagate in the metal film and leak back into reflected modes or are converted to heat by the absorption of the metal film. For a uniform film there will be no 633 nm radiation propagating into the air since the light is evanescent in this medium. The gold film was then heated periodically with a frequency doubled YAG laser (532 nm) modulated at a frequency of several kHz. The effect of the periodic heating produced by the green laser was to alter the local optical properties of the metal layer and the surrounding dielectric. This perturbation caused the SPs to be converted into light waves propagating in the air. This occurs because the k-vector of the local perturbation can change that of the SP so that the resulting k-vector corresponds to a propagating mode. The 633 nm light scattered by the green laser was detected with the photodetector, after passing through an interference filter which blocked the 532 nm radiation. The distribution of scattered light corresponds to closely to the SP distribution and is shown in Fig. 12.4 where see that there is a peak corresponding to the position of excitation and a 'hump' demonstrating the lateral propagation of the SPs. The dip between the two maxima arises from destructive interference between the two terms in the {} in 12.4 [10]. When the polariation of the incident 633 nm radiation was changed to be predominantly s-polarized very little signal was observed because of the field enhancement property discussed in the next section. Moreover, from the small amount of signal that could be seen there was no evidence of lateral displacement. The extreme sensitivity that makes SPs so sensitive as sensors arises from the large field enhancement at the

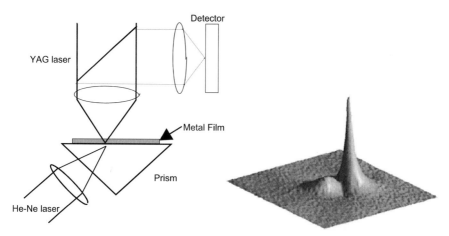

Fig. 12.4. (a) Schematic diagram of equipment used in photoscattering experiment, consisting of HeNe laser to generate surface plasmons and Nd doubled YAG laser to scatter them into propagating modes. Experimental results obtained with equipment shown in figure (b). Picture width: 90 × 80 μm

interface between n_2 and n_3. This is shown in Fig. 12.5 (a) for the case of gold at incident angle of 43.6 degrees. Interestingly, the field strength increases with depth in the metallic film. This is due to the large 'reflected' evanescent field, which decays away from the source and grows towards it.[1] There is an extremely large field at the interface between the metal layer and the final dielectric, n_3. The discontinuity in the absolute magnitude of the field at this interface arises because it is the tangential component of the E-field that is conserved across the interface. For s-polarized incident light, on the other hand, the field decays in the metal so the field at the interface is rather small, the field is continuous across the interface since the E-field vector lies in the plane of the interface. If we again consider an 'ideal' material with the same negative permittivity as gold and zero imaginary part the field distribution of Fig. 12.5 (b) results. This shows even greater p-field enhancement compared to gold and would be even more sensitive to surface properties. We note that the s-fields for 'real gold' and 'ideal gold' differ by about 2% indicating a relatively weak interaction with the metal layer. Fig. 12.6 (a) and 12.6 (b) compares the amplitude and phase of the reflection coefficients for a single gold layer in contact with the air and a similar structure with an additional intermediate layer corresponding to 20 nm of silicon dioxide, for clarity a relatively narrow range of incident angles is shown. We see that there is a considerable change in the position and shape of both the amplitude and phase features corresponding to SP excitation. In the case

---

[1] This viewpoint does not contravene conservation of energy as the field corresponding to the SPs is evanescent in medium 3. An interesting extension of this concept has been recently exploited by Pendry [11], where he predicts that it is possible, provided certain other conditions are met, to obtain an infinitely sharp focal spot by enhancing the evanescent wave components.

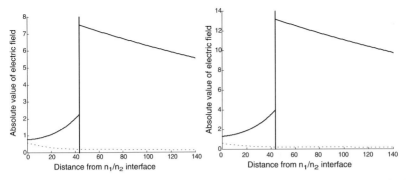

Fig. 12.5. (a) Modulus of the electric field distribution through the sample described in Fig. 12.2. Note the field increases increases throughout the sample. Angle of incidence from $n_1 = 43.6$ degrees. Solid line p-incident polarization, fine dashed line s-incident polarization. **(b)** Same case as Fig. 12.4 (a) except the metal layer is replaced with 'lossless' gold, that is $n_2 = \varepsilon^{1/2} = (-12.33)^{1/2}$

of the s-incident polarization the effect of the layer is much less distinct. Changes in the amplitude and phase of the reflection coefficient enable weak features to be observed using SP excitation. In Sect. 12.5.2 we will discuss how the sensitivity to interfacial conditions compares with total internal reflection microscopy. The most

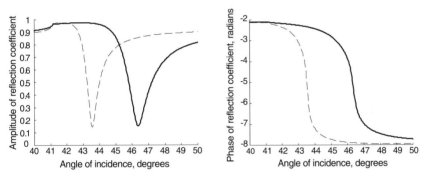

Fig. 12.6. (a) Comparison of modulus of reflection coefficient for bare gold layer as shown in Fig. 12.2 (*dashed line*) with same layer coated with a 20 nm layer of SiO_2 (*continuous line*). **(b)** Phase plot corresponding to Fig. 12.6 (a)

popular way to excite and detect SPs is to use the Kretschmann configuration, see Fig. 12.7. The metal film – usually gold or silver – is deposited on the hypotenuse of the prism and the light is incident along one of the other sides so it hits the metal film through the glass at an angle, θ, close to θ_p. The passage of the light through the glass increases the k-vector of the incident wave so that it matches that of the SP. The changes in the reflected light enable the properties of materials absorbed on the layer to be sensitively monitored. SPs find wide use as chemical and biosensors for application such as monitoring antibody–antigen reactions, cell attachment and toxicology (and indeed any areas where sensitivity to minute changes in the properties of an interface is required). Standard Kretschmann based SP techniques are thus extremely powerful, they do, however, lack the spatial resolution required to image highly localised regions on, for instance, at the cellular level. For this reason there has been substantial work to develop SP microscopy techniques that combine high lateral resolution with good surface sensitivity.

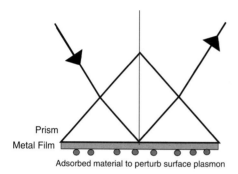

Prism

Metal Film

Adsorbed material to perturb surface plasmon

Fig. 12.7. Schematic diagram showing Kretschmann configuration used to excite surface plasmons on the hypotenuse of a prism

12.3   Surface Plasmon Microscopy – Kretschmann Prism Based Methods

In the mid-nineteen eighties the idea of combining the sensitivity of SPs with good lateral resolution took hold. The earliest work in the field was that of Yeatman and Ash [12] in the UK and Rothenhausler and Knoll [13] in Germany. Figure 12.8 shows the two wide field configurations based on the Kretschmann configuration. The sample is illuminated with a plane beam of light at an incident angle close to θ_p. In the 'bright field' configuration the light reflected from the sample is imaged onto the light sensitive detector where the image is recorded. Local variations in the SP propagation and local scattering will change the intensity of the reflected light allowing an SP image to be formed. Detection can also occur on the far side of the prism as also shown in Fig. 12.8; this effectively produces a 'dark field' configuration. In this case discontinuities in the sample scatter the SPs into propagating waves on the far side of the prism, rather like the pump beam in the experiment described in Sect. 12.2. This scattered light is then imaged onto a light sensitive detector. Clearly, in this case a continuous uniform film will not scatter any light so the image will appear dark. Figure 12.9 shows an alternative scanning configuration. In this case a weakly focused beam is incident on the sample and the reflectivity is monitored as function of scan position. In fact the imaging properties of this system are similar to that of the 'bright field' wide field system shown in Fig. 12.8, since the roles of the illumination and collection lenses are interchanged. This scanning configuration illustrates the trade-offs implicit in bright field Kretschmann based SP imaging. In

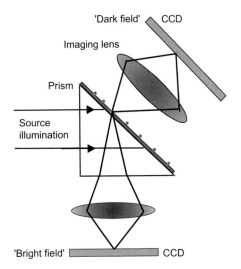

Fig. 12.8. Schematic diagram showing wide field 'bright field' and 'dark field' microscope systems based on the Kretschmann configuration

order to improve the lateral resolution one needs to focus the beam as tightly as possible, using a large numerical aperture. On the other hand, SPs are only excited over a narrow range of angles so that the numerical aperture must be limited to this range. As we showed in Sect. 12.2 the decay length of the SPs is proportional to gradient of the phase change of the reflectance function. A sharp phase change thus means a large decay length, which in turn restricts the maximum useful numerical aperture of the exciting beam and hence the lateral resolution. Yeatman [14] has shown how the improving the lateral resolution of the system results in reduction of sensitivity. Similarly the 'bright field' wide field system has lateral resolution limited by the decay length of the SPs. The lateral resolution of the prism based SP microscope

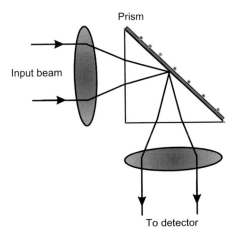

Fig. 12.9. Schematic diagram showing scanning system based on the Kretschmann configuration

has been studied in detail by Berger et al. [15]. When the observed contrast arises from the change in the k-vector of the SPs in different regions of the sample, it was demonstrated that the propagation distance or decay length of the SP determined the lateral resolution that could be obtained in a test structure. In essence, the SP may be considered to interact with the sample for a distance comparable with the decay length so that the localisation of the field will be of this order. Using the fact that the decay length of the SPs depends on the excitation wavelength, Berger et al. [15] were able to show that the lateral resolution could be improved by using a shorter wavelength with a correspondingly smaller decay length. The best resolution obtained was approximately two microns obtained at 531 nm, unfortunately when the decay length is this small the field enhancement at the interface and hence the sensitivity to the surface properties is greatly reduced. A similar approach to improving the lateral resolution has been recently presented by Giebel et al. [16], where a coating of aluminium was used, which has a relatively large imaginary part of the dielectric constant with a consequent reduction in decay length. Once again the improvement in lateral resolution is obtained with a reduction in sensitivity to surface properties. Despite this useful images monitoring cell attachment and motility were obtained.

Fig. 12.8 shows the 'dark field' system where SP imaging is obtained from the *scattering* on the far side of the prism. In this case if the scattering of the particle is imaged onto the CCD with a sufficiently high numerical aperture system then it possible at least in principle to obtain lateral resolution that exceeds that of the decay length of the SPs. Improved resolution using this approach does not appear to have been demonstrated in practical situations. This approach has the considerable disadvantage since it is necessary to probe on the far side of the sample, which means that the detection optics restricts access to the sample. 'Dark-field' surface plasmon microscopy can, however, be obtained in the reflection by blocking the zero order reflection in the Fourier plane as described by Grigorenko et al. [17], this will allow only scattered light to be detected. The method gives lateral resolution similar to the 'bright-field' method.

The difficulty with Kretschmann based configurations arises from two principal factors:

(i) the prism means the system is not readily compatible with a conventional microscope configuration, so that operation with other imaging modes is difficult and

(ii) the lateral resolution is poor compared to that achievable with optical microscopy, so that the range of applications will be restricted. In particular, applications in extremely important fields, such as cell biology, will be severely limited. For these reasons it is necessary to develop techniques capable of overcoming these limitations.

12.4 Objective Lens Based Surface Plasmon Microscopy: Amplitude Only Techniques

This section will discuss how surface plasmon microscopy can be carried out using a conventional objective lens. The principle behind the excitation of SPs from an objective lens is the same as the case for the prism based Kretschmann configuration. Essentially, the k-vector of the incident wave must match that of the SP. Clearly, this cannot be achieved with an air coupled objective, but an oil immersion lens may be used to achieve this. Consider an aplanatic objective with a source of illumination in the back focal plane. Each point incident on this plane maps to an incident plane wave propagating in a direction whose k-vector parallel to the plane of incidence is proportional to the radial position of the source in the back focal plane. This is shown schematically in Fig. 12.10. If the polarization of the beam incident on the back focal plane is linear, the polarization state of the incident radiation varies from pure p-polarized to pure s-polarized. In general the polarization is a combination of the two states, which can be determined by resolving with respect to the azimuthal angle, ϕ, as shown in Fig. 12.10. For an oil immersion lens with a numerical aperture greater than about 1.1 a fully illuminated objective lens will generate plane waves whose incident angle and polarization state can excite SPs on the sample. Kano et al. [18] have demonstrated both experimentally and theoretically that ex-

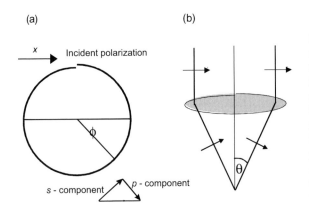

(a)   (b)

Fig. 12.10. Back focal plane distributions. (**a**) View of the back focal plane showing the relationship between azimuthal angle and polarization state. (**b**) View in the plane of incidence showing relation between radial position and incident angle. Also shows rotation of direction of E-field vector passing through the objective

citation through an oil immersion objective can lead to highly localized generation of SPs. In addition they demonstrated a system somewhat akin to the 'dark-field' system shown in Fig. 12.8, whereby SPs where excited on the sample through an oil immersion objective ($NA = 1.3$). The presence of local scatterers was detected on the far side of the sample by scattering into waves propagating away from the source. The rescattered SPs were collected with a dry objective on the far side of the sample. A simplified schematic of this system is shown in Fig. 12.11. This experiment demonstrated the potential of excitation using an objective lens as a means of exciting surface plasmons and showed that resolution comparable to the spot size is obtainable. On the other hand, the arrangement described is not practical for most imaging applications on account of the fact that like the prism based 'dark-field' arrangement detection takes place on the far side of the sample. Measuring the dis-

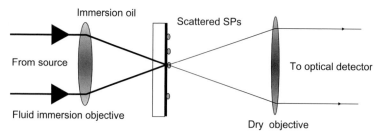

Fig. 12.11. Simplified schematic of 'dark-field' system of Kano and Kawata using fluid immersion excitation of surface plasmons

tribution of reflected light in the back focal plane is the basis of the technique of back focal plane ellipsometry, where the sample reflectivity as a function of incident angle and polarization state may be monitored. The advantage of this technique is, of course, that the focal spot is confined to a small submicron area allowing the properties to be measured in a highly localised region. This technique has been used with dry objectives by several authors to measure film thickness in semiconductors [19], to compensate for material variations in sample properties when measuring surface profile [20] and for extracting ellipsometric properties over a local region [21]. This concept has been recently extended by Kano and Knoll [22] using an oil immersion objective to measure the local thickness of Langmuir–Blodgett films. Figure 12.12 shows a calculated distribution in the back focal plane similar to that presented by Kano and Knoll. The results correspond to gold sample with the same parameters as those used to obtain Fig. 12.2. We note the presence the dark ring in the back focal plane distribution, this corresponds to the dip in the reflection coefficient shown in Fig. 12.2 (a). The strength of the dip is strongest in the horizontal direction where the incident light is pure p-polarized, the dip disappears in the vertical direction

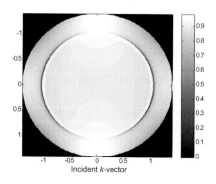

Fig. 12.12. Calculated back focal plane distribution, when surface plasmons are excited, showing partial ring corresponding to the dip in p-polarization reflection coefficient

where the incident light is entirely *s*-polarized. Measurement of the position of the dark ring allows the value of θ_p to be determined on a spatial scale corresponding that of the incident focal distribution. Localized measurements corresponding to 4 Langmuir Blodgett monolayers were detected reasonably easily with this technique.

These authors have more recently extended this approach a scanning microscope configuration [23] where the position of the ring is monitored as a function of scan position, thus allowing microscopic imaging. The lateral resolution obtained with this methods appears to be approximately 1.5 microns, which is rather lower than one would expect since the spot diameter is approximately a factor of three smaller than this. The difficulty in achieving the desired lateral resolution may, in part, arise from the fact that the experimental distributions in the back focal plane are rather prone to the presence of interference artifacts.

12.5   Objective Lens Based SP Microscopy: Techniques Involving the Phase of the Transmission/Reflection Coefficient

This subsection will discuss techniques in which the phase of the reflectance function rather than just the amplitude play a crucial role and discuss how the obtainable contrast and resolution benefits greatly from this approach. There has been quite a lot of recent interest in using the phase of the reflectance function in Kretschmann based SP imaging systems and brief discussion of some of these developments will help place our recent work on objective lens based techniques in context.

The variation of the phase of the reflectance function around θ_p provides an alternative quantity in addition to the amplitude of the reflection. A discussion of the use of the phase has been discussed by Grigorenko et al. [17]. They have shown that as the metal thickness is changed so that the minimum of the reflectivity is zero (corresponding to a thickness of approximately 47.7 nm of gold using the parameters given in Fig. 12.2) the gradient of the phase variation with incident angle increases without bound around θ_p. This means that even without these extreme conditions the phase of the reflectance function can be more sensitive to small changes compared to the amplitude. This sensitivity is borne out experimentally in the work of Notcovich et al. [24], who have imaged gas flows corresponding to a refractive index change of 10^{-6}.

12.5.1   Objective Lens Interferometric Techniques

In order to harness the advantages of phase sensitivity and objective lens imaging it is necessary to construct an interferometer using an oil immersion objective. In order to understand the technique it is appropriate to first describe the experimental set-up. Figure 12.13 shows a schematic diagram of the scanning heterodyne interferometer used in the experiments. The interferometer was illuminated with a 633 nm HeNe laser, one arm of the interferometer passed through an oil immersion objective

(Zeiss CP-Achromat 100×/1.25 oil), the other arm was formed by reflection from a reference mirror. Two Bragg cells, driven at slightly different frequencies (10 kHz), were used to effect the frequency shifting, so that the interference signal was detected at the beat frequency between the sample and reference arms. The use of two Bragg cells not only allows the interference signal to be detected at a very convenient frequency but also reduces the effects of spurious reflections in the Bragg cells [25]. The samples were mounted in an x, y, z piezo-scanning stage with resolution of 10 nm. The scan stages were controlled by a PC and the output from the heterodyne interferometer was converted to an amplitude and phase signal in a lock-in amplifier, which was, in turn, sent to the PC for display. One advantage of the heterodyne system is that is the heterodyne signal contains the interference term alone and no other processing is necessary as is the case with, say, phase stepping techniques. We now consider how we might expect an interferometer to behave in the presence

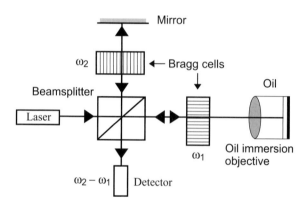

Fig. 12.13. Heterodyne interferometer configuration used for $V(z)$ measurements and high resolution surface plasmon imaging

of a sample that supports SPs. The contrast of the SP microscope can be explained by the so called $V(z)$ effect, where the output response, V, varies as a function of defocus, z. This effect has been used in recent publications to evaluate the pupil function of objective lens and to measure aberrations [26]. The effect is also very important in scanning acoustic microscopy where surface acoustic (Rayleigh) waves are excited by a large numerical aperture acoustic lens [27,28]. The resulting phase difference between ray paths corresponding to excitation and reradiation of surface waves and rays reflected normally from the sample provides a means of accurately measuring wave velocities. This effect also provides a powerful means to explain image contrast in the acoustic microscope, [29,30]. We will show that the situation when imaging structures that support SPs is analogous to the acoustic case, although there are important differences, in particular, with respect to the polarization of the incident beam.

In order to calculate the $V(z)$ response of a heterodyne interferometer we first need to calculate the field reflected to the back focal plane of the microscope objective, see Fig. 12.10. Consider linearly polarized light incident on the back focal plane of the microscope objective. The field amplitude passing through the objective is assumed to vary radially as $P(r)$, where the pupil function may include amplitude and phase variations (introduced by, for example, aberrations). For simplicity we normalise the aperture of the objective to unity. As mentioned in Sect. 12.4 a point excitation in the back focal plane can be considered to be transformed to a plane wave in a specific direction and polarization state after passing through the objective. Light passing through the objective at a normalized radial position r will be incident at an angle, $\theta = \arcsin(r \sin \theta_{max})$ where θ_{max} is the maximum angle of incidence produced by the objective lens. The position r in the back focal plane therefore maps directly to the sine of the incident angle at the focus of the objective. Similarly the polarization state of the light emerging from the objective varies as the azimuthal angle, ϕ.

Consider a plane wave emerging from a point (r, ϕ), in the back focal plane. The field reflected back along the original polarization direction due to this source will be given by:

$$E_{\text{reflected}} = P^2(\sin \theta) \left[R_p(\sin \theta) \cos^2 \phi + R_s(\sin \theta) \sin^2 \phi \right] \exp(i2n_1 k \cos \theta z) .$$

$$(12.5)$$

The pupil function appears squared since the light passes through the objective twice. $R_p(\sin \theta)$ and $R_s(\sin \theta)$ refer to the Fresnel reflection coefficients for p- and s- polarizations. The terms $\cos^2 \phi$ and $\sin^2 \phi$ refer to the fact that the proportion of each incident polarization changes with azimuthal angle. The p-incident light is the component resolved radially and the s-incident light is resolved along the tangent. The interference signal arises from the component of the field resolved along the direction of the original polarization direction; so it is necessary to resolve each reflected component a second time. The final term in the exponential refers to the phase shift due to defocus, z, k is the wavenumber in free space ($2\pi/\lambda_{\text{free}}$), where λ_{free} is the free space wavelength. Negative z refers to movement of the sample towards the objective.

The field reflected in the back focal plane is interfered with a frequency shifted reference beam E_o so that the frequency shifted output from the heterodyne interferometer is given by:

$$2\text{Re} \left\{ E_{\text{reflected}} E_o^* \right\}$$

The instantaneous interference signal from the heterodyne interferometer is a time varying signal proportional to the instantaneous change in intensity, and is thus real. The amplitude and phase of this signal can, of course, be represented as a complex phasor, this phase is equal to the phase difference between the reference and the signal beams. We can thus represent $V(z)$, omitting constant terms and assuming a uniform reference beam as:

$$V(z) = \iint_{\text{lens aperture}} P^2(\sin \theta) \left[R_p(\sin \theta) \cos^2 \phi + R_s(\sin \theta) \sin^2 \phi \right]$$
$$\times \exp(i2n_1 k \cos \theta z) \mathrm{d} \sin \theta \, \mathrm{d}\phi .$$

$$(12.6)$$

The interesting thing about this expression is that, as we will see, the modulus of the $V(z)$, $|V(z)|$, gives a considerable amount of information about the behaviour of the SPs. Despite this is it still necessary to detect the modulus of the *interferometer* output in order to get this information. Moreover, it is the phase behaviour of the reflectance functions that give $V(z)$ its interesting properties. In all the experimental and theoretical $V(z)$ curves presented in this article we only present $|V(z)|$; since the phase is not used modulus signs are omitted for brevity.

Figure 12.14 shows theoretical $V(z)$ curves calculated using (12.6). The excitation of SPs is represented by the phase change in R_p which induces a periodic ripple at negative defocus, whose periodicity depends on θ_p, whereas no such structure is visible in the $V(z)$ if it is formed from R_s alone. For a full aperture objective there will be equal contributions from s- and p- incident polarizations, which still gives the SP induced ripple albeit with reduced relative amplitude compared with pure p-incidence (Fig. 12.6 (b)). When the sample is defocused curvature is introduced to

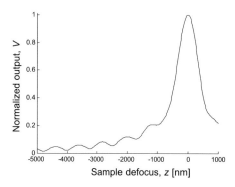

Fig. 12.14. Theoretical $V(z)$ curve corresponding to reflectance functions shown in Figs. 12.2 and 12.3. Note the periodic ripple at negative defocus

the reflected wavefront; the ripple arises from the deformation of the curved wavefront by the phase change in the reflection coefficient. This means that there are two principal components corresponding to the output signal: the reflection close to normal incidence and the reflection around θ_p. This effect has been explained for the case of the scanning acoustic microscope by Atalar [31]. These two contributions may be represented by ray paths A and B (Fig. 12.15); as the defocus changes the relative path length between the rays changes. Ray paths A and B do not interfere with each other directly because they return in different positions, however, they each interfere with a common reference beam [32]. This means that the strength of the interference signal will oscillate periodically, going through a complete cycle when the phase difference between A and B changes by 2π. The periodicity, Δz of the oscillations in the $|V(z)|$ signal can be obtained by noting that a defocus, z, introduces a phase shift in the wave travelling along the optical axis of $2kn_1z$; the phase difference in the wave exciting the surface plasmon is $2kn_1 \cos \theta$; Δz is obtained by setting the difference in these two phase shifts equal to 2π. The periodicity, Δz, is thus given by:

$$\Delta z = \frac{\lambda_{\text{free}}}{2n_1(1 - \cos \theta_p)} \,, \tag{12.7}$$

where λ_{free} is the free space wavelength. This expression is similar to the analogous expression for the $V(z)$ periodicity in the acoustic microscope. There is, however, one important difference between this effect and that observed in the case of the acoustic microscope. In the SP situation only the p-polarization excites surface waves, with the effect that the s-incident polarization contributes a background contribution to the $V(z)$ signal, reducing the relative amplitude of the ripples produced by SP excitation. Fig. 12.16 shows $V(z)$ curves obtained on a coverslip coated with

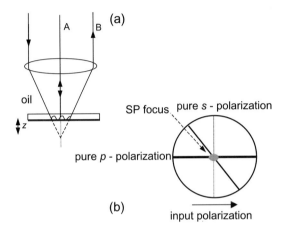

Fig. 12.15. Ray paths showing principal contributions contributing the ripple in the $V(z)$ curve. (**a**) Rays in plane of incidence showing normally reflected contribution (A) and surface plasmon contribution (B). (**b**) Schematic of sample showing surface plasmons excited over a ring, from which they can focus to the optical axis

44 nm of gold and a similar sample on which a further layer of 20 nm of silicon dioxide was deposited on the gold. A very thin layer of approximately 1 nm of chromium was deposited between the gold and the coverslip to improve adhesion. The presence of the dielectric perturbs θ_p, which, in turn, perturbs the periodicity of the ripple. The measured periodicity of the ripple for the bare metal film was 761 nm, which is extremely close to the value of 758 nm predicted from (12.7). The periodicity obtained on the dielectric coated sample was 676 nm compared to 672 nm predicted from a metal layer coated with 20 nm of silicon dioxide. The experiments indicate that the presence of the dielectric increase θ_p from 43.6 degrees to 46.4 degrees with a corresponding increase in the k vector of the excited SP. We have also inserted apertures in the back focal plane of the illumination optics to vary the proportion of s- and p-polarizations. These show that, as expected, when a narrow aperture is placed normal to the direction of the incident polarization, most the p-polarized light is suppressed and as expected the ripples in the $V(z)$ curve are substantially suppressed.

We now turn our attention to how the $V(z)$ curve can be related to imaging with the heterodyne interferometer. The $V(z)$ curve indicates that at focus there is very little contrast between different samples, but as the sample is defocused the different periodicities of the ripple patterns means that there is contrast between a bare metal film and a dielectric coated metal film. For instance, defocus A on Fig. 12.16, shows a region where the bare region will appear bright compared to the coated region.

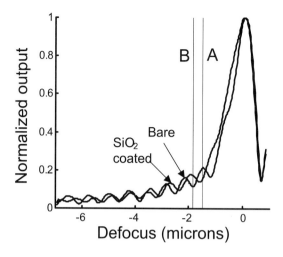

Fig. 12.16. Experimental $V(z)$ curves obtained on samples coated with gold (bare) and gold with an additional 20 nm layer of SiO_2 (coated)

For larger defocuses the contrast will reverse, at, for instance, defocus B. Although the $V(z)$ curves indicate that the microscope will detect contrast between a bare metal surface and a dielectric coating, they do not give any indication concerning the lateral resolution. When the sample is in focus the wavefront of the returning beam will make a substantial contribution to the output interference signal *except* where the phase changes rapidly over the region where the SPs are excited. For this reason the interferometer is not expected to give much SP contrast in focus.[2] It is thus necessary to defocus the sample to obtain good contrast.

The advantage in terms of contrast of operating with a negative defocus may, at first sight, appear to be a problem with the technique since defocus, of course, results in spreading of the focal spot. The SPs are, however, excited on a ring so that they propagate towards the optical axis where they come to a focus. This indicates that there will be a large increase in intensity over a highly localized diffraction limited region discussed in more detail in Sect. 12.5.2. The diameter of the spot will be small (although not equal) in both horizontal and vertical directions, since the polarization state, and hence the strength of SP excitation, varies with azimuthal angle. It is this self-focusing that leads to the possibility of high resolution imaging in the defocused state. The fact that the SPs are excited over a ring means that the SP microscope can be thought of as a 2π (radians) microscope by analogy with the 4π (steradians) microscope used in 3–D confocal microscopy [33].

A series of images were obtained on the structure shown schematically in Fig. 12.17, this consists of 2 μm stripes of bare metal followed by 6 μm of dielectric coated metal. A piezo-drive stage was used to scan in increments of 0.02 μm. The images shown in Fig. 12.18 (a), 12.18 (b) and 12.18 (c) were obtained by scanning

---

[2] It is worth noting that this effect is a consequence of the *detection* process, which will not be apparent with the amplitude methods described in Sect. 12.4. We will discuss later, however, how defocus will enhance the *generation* of the SPs even when the detection is sensitive to intensity.

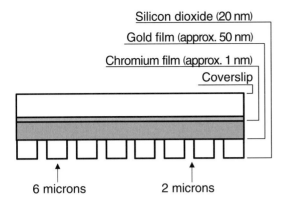

Silicon dioxide (20 nm)

Gold film (approx. 50 nm)

Chromium film (approx. 1 nm)

Coverslip

6 microns   2 microns

Fig. 12.17. Schematic of sample used for imaging experiment

the objective relative to the coverslip. Figure 12.18 (a) was obtained with the sample close to the focus and shows barely discernible contrast as expected from the $V(z)$ curves. Figure 12.18 (b) was obtained at a defocus of $-1.5\,\mu m$ (A on Fig. 12.16) and shows the stripes (bare metal) bright, whereas Fig. 12.18 (c) taken at $-1.85\,\mu m$ (B on Fig. 12.16) shows the stripes dark. Below each of the images horizontal line traces starting $1.6\,\mu m$ from the top left hand corner of the images are shown. The image corresponding to Fig. 12.18 (a) shows little contrast as expected. The traces corresponding to Figs. 12.18 (b) and 12.18 (c) show opposite contrast to each other and, more importantly, the transition across the stripes in both cases is less than $0.5\,\mu m$. Since the traces are horizontal and the stripes are not exactly vertical the actual transitions are at least 10% sharper than this value. These results clearly indicate that the resolution is limited by diffraction rather than the propagation length of the SPs. These effects are discussed in more detail in [34,35]. To fully understand

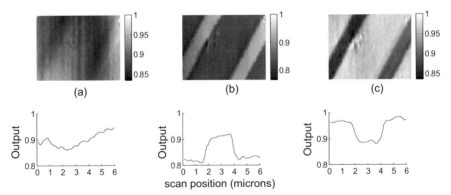

Fig. 12.18. Image of sample shown schematically in Fig. 12.17. (**a**) In focus-little contrast. (**b**) Defocus $-1.5\,\mu m$ – good contrast bare region appears bright. (**c**) Defocus $-1.85\,\mu m$ – good contrast coated region appears bright

the imaging performance of the interference based SP microscope in the presence of a structured sample it is necessary to carry out a vector diffraction analysis. We are currently examining the use of the so-called C-method of Chandezon [36] to model the imaging performance of the interferometric SP microscope so that the optimum achievable resolution may be quantified together with the effect of varying the input polarization state.

It is interesting to estimate the sensitivity of the interferometric $V(z)$ measurement to small variations in layer thickness. Consider a heterodyne interferometer operating in the shot noise limit. Provided the reference beam is sufficiently powerful the signal to noise ratio (SNR) can be readily shown to be given by:

$$\text{SNR} = \frac{\eta P}{h v B} \, . \tag{12.8}$$

Where η is the quantum efficiency, P is the optical power detected from the sample, h is Planck's constant, v is the optical frequency and B is the measurement bandwidth. Using values of $\eta = 0.8$, $P = 0.1$ mW and $B = 8$ kHz and a wavelength of 633 nm we obtain a signal to noise ratio of 3.2×10^9. This means that the number of resolvable voltage levels is approximately 5.6×10^4. We now look at the signal change on the $V(z)$ curve, such as that shown in Fig. 12.14, at a fixed defocus corresponding to approximately 1.8 microns as a layer of SiO_2 is deposited onto bare metal. This gives a signal change of approximately 1 part in 500 per nm. This indicates that the $V(z)$ should be capable of resolving changes in layer thickness of less than 10^{-2} nm. Clearly, in practice a system may give a noise performance a few dBs worse than the shot noise limit. On the other hand, using 1 point only of the $V(z)$ to measure a change in layer thickness is rather crude. We are confident therefore that the $V(z)$ technique can be used for measuring small fractions of a monolayer on localised regions.

12.5.2   Fluorescence Methods and Defocus

It may seem perverse to discuss fluorescence based methods in this section where the phase of the transmission and the reflection coefficient play a crucial role. In the interferometric technique discussed in Sect. 12.5.1 the role of the phase of the reflection coefficient in the *detection* process has been clearly explained. In this section we will discuss SP microscopy when the output signal is generated by a fluorescent process. Clearly, then the phase of the reflection and transmission coefficient do not have a role to play in the detection process, but this phase variation still plays an important role in the generation process. Once again this process is particularly apparent when the sample is defocused. Consider a fluorescence system, shown schematically in Fig. 12.19, a beam of radiation of wavelength λ_1 excites the sample and fluorescence at λ_2 is detected. In the conventional fluorescence process $\lambda_2 > \lambda_1$, if a pulsed laser (picosecond or femtosecond) is used the peak intensity can be sufficiently large that two photon fluorescence can arise from the excitation from the SPs. In this case the $\lambda_2 < \lambda_1$. There are several potential advantages in operating in two photon mode; from the instrumental point of view one of the most important

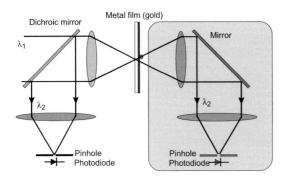

Fig. 12.19. Schematic diagram of surface plasmon fluorescence system

is that if the metal film is gold it will support SPs at typical pump wavelengths, while it will be largely transparent at the emission wavelength. In other words, the pump and emission wavelengths will be on different sides of the plasma resonance frequency. In this case the same objective can be used for both illumination and detection, so that the sub-system shown in the shaded box is unnecessary, the microscope can be single-ended and all observation may be performed in the reflection geometry. This is an advantage of systems described in Sect. 12.4 and 12.5.1 and one of the main problems with probe microscopy techniques. Another advantage of two photon emission, particularly when the sample is defocused, is that there will be very little emission except from the regions where the intensity is greatest, since the emission is proportional to the square of the intensity. This will be borne out in the computational results presented later in this section.

To assess the imaging performance of SP fluorescent microscopy, it is necessary to

(1) calculate the field distribution at the interface between n_2 and n_3, of Fig. 12.1.
(2) calculate of the fluorescence emission from the point object, allowing for the order of the fluorescence process (i.e. single photon, two photon etc).
(3) Account for the impulse response of the collection optics. This allows us to compare confocal and non-confocal operation.

The field distribution at the focus of a high numerical aperture objective can be calculated following the approach presented by many authors, for instance, Hopkins [37] and Richards and Wolf [38]. More recently, Török et al. [39] have extended this approach to account for focusing through media of different refractive indices. The case studied here is formally similar to the case described by Török et al. [39] except that the essential physics is rather different as will be explained in the following paragraphs.

In essence, as we mentioned earlier each point of illumination in the back focal plane is regarded as a point source which gives rise to a plane wave of appropriate polarization state and angle of incidence depending on the azimuthal angle and radial position respectively.

Following this approach leads to the following expression for the field distribution at the $z = 0$ plane corresponding to the interface between n_1 and n_2.

$$E_n = \int_0^{2\pi} \int_0^{\theta_{max}} A(\theta, \phi) f_n(\theta, \phi) \exp\left[in_1 k(x \sin\theta \cos\phi + y \sin\theta \sin\phi + z \cos\theta)\right] d\theta d\phi$$

(12.9)

where E_n denotes the x, y, z components of the electric field vector.

For E_x $\longrightarrow$ $f_x(\theta, \phi) = \cos^2\phi \cos\theta \tau_p(\theta) + \sin^2\phi \tau_s(\theta)$

For E_y $\longrightarrow$ $f_y(\theta, \phi) = \cos\phi \sin\phi \left[\cos\theta \tau_p(\theta) - \tau_s(\theta)\right]$

For E_z $\longrightarrow$ $f_z(\theta, \phi) = -\cos\phi \sin\theta \tau_p(\theta)$

$$E_i = \int_0^{2\pi} \int_0^{\theta_m} A(\theta, \phi) f(\theta, \phi) \exp\left[ik(x \sin\theta \cos\phi + y \sin\theta \sin\phi + z \cos\theta)\right] \sin\theta \, d\theta d\phi .$$

Where k is the wavevector in free space, $A(\theta, \phi)$ represents the source of the excitation in the back focal plane, for an aplanatic system an additional factor of $\sqrt{\cos\theta}$ should be incorporated to account for the change in area subtended by an element of the incident beam as is changes direction passing through the objective. Other polarization states, such as circular, may be readily derived by appropriate summations of the Jones matrix in the back focal plane. $\tau_p(\theta)$ and $\tau_s(\theta)$ represent the transmission coefficients for p- and s-incident polarization as a function of incident angle θ.

Equation (12.9) is similar to that developed by Török et al. [39], there are certain issues that need clarifying to appreciate how the expression is used for the problem of SP imaging. Imaging through a stratified medium, as addressed by Török et al. [39] is of considerable importance in conventional two photon microscopy, since the aberration of the focused beam as the light passes through regions of different refractive index has a powerful influence on the field distribution and the resulting fluorescent yield. Our problem looks formally similar; the key physical difference here is that the layer through which the light travels prior to reaching the sample is extremely thin compared to the optical wavelength, so aberrations from this source are less important (although accounted for). The form of the transmission coefficients $\tau_s(\theta)$ and, particularly $\tau_p(\theta)$ are crucial in our case and they have a different meaning from the interface transmission coefficients as described in the matrix formulation of Török et al. [39]. In (12.9) the coefficients are the total electric field transmission coefficients, between n_1 and n_3, thus accounting for multiple reflections at the interfaces as well as the propagation within n_2.

Once the field distribution is calculated the fluorescent yield can be calculated from (12.10) below:

$$I = \alpha \left\{ |E_x|^2 + |E_y|^2 + |E_z|^2 \right\}^n ,$$

(12.10)

where α is a constant of proportionality related to the efficiency of the fluorescence process, the power n is the order of the process, i.e. 1 for single photon, 2 for two photon etc. To get an idea how the fluorescence system will perform we can make some strong simplifying assumptions. The light emitted is assumed to be unpolarized and uniformly distributed. The model for the fluorescent yield is a significant approximation since we are considering the fluorescent process to be independent of the polarization state of the light incident on the fluorescent particle. This is the same assumption as used in the early work of van der Voort and Brakenhoff [40], where they considered 3–D image formation in fluorescent confocal microscopy. Higdon et al. [41] also compare this model with other polarization dependent models for multiphoton fluorescence microscopy. The point spread function of the microscope can be regarded as the product of the point spread function for both excitation and detection; this leads to a further sharpening of the response for a confocal system. More details of this modelling process are given in [42].

In the $V(z)$ calculation the reflection coefficients play a vital role in the determination of the imaging properties. In the fluorescence based system the transmission coefficient play a similarly important role, see Fig. 12.8. Figures 12.20 and 21 show the amplitude and phase of the transmission coefficients corresponding to the amplitude and phase of the reflection coefficients shown in Figs. 12.2 and 12.3. The amplitude transmission coefficient for incident p-polarization shows a large peak corresponding to θ_p the peak value at this point arises from an intensification of the field between n_2 and n_3; as depicted in Fig. 12.5 (a). Since the waves are evanescent, however, there is no energy coupled into the bottom layer. The phase of the transmission coefficient for p-incident polarization also shows a continuous phase shift similar (albeit through approximately π radians) to that observed for the reflection coefficient indicating lateral transfer of (stored) energy excited. We will show that this phase shift leads to self-focusing of the SPs on the sample surface. In the calculations that follow the transmission coefficients of Fig. 12.20 and 21 are used; the results also concentrate on the two photon response where $\lambda_1 = 2\lambda_2 = 2$ wavelength units. Consider the case when the light incident in the back focal plane is linearly

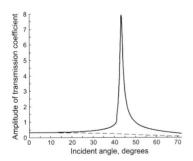
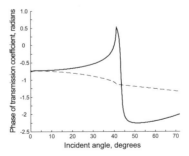

Fig. 12.20. (**a**) Modulus of transmission coefficient for parameters corresponding to reflectance functions of Figs. 12.2 and 12.3. *Solid line* – incident *p*-polarization, *dashed line* – incident *s*-polarization. Note that the peak in the transmission coefficient corresponds to the field enhancement shown in Fig. 12.5 (a). (**b**) Phase of transmission coefficient corresponding to modulus shown in Fig. 12.20 (a)

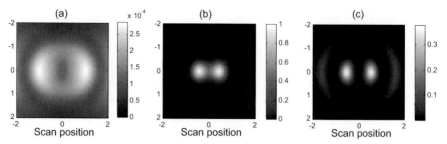

Fig. 12.21. Non-confocal point spread function for incident linear polarization. (**a**) Defocus +2 pump wavelengths (4 wavelength units), (**b**) in focus, (**c**) defocus −2 pump wavelengths (4 wavelength units)

polarized. Figure 12.21 shows the two photon non-confocal responses for a point object for different defocuses of 2 pump wavelengths, that is 4 wavelength units. We note that in both cases there is a minimum at the centre of the scan. This arises because the dominant z-components of the field cancel here; the x- and y-components reinforce but these are relatively small. This leads to the two spot distribution similar to that predicted by Kano et al. [18] for annular excitation. It is important also to note the effect of defocus, when the sample is below the focal plane (positive defocus) the surface waves propagate away from the centre and there is a redistribution of energy away from the lens axis. On the other hand, we can see that when the sample is moved above the focal plane the distribution remains tightly confined and self-focusing of the surface waves is evident. The negative defocus image is rather sharper than the in focus image. It is clearly not desirable to have a distribution with a minimum in the centre, this arises from the cancellation of the z-components on the optical axis. For fluorescence imaging it is therefore appropriate to consider modifying the input polarization so that the z- components reinforce at the centre of the distribution. If the input polarization at the back focal plane is radially polarized, so that the polarization is always parallel to the radius, then only p-incident light will emerge from the objective, there will be a reinforcement of the field at the optical axis. This reinforcement can also be achieved with a distribution that it is easier to achieve in practice, where the incident polarization is linear, but the phase is inverted by 180 degrees in one semicircle of the back focal plane. Figure 12.22 shows the two photon response for 'flipped' incident polarization for non-confocal detection. We see that for positive defocus the energy propagates away from the optical axis making it unsuitable for imaging. The in focus response shows a sharply peaked response, the defocused response although having a rather lower intensity is even more confined in space, albeit by a relatively small amount. In both cases the sidelobes are rather high along the direction of SP propagation, a situation similar to that encountered in 4π microscopy [33]. The fact that the intensity is maximum at the optical axis means that the resolution may be further enhanced (at the expense of some loss of signal) by detecting the light emitted from the sample using a confocal pinhole. This can be seen (see Fig. 12.23) to sharpen up the mainlobe considerably as well as reducing any sidelobes. For the non-confocal two photon re-

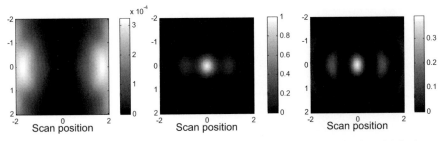

Fig. 12.22. Two photon non-confocal response for 'flipped' input polarization. (**a**) Defocus +2 pump wavelengths (4 wavelength units), (**b**) in focus, (**c**) defocus −2 pump wavelengths (4 wavelength units)

sponse the FWHM (full width half maximum) of the response is approximately 0.9 wavelength units along the direction of SP propagation (with large sidelobes) and approximately 1.24 wavelengths normal to this direction. For confocal detection, on the other hand the FWHM is 0.66 wavelengths and 0.76 wavelengths respectively, with the sidelobes even in the direction of SP propagation severely suppressed. For excitation at 800 nm and detection at 400 nm, lateral resolution approaching 100 nm may be expected. The lateral resolution is not sensitive to the numerical aperture of the objective lens, provided it covers sufficient angle to excite SPs. The two photon microscope thus offers the possibility of excellent sensitivity combined with lateral resolution equivalent to the very best microscopes. Up to now we have considered

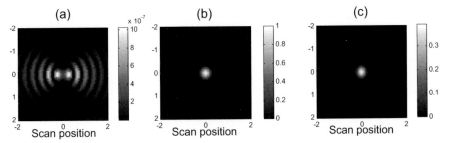

Fig. 12.23. Two photon confocal response for 'flipped' input polarization. (**a**) Defocus +2 pump wavelengths (4 wavelength units), (**b**) in focus, (**c**) defocus −2 pump wavelengths (4 wavelength units)

illumination with a full aperture, it is interesting to consider what happens when the sample is illuminated with an annular aperture with 'flipped' linear polarization. The two photon non-confocal in focus response is shown in Fig. 12.24. The response in

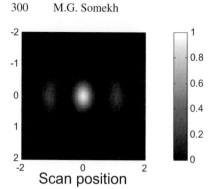

Fig. 12.24. Non-confocal two photon response for annular aperture

form, if not amplitude, is extremely similar to Fig. 12.22, furthermore, for the case of annular excitation, the form does not vary significantly with defocus over the range of practical defocuses. This shows that selecting the SPs either by defocus or by restricting the range of input spatial frequencies gives similar results. Although the shape of the point spread function does not change with defocus for excitation through an annular aperture, there is a very strong dependence of the amplitude of the signal with defocus. We see in Fig. 12.25 that the two photon response reaches a maximum at a defocus of approximately -5 pump wavelengths (i.e. 10 wavelength units) with the yield being approximately 5.5 times larger than the in focus value. This arises from the fact that the phase variation of the defocused beam matches that of the k-vector of SPs better as the sample is defocused. As the sample is defocused further two factors limit the magnitude of the response, firstly the SP generation becomes less efficient and secondly, the attenuation of the SPs as they propagate to the optical axis reduces the field strength. Another technique closely related to SP

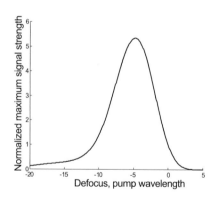

Fig. 12.25. Variation of the peak two photon response with defocus for 'flipped' excitation through an annular aperture

fluorescence is total internal reflectance fluorescence microscopy [43,44]. This technique can in simple terms be thought of a similar to the techniques described above with the metal layer removed. Figs. 12.26a,b show the transmission coefficient for p-polarized light for the same case as shown in Figs. 12.20a,b with the metal layer, which supports the SPs, removed. The transmitted field amplitudes show quite such a large peak just above the critical angle, we also note from Fig. 12.26 that even though there is a phase variation, this is considerably weaker than that observed when the metal layer is present. The mean gradient is small indicating that significant energy is only transferred for short distances, moreover, the gradient does not remain constant so that the displaced field will be distorted. Modelling, the situation with the annular lens we see that when the sample is moved above the focus there is only a tiny increase in signal, of approximately 3%; this indicates that the self-focusing and improved generation efficiency are extremely weak compared to the case when SPs are excited. It seems, therefore, that compared to total internal

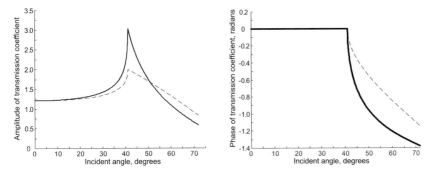

Fig. 12.26. (a) Modulus of transmission coefficient corresponding to situation in Fig. 12.20 with gold layer removed. *Solid line p-*incident polarization, *fine dashed line s-*incident polarization. **(b)** Phase of transmission coefficient corresponding to modulus curve in Fig. 12.26 (a). *Solid line p-*incident polarization, *fine dashed line s-*incident polarization

reflection methods there is greater field enhancement and also far greater ability to further increase by defocusing the signal with SP microscopy. On the other hand, the penetration depth of SPs is fixed by their k-vector, the penetration depth in TIR microscopy may be tuned, to some extent, by controlling the angle of incidence.

12.5.3 Discussion

The numerical simulations presented in Sect. 12.5.2 show that defocusing the sample results in a larger excitation of the SPs, because by defocusing the phase variation of the incident beam matches the k-vector of the SPs resulting in more efficient excitation. The self-focusing of the SPs means that there is a diffraction limited focus concentrated on the lens axis. In the case the $V(z)$ method discussed in Sect. 12.5.1 the enhanced efficiency of generation is further reinforced by a similar improvement in the efficiency of the detection process. Interferometric detection ensures that the detection process is sensitive to the phase gradient of the reflected light.

12.6   SP Microscopy in Aqueous Media

In the experimental and theoretical results presented up to this point the final dielectric, n_3 in Fig. 12.1 (a), and n_4 in Fig. 12.1 (b) have consisted of air or a material with refractive index 1. When imaging biological samples the tissue or the culture medium will have a refractive index close to that of water $n \approx 1.33$. This has significant effect on performance of SP microscopy. This is illustrated very clearly in the graph of reflection coefficient for p-polarized light versus angle shown in Fig. 12.27, for the parameters shown in Figs. 12.2 and 12.3, except that the index of the final medium n_3 is equal to 1.33. We can see that the basic shape of both the amplitude and the phase of the reflection coefficient is unchanged. The angle necessary to excite SPs, however, is now close to 70 degrees; this means that the numerical aperture necessary to excite SPs is greater than that of commonly available objectives.[3] We discuss two solutions to this problem (i) use an objective coupled with a high refractive index immersion fluid and (ii) alter the structure of the coverslip so that the angle for surface wave excitation is reduced.

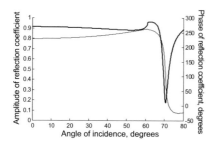

Fig. 12.27. Reflectance function for p-polarization corresponding, with aqueous backing layer $n_3 = 1.33$. Coupling oil, $n_1 = 1.52$

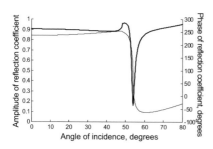

Fig. 12.28. Reflectance function for the same case as Fig. 12.27 except $n_1 = 1.78$

---

[3] 1.4 NA is commonly available, some manufacturers (e.g. Zeiss) are making a 1.45 NA objective.

(i) High refractive index immersion fluid. In principle this is the most obvious solution. Figure 12.28 shows the p-reflection coefficient for the same structure as
the shown in Fig. 12.27 except that that top medium is now replaced by a refractive index of 1.78 compared to 1.52 in the previous case. We can now see
that the value of θ_p is now approximately 54 degrees, and when the incident
angle is close to 58 degrees the features relating to SP excitation are not longer
apparent. This means that the required numerical aperture to excite SPs in this
situation is, in practice, around 1.5. The field distribution enhancement at the
interface between n_2 and n_3 in this case is approximately 6, so that only slightly
less sensitivity compared to the case of air backing may be expected. Recently,
objectives using as diiodomethane as couplant with quoted NAs of 1.65 (Olympus Apo 100×/1.65 Oil HR) have become available. These objectives should, in
principle, excite SPs with aqueous backing. The present author and coworkers
have used such an objective. Our results indicate clearly it is possible to excite
SPs with aqueous backing using this objective. This was demonstrated by comparing the $V(z)$ curves obtained with a 44 nm gold layer and a 150 nm gold layer
where the SP excitation is suppressed. On the other hand, we were not able to
produce good quality $V(z)$ curves such as those presented in Fig. 12.16. This
may be due to the fact that the $V(z)$ is very sensitive to the pupil function of the
lens and the presence of significant aberrations or inappropriate apodization of
the pupil, while not preventing SP excitation will render the quantitative interpretation of the $V(z)$ curve invalid; this has been discussed for the acoustic case
by Bertoni and Somekh [45]. Indeed, as we mentioned in Sect. 12.5.1 one of
the main uses of the $V(z)$ response is to measure aberrations since the axial response is particularly sensitive to this. A more mundane practical reason to seek
other methods to examine samples with aqueous backing is that the immersion
fluid is rather unpleasant to work with, possibly toxic [44] and has a particularly unpleasant smell. Furthermore, to increase the refractive index sulphur is
dissolved in the diiodomethane, with the consequence that there is a tendency
to deposit sulphur liberally around the experimental rig. Apart from the obvious
inconvenience this causes it also means that obtaining consistent quantitative
results over an extended period is difficult.

(ii) Modified coverslip. This involves generating surface waves with modified properties [46,47]. We now examine ways in which a conventional immersion fluid
may be used with aqueous backing. The critical angle between a medium of
index 1.52 and 1.33 is approximately 61 degrees. Even if a surface wave is excited at at this angle then a numerical aperture greater than the index of the
aqueous medium is required. The question is whether it is possible to excite a
surface wave with useful properties at an angle as close as possible to this 61
degree incident angle. This may be achieved by using a structure with two layers sandwiched between upper and lower dielectrics as shown in Fig. 12.1 (b).
Following the notation of Fig. 12.1 (b) n_1 is 1.52, n_2 is the 43.5 nm layer of gold
used throughout this article, n_3 is an 440 nm additional layer of quartz (refractive index= 1.457, which is intermediate between n_1 and n_4), the final layer n_4
is 1.33. Figure 12.29 shows the amplitude and phase of the reflection coefficient

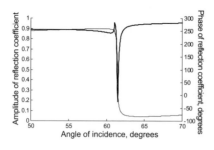

Fig. 12.29. Reflectance function for incident *p*-polarization for 4 layer system with aqueous backing, with additional layer of 440 nm of SiO$_2$. Otherwise parameters are as for Fig. 12.27. (See text for full details). Note incident angles are shown only between 50 degrees and 70 degrees

for *p*-incident radiation. This is shown only between incident angles 50 to 70 degrees, because there is little structure of present interest outside this range. We can see immediately that there is a phase change through approximately 2π radians which corresponds to excitation of a surface wave and dip in the amplitude of the reflection coefficient similar to that observed for SP excitation. Interestingly, surface wave excitation occurs at an angle only slightly greater than 61 degrees so that this wave should be excited at least in principle using a conventional high quality oil immersion objective. Figure 12.30 shows the modulus of the electric field distribution corresponding to excitation at 61.45 degrees, where we can see that the field in the metal layer grows in the metal layer prior to setting up a standing wave the intermediate layer. Importantly, for the purposes of sensitive microscopy there is a large field enhancement at the interface between n_3 and the aqueous backing n_4.

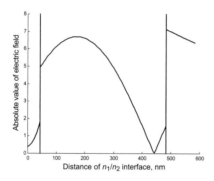

Fig. 12.30. Electric field distribution through the layered structure of Fig. 12.29, for incident angle of 61.45 degrees

12.7   Discussion and Conclusions

This chapter has discussed some of the technical issues concerning the development of SP microscopy in the far field. One of the key issues has been how diffraction limited rather than attenuation limited resolution may be achieved. We have discussed this in the context of the so-called $V(z)$ technique, which relies on detection of the phase difference between directly reflected light and light involved in conversion to SPs. We show that this technique is very sensitive to the properties of the sample as well as capable of giving excellent lateral resolution surpassing other techniques used to obtain SP images in the far field. We attribute the good lateral resolution of the technique to the fact that SPs come to a diffraction limited focus on the optical axis. We also describe the operation of SP fluorescence microscopy where we show how the excited SPs do in fact come to a diffraction limited focus and, furthermore, we show how defocusing increases the efficiency with which these waves are generated. We expect the techniques described to have considerable application in both material science and, more particularly, life science where the interfacial conditions play a significant role. In the life sciences the techniques we describe will expected to have major impact on the studies of cell membranes where strong field enhancement will give excellent sensitivity to local variations in mechanical properties as well as changes in membrane potential. Prism based Kretschmann techniques are already being used for cell signaling studies as well as studies of cell attachment; their applications are, of course, severely restricted by the limited lateral resolution, which the techniques described here will overcome.

It must be borne in mind that this article should be read as a prospectus for the potential of diffraction limited SP microscopy techniques and we will devote the rest of the article to a brief discussion of how the ideas discussed here may be further extended.

Second harmonic imaging in transmission has recently been described in live cell using a short pulsed laser [48]. This has been used to image neuroblastoma cells in suspension, the technique was compared with two photon imaging where small differences were observed in isolated regions of the cell. Extension of this technique to second harmonic SP generation using the configuration shown in Fig. 12.19 is a very attractive option. Firstly second harmonic imaging of cells in suspension requires that imaging is performed in transmission, the SP second harmonic imaging will allow a single ended reflection geometry. The break in symmetry at the interface of the cell membrane, especially when it is bound to the sample means that a strong second harmonic signal is expected. The technique will also be expected to be a more sensitive measure of membrane potential than two photon techniques, giving much better dynamic range.

Up to this point all the high resolution SP imaging techniques we have discussed have been scanning methods. Widefield imaging has been demonstrated with the Kretschmann prism based techniques but with poor lateral resolution. It is possible to combine the widefield imaging with high lateral resolution. We presented results in Sect. 12.5.1, which showed how a scanning heterodyne interferometer can be used to give excellent spatial resolution for SP imaging. If the transfer function of the scanning heterodyne interferometer can be produced in a widefield configuration it should be possible, at least in principle, to produce a high resolution widefield SP microscope. We have recently developed a wide field confocal interferometer system using speckle illumination [49]. The sample is illuminated through a diffuser, this gives a random distribution of spatial frequencies on the sample covering those of the lens objective. The speckle pattern illuminating the sample is interfered with a similar speckle pattern reflected from a reference surface. The diffuser is then rotated to average the speckle noise; under these conditions may be readily shown that the interference term is precisely the same as that of the heterodyne interferometer. To extract the interference term we have employed different phase stepping strategies including parallel phase stepping in a single image [50]. The result is that it is quite possible to obtain high lateral resolution SP microscopy with the resolution reported in Sect. 12.5.1. The difficulty lies in the fact that, while it is easy to extract a tiny variation on a large background with the heterodyne interferometer since the frequency of the interference term differs from that of the background, this is less simple is a phase stepping system because of the limited dynamic range of a CCD camera. The challenge is thus to develop an optical system where the variations produced by the SP signal are a larger proportion of the overall signal or to develop pixellated detectors capable of extracting a time varying signal on a relatively large background.

This chapter has discussed some the recent work in SP and surface wave microscopy, and has shown that high lateral resolution combined with extreme sensitivity to surface properties should be expected to become an important part of the microscopist's armoury especially in the biological sciences.

Acknowledgements

The author wishes to thank the Paul Instrument fund for financial support and is very happy to acknowledge the major contribution of his colleagues Drs. T.S. Velinov, Shugang Liu and Chung Wah See.

References

1. H. Raether: *Surface plasmons on smooth and rough surfaces and on gratings* (Springer-Verlag, Berlin 1998)
2. M.T. Flanagan, R.H. Pantell: Electron. Lett. **20**, 969 (1984)
3. R.J. Green, S. Tasker, J. Davies, M.C. Davies, C.J. Roberts, S.J.B. Tendler: Langmuir **13**, 6510 (1997)
4. VN. Konopsky : Opt. Commun. **185**, 83 (2000)
5. S.I. Bozhevolnyi: Phys. Rev. B **54**, 8177 (1996)
6. B. Hecht, H. Bielefeldt, L. Novotny, Y. Inouye, D.W. Pohl: Phys. Rev. Lett. **77**, 1889 (1996)
7. R.M.A. Azzam, N.M. Bashara: *Ellipsometry and polarized light* (Elsevier, Amsterdam 1986)
8. M. Born, E. Wolf: *Principles of Optics* (Pergamon, Oxford 1999)
9. L.M. Brekhovskikh: *Waves in layered media* (Academic Press, New York, London 1960)
10. T. Velinov, M.G. Somekh, S. Liu : Appl. Phys. Lett. **75**, 3908 (1999)
11. J.B. Pendry: Phys. Rev. Lett. **85**, 966 (2000)
12. E.M. Yeatman, E.A. Ash: Electron. Lett. **23**, 1091 (1987)

13. B. Rothenhausler, W. Knoll: Nature **332**, 615 (1988)
14. E.M. Yeatman: Biosens. Bioelectron. **11**, 635 (1996)
15. C.E.H. Berger, R.P.H. Kooyman, J. Greve: Rev. Sci. Instrum. **65**, 2829 (1994)
16. K.F. Giebel, C. Bechinger, S. Herminghaus, M. Riedel, P. Leiderer, U. Weiland, M. Bastmeyer: Biophys. J. **76**, 509 (1999)
17. A.N. Grigorenko, A.A. Beloglazov, P.I. Nikitin, C. Kuhne, G. Steiner, R. Salzer: Opt. Commun. **174**, 151 (2000)
18. H. Kano, S. Mizuguchi, S. Kawata: J. Opt. Soc. Am. B **15**, 1381 (1998)
19. J.T. Fanton, J. Opsal, D.L. Willenborg, S.M. Kelso, A. Rosencwaig: J. Appl. Phys. **73**, 7035 (1993)
20. C.W. See, M.G. Somekh, R. Holmes: Appl. Optics **35**, 6663 (1996)
21. S.V. Shatalin, R. Juškaitis, J.B. Tan, T. Wilson: J. Microsc. – Oxford **179**, 241 (1995)
22. H. Kano, W. Knoll: Opt. Commun. **153**, 235 (1998)
23. H. Kano, W. Knoll: Opt. Commun. **182**, 11 (2000)
24. A.G. Notcovich, V. Zhuk, S.G. Lipson: Appl. Phys. Lett. **76**, 1665-1667 (2000)
25. A.L. Migdall, B. Roop, Y.C. Zheng, J.E. Hardis, G.J. Xia: Appl. Optics **29**, 5136 (1990)
26. H. Zhou, C.J.R. Sheppard: J. Mod. Optics **44**, 1553 (1997)
27. A. Atalar: J. Appl. Phys. **49**, 5130 (1978)
28. W. Parmon, H.L. Bertoni: Electron. Lett. **15**, 684 (1979)
29. C. Ilett, M.G. Somekh, G.A.D. Briggs: P. Roy. Soc. Lond. A **393**, 171 (1984)
30. M.G. Somekh, H.L. Bertoni, G.A.D. Briggs, N.J. Burton: P. Roy. Soc. Lond. A **401**, 29 (1985)
31. A. Atalar: J. Appl. Phys. **50**, 8237 (1979)
32. M.J. Offside, M.G. Somekh, C.W. See: Appl. Phys. Lett. **55**, 2051 (1989)
33. S.W. Hell, S. Lindek, E.H.K. Stelzer: J. Mod. Optics **41**, 675 (1994)
34. M.G. Somekh , S. Liu, T.S. Velinov: Opt. Lett. **25**, 823 (2000)
35. M.G. Somekh, S.G. Liu, T.S. Velinov, C.W. See: Appl. Optics **39**, 6279 (2000)
36. L.F. Li, J. Chandezon, G. Granet, J-P. Plumey: Appl. Optics **38**, 304 (1999)
37. H.H. Hopkins: P. Roy. Soc. Lond. A **55**, 116 (1943)
38. B. Richards, E. Wolf: P. Roy. Soc. Lond. A **253**, 358 (1959)
39. P. Török, P. Varga, Z. Laczik, G.R. Booker: J. Opt. Soc. Am. A **12**, 325 (1995)
40. H.T.M. van der Voort, G.J. Brakenhoff: J. Microsc. – Oxford **158**, 43 (1989)
41. P.D. Higdon, P. Török, T. Wilson: J. Microsc. – Oxford **193**, 127 (1999)
42. M.G. Somekh: J. Microsc. – Oxford **206**, 120 (2002)
43. D. Axelrod, T.P. Burghardt, N.L. Thompson: Annu. Rev. Biophys. Bio. **13**, 247 (1984)
44. D. Toomre, D.J. Manstein: Trends Cell Biol. **11**, 298 (2001)
45. H.L. Bertoni, M.G. Somekh: IEEE T. Ultrason. Ferr. **33**, 91 (1986)
46. G. Kovacs: 'Optical excitation of surface plasmon-polaritons in layered media', In: *Electromagnetic surface modes*, ed. by A.D. Boardman (John Wiley & Sons, New York 1982) pp. 20–56
47. W. Knoll: Annu. Rev. Phys. Chem. **49**, 569 (1998)
48. P.J. Campagnola, M.D. Wei, A. Lewis, L.M. Loew: Biophys. J. **77**, 3341 (1999)
49. M.G. Somekh, C.W. See, J. Goh: Opt. Commun. **174**, 75 (2000)
50. N.B.E. Sawyer, S.P. Morgan, M.G. Somekh, C.W. See, X.F. Cao, B.Y. Shekunov, E. Astrakharchik: Rev. Sci. Instrum. **72**, 3793 (2001)

13   Optical Coherence Tomography

Stephen A. Boppart

13.1   Introduction

Optical coherence tomography (OCT) is an emerging imaging technique for a wide range of biological, medical, and material investigations [1,2]. Advances in the use of OCT for microscopy applications have provided researchers with a novel means by which biological specimens and non-biological samples are visualized. OCT was initially developed for imaging biological tissue because it permits the imaging of tissue microstructure *in situ*, yielding micron-scale image resolution without the need for excision of a specimen for tissue processing. OCT is analogous to ultrasound B-mode imaging except that it uses low-coherence light rather than high-frequency sound and performs imaging by measuring the backscattered intensity of light from structures in tissue. OCT can image tissue or specimens in cross-section, as is commonly done in ultrasound, or in *en face* sections, as in confocal and multi-photon microscopy. The OCT image is a gray-scale or false-color multi-dimensional representation of backscattered light intensity that represents the differential backscattering contrast between different tissue types on a micron scale. Because OCT performs imaging using light, it has a one to two order-of-magnitude higher spatial resolution than ultrasound and does not require contact with the specimen or sample. The use of light also enables the spectroscopic characterization of tissue and cellular structures.

OCT was originally developed and demonstrated in ophthalmology for high-resolution tomographic imaging of the retina and anterior eye [3–5]. Because the eye is transparent and is easily accessible using optical instruments and techniques, it is well suited for diagnostic OCT imaging. OCT is promising for the diagnosis of retinal disease because it can provide images of retinal pathology with micron-scale resolution. Clinical studies have been performed to assess the application of OCT for a number of macular diseases [4,5]. OCT is especially promising for the diagnosis and monitoring of glaucoma and macular edema associated with diabetic retinopathy because it permits the quantitative measurement of changes in the retina or the retinal nerve fiber layer thickness. Because morphological changes often occur before the onset of physical symptoms, OCT can provide a powerful approach for the early detection of these diseases.

Following the application of OCT in ophthalmology, OCT has been applied for imaging in a wide range of nontransparent tissues [6–10]. In tissues other than the eye, the imaging depth is limited by optical attenuation due to scattering and absorption. Ophthalmic imaging is typically performed at 800 nm wavelengths. Because

Török/Kao (Eds.): Optical Imaging and Microscopy, Springer Series in Optical Sciences
Vol. 87 – © Springer-Verlag, Berlin Heidelberg 2003

optical scattering decreases with increasing wavelength, OCT imaging in nontransparent tissues is possible using 1.3 μm or longer wavelengths. In most tissues, imaging depths of 2–3 mm can be achieved using a system detection sensitivity of 100 to 110 dB. Imaging studies have also been performed in virtually every organ system to investigate applications in cardiology [11–14], gastroenterology [15,16], urology [17,18], neurosurgery [19], and dentistry [20], to name a few. High-resolution OCT using short coherence length, short pulsed light sources has been demonstrated and axial resolutions of less than 2 μm have been achieved [21–24]. High-speed OCT at image acquisition rates of 4 to 8 frames per second have been achieved, depending on the image size [25–27]. OCT has been extended to perform Doppler imaging of blood flow [28–30] and birefringence imaging to investigate laser intervention [31–33]. Different imaging delivery systems including transverse imaging catheter/endoscopes and forward imaging devices have been developed to enable internal body OCT imaging [34–36]. Most recently, OCT has been combined with catheter/endoscope-based delivery to perform *in vivo* imaging in animal models and human patients [37–41].

13.2   Principles of Operation

OCT is based on optical ranging, the high-resolution, high dynamic-range detection of backscattered light as a function of optical delay. In contrast to ultrasound, the velocity of light is extremely high. Therefore, the time delay of reflected light cannot be measured directly and interferometric detection techniques must be used. One method for measuring time delay is to use low-coherence interferometry or optical coherence domain reflectometry. Low-coherence interferometry was first developed in the telecommunications field for measuring optical reflections from faults or splices in optical fibers [42]. Subsequently, the first applications in biological samples included one-dimensional optical ranging in the eye to determine the location of different ocular structures [43,44].

The time delay of reflected light is typically measured using a Michelson interferometer (Fig. 13.1). Other interferometer designs, such as a Mach–Zehnder interferometer, have been implemented to optimize the delivery of collection of the OCT beam [45,46]. The light reflected from the specimen or sample is interfered with light that is reflected from a reference path of known path length. Interference of the light reflected from the sample arm and reference arm of the interferometer can occur only when the optical path lengths of the two arms match to within the coherence length of the optical source. As the reference arm optical path length is scanned, different delays of backscattered light from within the sample are measured. The interference signal is detected at the output port of the interferometer, electronically band-pass filtered, demodulated, digitized, and stored on a computer. The position of the incident beam on the specimen is scanned in the transverse direction and multiple axial measurements are performed. This generates a two-dimensional (2-D) data array that represents the optical backscattering through a cross-sectional plane in the specimen (Fig. 13.2). The logarithm of the backscatter intensity is then mapped to a false-color or gray-scale and displayed as an OCT image. The interferometer in an OCT instrument can be implemented using a fiber optic coupler and beam-scanning can be performed with small mechanical galvanometers in order to yield a compact and robust system (Fig. 13.2).

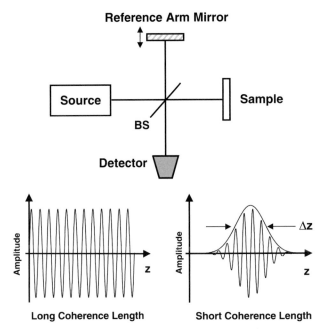

Fig. 13.1. Low-coherence interferometry. Using a short coherence length light source and a Michelson-type interferometer, interference fringes are observed only when the path lengths of the two interferometer arms are matched to within the coherence length of the light source

In contrast to conventional microscopy, the axial resolution in OCT images is determined by the coherence length of the light source. The axial point spread function of the OCT measurement, as defined by the signal detected at the output of the interferometer, is the electric-field autocorrelation of the source. The coherence length of the light is the spatial width of the field autocorrelation and the envelope of the field autocorrelation is equivalent to the Fourier transform of its power spectrum. Thus, the width of the autocorrelation function, or the axial resolution, is inversely proportional to the width of the power spectrum. For a source with a Gaussian spectral distribution, the axial resolution Δz is given:

$$\Delta z = \frac{2 \ln 2}{\pi} \cdot \frac{\lambda^2}{\Delta \lambda} , \qquad (13.1)$$

where Δz and $\Delta \lambda$ are the full-widths-at-half-maximum of the autocorrelation function and power spectrum respectively and λ is the center wavelength of the op-

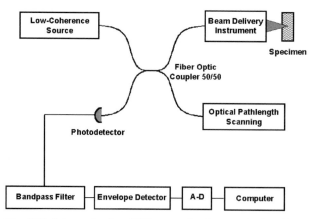

Fig. 13.2. Schematic of an OCT system implemented using fiber optics. The Michelson interferometer is implemented using a fiber-optic coupler. Light from the low-coherence source is split and sent to a sample arm with a beam delivery instrument and a reference arm with an optical pathlength scanner. Reflections from the arms are combined and the output of the interferometer is detected with a photodiode. The signals are demodulated, processed by a computer, and displayed as an image

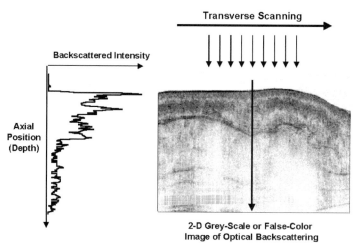

Fig. 13.3. OCT image formation. An OCT image is based on the spatial localization of variations in optical backscatter from within a specimen. Images are acquired by performing axial measurements of optical backscatter at different transverse positions on the specimen and displaying the resulting two-dimensional data set as a gray-scale or false-color image

tical source. Figure 13.4 illustrates the dependence of the axial resolution on the bandwidth of the optical source. To achieve high axial resolution requires broad bandwidth optical sources. Curves are plotted in Fig. 13.4 for three commonly used wavelengths. Higher resolutions are achieved with shorter wavelengths, however,

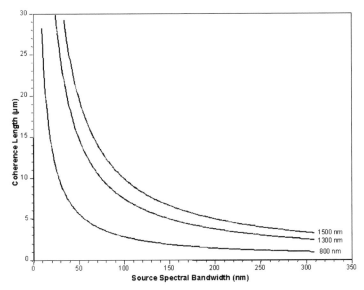

Fig. 13.4. Dependence of coherence length (axial resolution) on optical source bandwidth. Curves are plotted for 800 nm, 1300 nm, and 1500 nm, three common wavelengths used in OCT. High axial imaging resolution is achieved with broad spectral bandwidths and shorter wavelengths. Shorter wavelengths, however, are more highly absorbed in biological tissue, decreasing imaging penetration depth

shorter wavelengths are more highly scattered, resulting in less imaging penetration.

The transverse resolution in an OCT imaging system is determined by the focused spot size in analogy with conventional microscopy and is given by

$$\Delta x = \frac{4\lambda}{\pi} \cdot \frac{f}{d} , \qquad (13.2)$$

where d is the beam diameter incident on the objective lens and f is the focal length of the objective. High transverse resolution can be obtained by using a large numerical aperture and focusing the beam to a small spot size. The transverse resolution is also related to the depth of focus or the confocal parameter $2z_R$ (two times the Raleigh range),

$$2z_R = \frac{\pi \Delta x^2}{2\lambda} . \qquad (13.3)$$

Thus, increasing the transverse resolution results in a reduced depth of field. Typically, the confocal parameter or depth of focus is chosen to match the desired depth of imaging. High transverse resolutions may be utilized in OCT, however, the short depth of field requires spatially tracking the focus in depth along with the axial OCT scanning.

Finally, the detection signal-to-noise ratio (SNR) is given by the optical power backscattered from the sample (P_{SAM}) divided by the noise equivalent bandwidth (NEB),

$$SNR = 10 \ \log \left(\frac{\eta}{\hbar \omega} \frac{P_{SAM}}{NEB} \right) , \tag{13.4}$$

where η is the quantum efficiency of the detector, ω is radian frequency of the source, and $\hbar$ is Plank's constant divided by 2π.

Depending upon the desired signal to noise performance, incident powers of 5–10 mW are required for OCT imaging. Typically, the acquisition of 250–500 square pixel images at several frames per second can be achieved at a signal-to-noise ratio of 100–110 dB with 5–10 mW of incident optical power. If lower data acquisition speeds or signal-to-noise can be tolerated, power requirements can be reduced accordingly.

13.3   Technological Developments

Since the inception of OCT in the early 1990's, there have been rapid technological developments aimed at improving the imaging resolution, acquisition rate, and methods of beam delivery to the tissue or sample. Investigators have also explored other imaging methods using the principles of OCT to extract information from the tissue or sample. Some of these methods have included acquiring optical Doppler signals from moving scatterers or structures, obtaining images based on the polarization state of the incidence and returned light, and extracting spectroscopic information based on the local absorption or scattering properties of the tissue.

13.3.1   Optical Sources for High-Resolution Imaging

The majority of OCT imaging systems to date have used superluminescent diodes (SLDs) as low-coherence light sources [47,48]. Superluminescent diodes are commercially available at several wavelengths including 800 nm, 1.3 μm, and 1.5 μm and are attractive because they are compact, have high efficiency, and low noise. However, output powers are typically less than a few milliwatts, which limit fast real-time image acquisition rates. Additionally, the available bandwidths are relatively narrow, permitting imaging with 10–15 micron resolution. Recent advances in short-pulse solid-state laser technology make these sources attractive for OCT imaging in research applications. Femtosecond solid-state lasers can generate tunable, low-coherence light at powers sufficient to permit high-speed OCT imaging. Short pulse generation has been achieved across the full wavelength range in titanium:sapphire ($Ti:Al_2O_3$) from 0.7 μm to 1.1 μm and over more limited tuning ranges near 1.3 µm and 1.5 µm in chromium:forsterite (Cr^{4+}:Mg_2SiO_4) and chromium:ytterbium-aluminum-garnet (Cr^{4+}:YAG) lasers, respectively. OCT imaging with resolutions of 1.5 µm and 5 µm has been demonstrated at 800 nm and 1.3 µm respectively using Ti:Al_2O_3 and Cr^{4+}:Mg_2SiO_4 sources [23,24]. A comparison between the power spectra and autocorrelation functions of a 800 nm center wavelength SLD and a short pulse (≈ 5.5 fs) titanium:sapphire laser is shown in Fig. 13.5. An order-of-magnitude improvement in axial resolution is noted for the short pulse titanium:sapphire laser source [24]. More compact and convenient sources such as rare-earth-element-doped fiber sources, are currently under investigation [49–52]. The titanium:sapphire (Ti:Al_2O_3) laser technology is routinely used in multi-photon microscopy (MPM) applications for its high peak intensities to enable multi-photon absorption and subsequent emission of fluorescence from exogenous fluorescent contrast agents [53,54]. Combined OCT and multi-photon microscopy has been used to provide complementary image data using a single optical source [55].

13.3.2   Spectroscopic OCT

OCT techniques can be used to extract spatially-distributed spectroscopic information from within the tissue specimen or material sample [56]. In standard OCT imaging, the amplitude of the envelope of the field autocorrelation is acquired and used to construct an image based on the magnitude of the optical backscatter at each position. Spectroscopic OCT information can be obtained by digitizing the full interference signal and performing digital signal processing using a Morlet wavelet transform. The Morlet wavelet transform is used rather than the short-time Fourier transform because it reduces windowing artifacts [57,58]. Spectroscopic data can be extracted from each point within the specimen. The spectral center of mass can be calculated and compared to the original spectrum from the laser source. Shifts from the center of mass are displayed on a 2-D image using the multi-dimensional hue-saturation-luminence (HSL) color space. At localized regions within the tissue, a color is assigned with a hue that varies according to the direction of the spectral shift (longer or shorter wavelength) and a saturation that corresponds to the magnitude of that shift. The luminance is held constant.

In scattering media, longer wavelengths of light are scattered less. Therefore, for a homogeneous scattering sample, one would expect shorter wavelengths to be scattered near the surface and a smooth color-shift to occur with increasing depths as longer wavelengths are scattered. In more heterogeneous samples, such as tissue, scattering objects such as cells and sub-cellular organelles produce variations in the spectroscopic OCT data (Fig. 13.6) [56]. Although images can indicate changes in the spectroscopic properties of the tissue, further investigation is needed to determine how the spectroscopic variations relate to the biological structures and how this information can be used for diagnostic purposes.

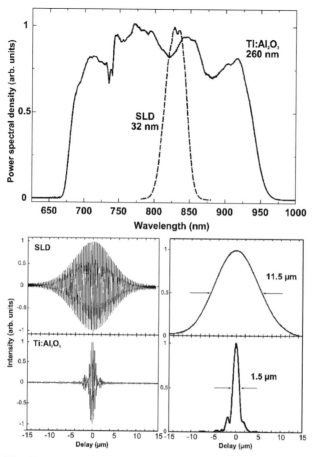

Fig. 13.5. High-resolution OCT imaging. A comparison of optical output spectrum (*top*) with interference signals (*bottom left column*) and envelopes (*bottom right column*) for a Kerr-lens mode-locked titanium:sapphire (Ti:Al$_2$O$_3$) laser versus a superluminescent diode (SLD) is shown. The broad optical bandwidth of the titanium:sapphire laser (260 nm) permits a free-space axial resolution of 1.5 μm. In comparison, the superluminescent diode with a 32 nm spectral bandwidth permits an axial resolution of 11.5 μm. Figure reprinted with permission [24]

13.3.3 Real-Time OCT Imaging

The short-pulse solid-state laser technology not only provides broad spectral bandwidths for high-resolution OCT imaging, but also higher output powers to enable fast real-time OCT imaging. Higher incident powers are required to maintain equivalent signal-to-noise ratios when scanning at a faster rate. Linearly translating a reference arm mirror provides axial scan frequencies of approximately 100 Hz, depending on the mirror size and the translating galvanometer, but is problematic at higher rates. Several investigators have utilized rotating glass cubes, piezoelectric

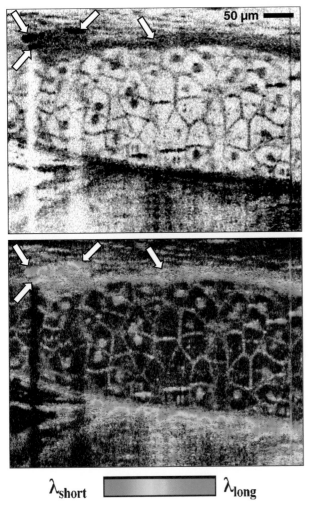

λ_{short} λ_{long}

Fig. 13.6. Spectroscopic OCT imaging. Conventional OCT imaging (*top*) and spectroscopic OCT imaging (*bottom*) of *in vivo* mesenchymal cells in a *Xenopus laevis* (African frog) tadpole. Conventional OCT images represent the optical backscatter intensity while spectroscopic OCT images represent local changes in the absorption or scattering properties of the incident optical spectrum. Melanocytes (*arrows*) appear red because the melanin within these cells absorb shorter wavelengths of light. The color scale represents the shift of the center of gravity of the optical spectrum for each pixel in the image. Figure reprinted with permission [56]

modulators, and multi-pass optical cavities to increase axial scan rates while maintaining scan ranges of 1–2 mm [25,59–61]. An optical delay line based on the principles used in femtosecond pulse shaping has been demonstrated for OCT [26]. This delay line spectrally disperses the reference arm beam with a grating. The dispersed beam is then focused by a lens on to a rotating mirror mounted on a galvanometer. The mirror, located in the Fourier-transform plane of the lens, imparts a wavelength-dependent phase shift on the light. Subsequently, when re-coupled back into the interferometer, this phase-shift is equivalent to a time-delay in the time domain. The use of high-speed resonant galvanometers has permitted axial scan rates as high as 3 kHz over scan ranges of several millimeters. Depending on the image pixel size (number of axial scans within each image), this scan rate can provide video-rate OCT imaging (30 frames per second for frames with 100 axial scans each). An example of high-speed *functional* OCT imaging is shown in Fig. 13.7. Images of a beating *Xenopus laevis* (African frog) tadpole heart were acquired in 30 s and in 250 ms (256 × 256 pixel images) [62]. The image acquired over 30 s shows multiple motion artifacts from the beating heart. At an acquisition rate of 250 ms per image (4 frames per second), motion artifacts are significantly reduced and cardiac morphology can be observed. Functional cardiac parameters could be measured and high-speed processes such as chamber contracting and valve function could be visualized at acquisition rates of 4 to 8 frames per second [62].

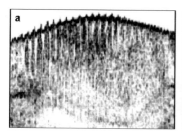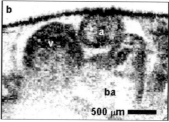

Fig. 13.7. Low- and high-speed OCT images of a beating *Xenopus laevis* (African frog) tadpole heart. (**a**) Image acquired over 30 s reveals multiple motion artifacts. (**b**) High-speed imaging at 4 frames per second (250 ms per image) significantly reduces motion artifacts, allowing the ventricle (*v*), atrium (*a*), and bulbous arteriosus (*ba*) to be visualized. High-speed imaging permits quantification of functional parameters. Figure reprinted with permission [62]

Depth-priority scanning, as described above and commonly implemented in OCT, is performed by rapidly varying the optical delay in the reference arm and collecting a single axial scan before translating the beam laterally and repeating this depth-scanning. An alternative method for generating OCT images is with transverse-priority scanning [63,64]. For this method, the optical pathlength in the reference arm is held momentarily constant, apart from a low-amplitude, high-frequency modulation to enable heterodyne detection. The OCT beam in the sample arm is then scanned laterally across the specimen in both the X and Y planes while the amplitude of the interference signal is recorded. The result, in contrast to a cross-sectional plane in depth, is a horizontal (*en face*) plane at a given depth within the specimen. This is equivalent to optical sectioning in confocal and multi-photon microscopy.

Three-dimensional (3-D) OCT imaging using transverse-priority scanning can be obtained by stepping the position of the reference arm mirror after each *en face* image is acquired [65].

The OCT images produced from transverse-priority scanning are readily correlated with confocal or multi-photon microscopy images, which are typically acquired with a similar scanning method. This method can also utilize higher numerical aperture objective lenses to provide high transverse resolutions since a large depth of focus is not needed as in depth-priority OCT scanning. The combination of OCT with high-numerical aperture objectives has been termed optical coherence microscopy (OCM) [66].

13.3.4   Optical Coherence Microscopy

Optical coherence microscopy combines the principles of confocal microscopy with the principles of low-coherence interferometry [66,67]. High detection sensitivity and high contrast from rejection of out-of-focus light is achieved, resulting in improved optical sectioning capabilities deep within highly-scattering specimens. The axial point-spread function (PSF) in OCM is narrower, given by the product of the field autocorrelation PSF and the microscope objective PSF, and enhances the optical sectioning capabilities of the microscopy technique. Figure 13.8 shows experimental measurements of the axial point-spread function with a 20× microscope objective (NA= 0.4) [66]. This data was obtained by scanning a mirror through the focal plane, with and without coherence gating.

Considering a single-backscatter model for photons, two limits define when OCM can outperform confocal microscopy. A parametric plot of these theoretical limits is shown in Fig. 13.9 [66]. The first limit, for shorter mean-free-paths, defines when the coherence gating of OCM enhances the confocal microscopy image. This occurs only when the collected scattered light from outside of the focal plane dominates the light collected from within the focal plane. The second limit, for longer mean-free-paths, is determined by the maximum depth at which single-backscattered photons can be detected given quantum detection and the tissue laser damage thresholds. For approximately 5–15 scattering mean-free-paths, the use of OCM enables imaging of structures that would otherwise be obscured by scattering in conventional confocal microscopy.

While the use of OCM can improve the optical sectioning capability of confocal microscopy, further enhancement is likely with MPM. For several years investigators have recognized that OCT and MPM can utilize a single laser source for multi-modality imaging. Recently, a microscope that integrates OCM and two-photon-excitation fluorescence imaging has been used to image cells in *Drosophila* embryos (Fig. 13.10) [55]. OCT and MPM provide complementary image data. OCT can image deep through transparent and highly-scattering structures to reveal the 3-D structural information. OCT, however, cannot detect the presence of a fluorescing particle. In a complementary manner, MPM can localize fluorescent probes in three-dimensional space. MPM can detect the fluorescence, but not the microstructure nor

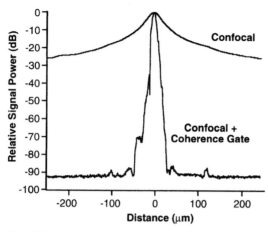

Fig. 13.8. Comparison of confocal and optical coherence microscopy point-spread functions. Experimental confocal microscopy data was obtained using a 20× microscope objective (NA= 0.4). A mirror was scanned through the focal plane and reflected signal power was measured with and without coherence gating. Figure reprinted with permission [66]

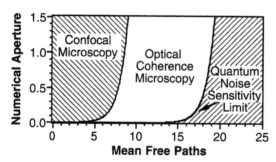

Fig. 13.9. Theoretical parametric plot of limits in optical coherence microscopy. Assuming a single-backscatter model for photons, this plot illustrates that optical coherence microscopy can enhance detection in highly-scattering media, beyond the limits of confocal microscopy alone. The right-most limit is bounded by the quantum noise sensitivity. Figure reprinted with permission [66]

the location of the fluorescence relative to the microstructure. Hence, the development of an integrated microscope capable of OCT and MPM uniquely enables the simultaneous acquisition of microstructural data and the localization of fluorescent probes.

13.3.5   Beam Delivery Systems

The OCT imaging technology is modular in design. This is most evident with the various optical instruments through which the OCT beam can be delivered to the tissue or sample. Because OCT is fiber-optic based, single optical fibers can be used

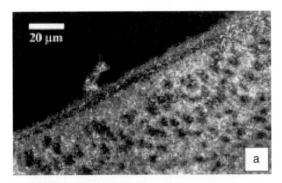

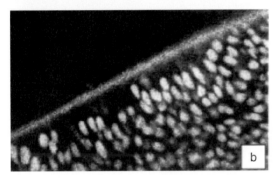

Fig. 13.10. Combined optical coherence and two-photon microscopy imaging. *En face* images of cells in a live *Drosophila* embryo were acquired simultaneously using (**a**) optical coherence microscopy and (**b**) two-photon microscopy. The embryo was transfected with genes to express green fluorescent protein in its cell nuclei. In the OCM image (**a**), the nuclei appear as non-scattering structures. Figure reprinted with permission [55]

to deliver the OCT beam and collect the reflected light. The OCT technology can readily be integrated into existing optical instruments such as research and surgical microscopes [55,68], ophthalmic slit-lamp biomicroscopes [3], catheters [37], endoscopes [38–41,69], laparoscopes [35], and hand-held imaging probes [35].

Imaging penetration is determined by the optical absorption and scattering properties of the tissue or specimen. The imaging penetration for OCT ranges from tens of millimeters for transparent tissues such as the eye to less than 3 mm in highly-scattering tissues such as skin. To image highly-scattering tissues deep within the body, novel beam-delivery instruments have been developed to relay the OCT beam to the site of the tissue to be imaged. An OCT catheter has been developed for insertion into biological lumens such as the gastrointestinal tract [37]. Used in conjunction with endoscopy, the 1 mm diameter catheter can be inserted through the working channel of the endoscope for simultaneous OCT and video imaging [38–40]. Minimally-invasive surgical procedures utilize laparoscopes which are long, thin, rigid optical instruments to permit video-based imaging within the abdominal cavity. Laparoscopic OCT imaging has been demonstrated by passing the OCT beam through the optical elements of a laparoscope [35]. Deep solid-tissue imaging is possible with the use of fiber-needle probes [36]. Small (400 µm diameter) needles housing a single optical fiber and micro-optical elements have been inserted into solid tissues and rotated to acquire OCT images. Recently, microfabricated micro-electro-optical-mechanical systems (MEOMS) technology has been used to miniaturize the OCT beam scan mechanism [69].

13.4   Applications

In parallel with the development of the OCT technology has been the investigation of a wide range of applications in the fields of biology, medicine, and material science. A brief overview of these fields with representative images follows.

13.4.1   Developmental Biology

OCT has been demonstrated in the field of developmental biology as a method to perform high-resolution, high-speed imaging of developing morphology and function. Cellular-level imaging is possible, providing a non-invasive technique for visualizing cellular processes such as mitosis and migration. Imaging studies have been performed on several standard biological animal models commonly employed in developmental biology investigations including *Rana pipiens* (Leopard frog), *Xenopus laevis* (African frog) [70], and *Brachydanio rerio* (zebra fish) embryos and eggs, and the murine (mouse) model [62,71–74].

A series of cross-sectional images acquired *in vitro* from the dorsal and ventral sides of a Stage 49 (12 day) *Rana pipiens* (Leopard frog) tadpole are shown in Fig. 13.11 [71]. These images were acquired using a SLD source operating at 1.3 μm center wavelength. Axial and transverse resolutions were 16 μm and 30 μm, respectively. Features of internal architectural morphology are clearly visible in the images. The image of the eye differentiates structures corresponding to the cornea, lens, and iris. The corneal thickness is on the order of 10 μm and can be resolved due to the differences in index of refraction between the air, cornea, and aqueous humor. By imaging through the transparent lens, the incident OCT beam images several of the posterior ocular layers including the ganglion cell layer, retinal neuroblasts, and choroid. Other identifiable structures include several different regions of the brain as well as the ear vesicle. The horizontal semicircular canal and developing labyrinths can be observed. Internal morphology not accessible in one orientation due to the specimen size or shadowing effects can be imaged by re-orienting the specimen and scanning in the same cross-sectional image plane. The images in the middle column in Fig. 13.11 were acquired with the OCT beam incident from the ventral side to image the respiratory tract, ventricle of the heart, internal gills, and gastrointestinal tract.

These images can be compared with corresponding histology. Histological images are acquired by euthanizing the specimen, immersing the specimen in a chemical fixative, and physically sectioning thin (2–5 micron-thick) slices using a microtome. The slices are placed on a microscope slide, selectively stained to highlight particular features, and commonly viewed with light microscopy. The correlations between OCT and histology images are strong, suggesting that OCT images can

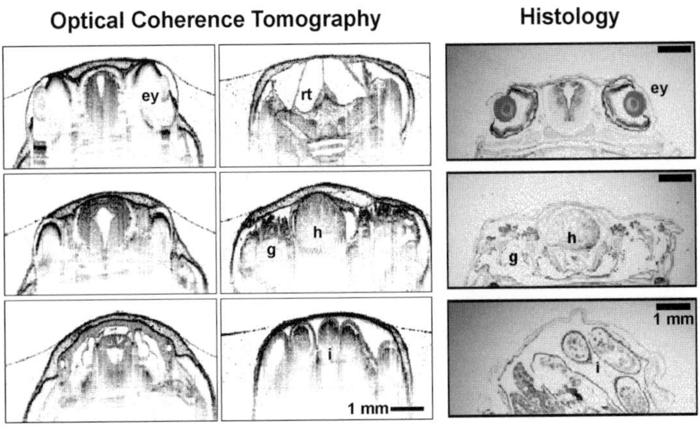

Optical Coherence Tomography   **Histology**

Fig. 13.11. OCT of developing biology. OCT images and corresponding histology acquired from a Stage 49 (12 day) *Rana pipiens* (Leopard frog) tadpole. OCT images in the left and middle columns were acquired with the OCT beam incident from the dorsal and ventral sides of the tadpole, respectively. Structures including the eye (*ey*), respiratory tract (*rt*), gills (*g*), heart (*h*) and gastrointestinal tract (*i*) are observed. The corresponding histology in the right column correlates strongly with the structures observed using OCT. Figure reprinted with permission [71]

accurately represent the *in vivo* specimen morphology. The potential exists to re-peatedly image specimens to quantify organo- and morphogenesis throughout de-velopment. Technologies such as OCT are likely to become increasingly important in functional genomics, relating genotype to the morphology (phenotype) and func-tion in living specimens.

OCT images represent the optical backscatter intensity from regions within the tissue or sample. Because OCT relies on the inherent optical scattering changes to produce image contrast, no exogenous contrast agents or fluorophores are necessary. This permits long-term sequential imaging of development *in vivo* without loss of specimen viability. Repeated imaging of the *Xenopus laevis* hind limb patterning and formation has been demonstrated (Fig. 13.12) without loss of specimen via-bility and without developmental abnormalities [75]. In this example, formation of the cartilaginous skeletal system and the developing vascular system was observed. By imaging subtle differences in backscattering intensity, morphological features millimeters deep within specimens can be clearly delineated.

Previous OCT images have characterized morphological features within bio-logical specimens. These structures are static even though they may have been ac-quired from *in vivo* specimens. *In vivo* imaging in living specimens, particularly in larger organisms and for medical diagnostic applications, must be performed at high speeds to eliminate motion artifacts within the images. Functional OCT imaging is the quantification of *in vivo* images which yield information characterizing the func-tional properties of the organ system or organism. Studies investigating normal and abnormal cardiac development have been frequently limited by an inability to access

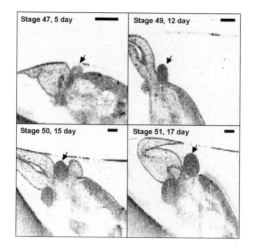

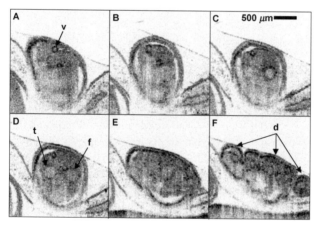

Fig. 13.12. Tracking biological development over time. Top sequence illustrates hind limb development (arrows) in a *Xenopus* tadpole over several weeks. Images were acquired at days 5, 12, 15, and 17. The bottom sequence represents multiple OCT images of a single hind limb acquired at day 28. Images were acquired proximal-to-distal along the length of the hind limb, revealing the vasculature (*v*), the tibia (*t*), the fibula (*f*), and multiple digits (*d*) [75]

cardiovascular function within the intact organism. OCT has been demonstrated for the high-resolution assessment of structure and function in the developing *Xenopus laevis* (African frog) cardiovascular system (Fig. 13.7) [62]. The morphology of the *in vivo* cardiac chambers can be clearly delineated. Image acquisition rates are fast enough to capture the cardiac chambers in mid-cycle. With this capability, images can be acquired at various times and displayed in real-time to produce a movie illustrating the dynamic, functional behavior of the developing heart [27,62]. OCT, like technologies such as computed tomography and magnetic resonance imaging, provides high-speed *in vivo* imaging, allowing quantitative dynamic activity, such as ventricular ejection fraction, to be assessed.

The field of functional genomics involves the study of the genome and how the expression of genes relates to morphology and function. With advanced molecular biology techniques available to site-specifically modify the genomes of animal models, imaging techniques are needed to identify changes in both morphology and function in these genetically modified specimens. The use of OCT for identifying normal and abnormal development is shown in Fig. 13.13 [71]. These images were from *Xenopus* specimens. The abnormal morphology observed was believed to be the result of spontaneous developmental abnormalities, not the result from a mutagenic agent. In addition to identifying abnormalities in morphology, the corresponding histology illustrates the difficulty often associated with histological processing of many developmental biology specimens. Routine histology can be time consuming, costly, and often prone to processing artifacts. OCT has the potential to image throughout development, without having the sacrifice specimens at single time points and reducing the number of mutant specimens available for later studies.

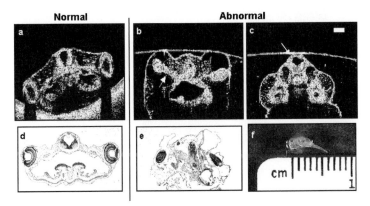

Fig. 13.13. Identification of normal and abnormal development. Normal *Xenopus* specimen (**a**) with corresponding histology (**d**) can be compared to an abnormal specimen (**b**, **e**). Arrow in (**b**) indicates longitudinal section of the left optic nerve and ocular muscles. A subtle neural tube defect is shown by the arrow in (**c**). Bar represents 500 μm. Figure reprinted with permission [71]

13.4.2 Cellular Imaging

Although previous studies have demonstrated *in vivo* OCT imaging of tissue morphology, most have imaged tissue at ~10–15 μm resolutions, which does not allow differentiation of cellular structure. The ability of OCT to identify the mitotic activity, the nuclear-to-cytoplasmic ratio, and the migration of cells has the potential to not only impact the fields of cell and developmental biology, but also impact medical and surgical disciplines for the early diagnostics of disease such as cancer.

The *Xenopus laevis* (African frog) tadpole has been used to demonstrate the feasibility of OCT for high-resolution *in vivo* cellular and subcellular imaging [24,73]. Many of the cells in this common developmental biology animal model are rapidly dividing and migrating during the early growth stages of the tadpole, providing an opportunity to image dynamic cellular processes. Three-dimensional volumes of high-resolution OCT data have been acquired from these specimens throughout development [73]. From this 3-D data set, cells undergoing mitosis were identified and tracked in three dimensions (Fig. 13.14). In a similar manner, 3-D data sets were acquired to track single melanocytes (neural crest cells) as they migrated through the living specimens. The ability of OCT to characterize cellular processes such as mitosis and migration is relevant for cancer diagnostics and the investigation of tumor metastasis in humans.

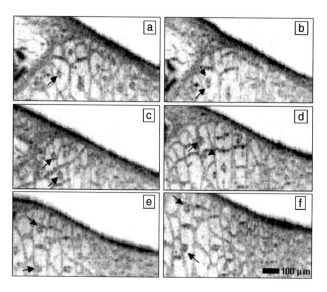

Fig. 13.14. Tracking cell mitosis. OCT imaging of a mesenchymal parent cell in (**a**) undergoes mitosis, dividing into two daughter cells (**b–f**). Following cell division, daughter cells migrate apart and increase in size, preparing for subsequent cell division. Images were extracted from 3-D data sets acquired at 10 min intervals. Figure reprinted with permission [73]

An example of cellular-level OCT imaging in these specimens using a state-of-the-art femotosecond laser is shown in Fig. 13.15 [24]. This composite image (0.83 × 1 mm, 1800 × 1000 pixels) was acquired using a titanium:sapphire laser with an extremely broad bandwidth (~260 nm). The axial and transverse resolutions for this image are 1 μm and 5 μm, respectively. Because the high transverse resolution reduced the depth of focus to 49 μm, separate OCT images were first acquired with the focus at different depths within the specimen. These images were then assembled to produce the composite image shown in Fig. 13.15. This type of image construction is similar to C-mode ultrasound. Cellular features including cell membranes, nuclei, and nuclear morphology are clearly observed up to nearly 1 mm into the specimen. The high-resolution, optical-sectioning ability of this system enables 3-D imaging of a single cell, as shown in Fig. 13.16 [76]. Sections were acquired at 2 micron intervals. Sub-cellular organelles are observed in several of the sections. The images presented here approach the resolutions common in confocal and multi-photon microscopy, but were obtained at greater depths in highly-scattering specimens.

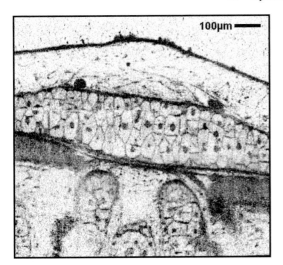

Fig. 13.15. Ultrahigh resolution OCT imaging. Using a state-of-the-art titanium:sapphire laser, broad spectral bandwidths enable high axial resolutions. This composite OCT image was acquired with axial and transverse resolutions of 1 and 5 microns, respectively. The high transverse resolution corresponds to a confocal parameter (depth of field) of 49 μm. Therefore, multiple images were acquired at increasing depths, then combined to produce this composite image. Figure reprinted with permission [24]

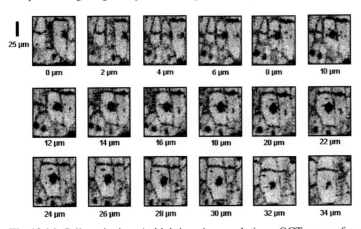

Fig. 13.16. Cell sectioning. At high imaging resolutions, OCT can perform optical sectioning similar to confocal and multi-photon microscopy. Multiple sections at 2 μm intervals are shown for a single *Xenopus* mesenchymal cell. These sections reveal sub-cellular organelles and can be assembled to produce a 3-D data set. Figure reprinted with permission [76]

13.4.3 Medical and Surgical Microscopy – Identifying Tumors and Tumor Margins

OCT has been used to differentiate between the morphological structure of normal and neoplastic tissue for a wide range of tumors [19,77–79]. The use of OCT to iden-

328   Stephen A. Boppart

tify tumors and tumor margins *in situ* will represent a significant advancement for medical or image-guided surgical applications. OCT has been demonstrated for the detection of brain tumors and their margins with normal brain parenchyma, suggesting a role for guiding surgical resection [19]. A hand-held surgical imaging probe was constructed for this application. The compact and portable probe permits OCT imaging within the surgical field while the OCT instrument can be remotely located in the surgical suite.

Figure 13.17 shows an *in vitro* specimen of outer human cerebral cortex with metastatic melanoma. The OCT images in Figs. 13.17a and 13.17b were acquired through the tumor. These original images were threshold-segmented to identify regions of high backscatter within the tumor. The original images were then overlaid with the segmented data and shown in Figs. 13.17c and 13.17d. The OCT images show increased optical backscatter in the region of the tumor (*white arrows*). Smaller tumor lesions also appear within the image. A shadowing effect is observed below each tumor site due to the increased optical backscatter and the subsequent loss of optical power penetrating beneath the tumor. In Figs. 13.17a and 13.17c, the boundary of the tumor can be identified. In Figs. 13.17b and 13.17d, the tumor is identified below the surface of normal cortex. The histology in Figs. 13.17e and 13.17f confirms the presence and relative sizes of the metastatic tumors.

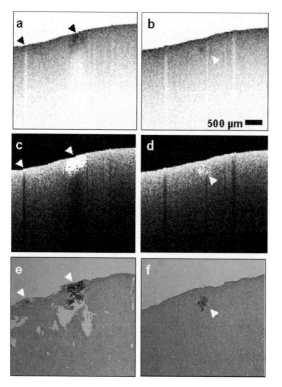

Fig. 13.17. Tumor detection. OCT imaging of metastic melanoma in *in vitro* human cortical brain tissue. Original OCT images (**a, b**) were threshold-segmented (**c, d**) to highlight backscattering tumor. Comparison with histology (**e, f**) is strong. Figure reprinted with permission [19]

The image resolutions used to acquire the images in Fig. 13.17 were as high as 16 μm, higher than current ultrasound, CT, or MRI intraoperative imaging techniques. This allowed the tumor-cortex interface and the extent of tumor below the surface to be defined with high resolution. At higher imaging resolutions, it may be possible to image individual tumor cells that have migrated away from the central tumor. OCT represents a new high-resolution optical imaging technology that has the potential for identifying tumors and tumor margins on the micron scale and in real-time. OCT offers imaging performance not achievable with current imaging modalities and may contribute significantly toward the surgical resection of neoplasms.

13.4.4   Image-Guided Surgery

The repair of vessels and nerves is necessary to restore function following traumatic injury [80]. Although the repair of these sensitive structures is performed with the aid of surgical microscopes and loupes to magnify the surgical field [81], surgeons are limited to the *en face* view that they provide. A technique capable of subsurface, three-dimensional, micron-scale imaging in real-time would permit the intraoperative monitoring of microsurgical procedures. The capabilities of OCT for the intraoperative assessment of microsurgical procedures have been demonstrated [68]. High-speed OCT imaging was integrated with a surgical microscope to performed micron-scale three-dimensional imaging on microsurgical specimens.

The ability of OCT to assess internal structure and vessel patency within an arterial anastomosis (surgical repair) is shown in Fig. 13.18 [68]. Cross-sectional OCT images (2.2×2.2 mm, 250×600 pixels) and 3-D projections of a 1 mm diameter rabbit inguinal artery are illustrated. Two-dimensional cross-sectional images in Figs. 13.18b–e were acquired from the locations labeled in Fig. 13.18a. Figures 18b and 18e were acquired from each end of the anastomosis. The patent lumen is readily apparent. This is in contrast to what is observed in Fig. 13.18d where at the site of anastomosis, the lumen had been occluded by tissue. By assembling a series of cross-sectional 2-D images, a 3-D data set was produced. From this data set, arbitrary planes can be selected and corresponding sections displayed. Figure 13.18f is a longitudinal section from the 3-D data set that confirms the occlusion within the anastomosis site. These results have shown how 2-D OCT images and 3-D OCT projections can provide diagnostic feedback to assess microsurgical anastomoses. This previously unavailable diagnostic ability offers the potential to directly impact and to improve patient outcome by incorporating high-speed, high-resolution intraoperative image-guidance during microsurgical procedures.

Surgical intervention requires adequate visualization to identify tissue morphology, precision to avoid sensitive tissue structures, and continuous feedback to monitor the extent of the intervention. OCT may provide the technological advancements to improve the operative procedure. The feasibility of OCT to perform image-guided surgical intervention has been investigated [82,83]. High-power argon laser (514 nm wavelength) ablation has been used as the representative interventional surgical technique. Figure 13.19 illustrates the use of OCT for real-time monitoring of

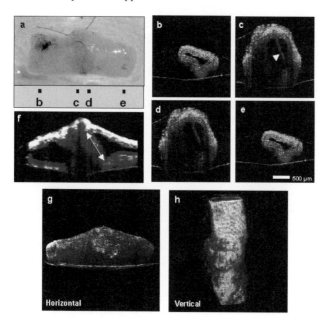

Fig. 13.18. Microsurgical guidance of arterial anastomosis. Labeled vertical lines in (**a**) refer to cross-sectional imaging locations (**b-e**) from an anastomosis of an *in vitro* rabbit artery. An obstructive flap is identified in (**c**). By resectioning the 3-D data set, a complete obstruction is observed in (**f**). Horizontally- and vertically-rotated 3-D projections are shown in (**g**) and (**h**), respectively. Figure reprinted with permission [68]

surgical intervention, including dosimetry of laser ablation [83]. A surgical ablation threshold is documented by a comparison of thermal injury in rat kidney and liver for thermal energy doses below and above the ablation threshold. Two sequences with corresponding histology are shown in Fig. 13.19. The first exposure of 1 W on *in vitro* rat kidney is halted after 1 s, prior to the ablation threshold. The second exposure of 1 W on *in vitro* rat liver is allowed to continue for 7 s, past the ablation threshold, resulting in ejection of tissue and crater formation. The corresponding histology for the sub-threshold lesion indicates a region of coagulated tissue (*arrows*). In contrast, the above-threshold lesion histology shows marked tissue ablation and fragmentation within the lesion crater. Below the crater extends a zone of coagulated tissue (*arrows*) which is not observed in the OCT image due to the poor penetration through the carbonized crater wall.

These examples demonstrate the use of OCT for guiding and monitoring surgical intervention. Although a continue-wave argon laser was used as an interventional device, the argon laser is only a representative technique for a wide range of instruments and techniques including scalpels, electrosurgery, radiofrequency, microwaves, and ultrasound ablation [84]. OCT imaging was performed at 8 frames per second, fast enough to capture dynamic changes in the optical properties of the tissue during thermal ablation. These image sequences provided interesting insight into ablation mechanisms for a variety of tissue types. OCT can monitor the extent of thermal injury below the surface of the tissue by imaging the changes in optical backscatter. OCT imaging can therefore provide empiric information for dosimetry to minimize the extent of collateral injury. The use of OCT for guiding surgical inter-

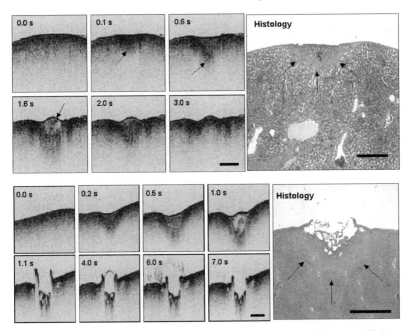

Fig. 13.19. Image-guided surgery. Top sequence: Argon laser (514 nm, 1 W, 3 s exposure) thermal injury to *in vitro* rat kidney. Exposure was stopped prior to membrane rupture of kidney. Arrows in the corresponding histology indicate zone of thermal damage. Bottom sequence: Above-threshold argon laser ablation (1 W, 7 s exposure) of rat liver. Arrows in histology image illustrate regions of thermal damage surrounding a region of ablated tissue. Bar represents 1 mm. Figures reprinted with permission [83]

ventions has the potential to improve intraoperative monitoring and more effectively control interventional procedures.

13.4.5   Materials Investigations

While the majority of OCT applications have been in the fields of biology and medicine. OCT has also been demonstrated in non-biological areas of materials investigation [85,86], optical data storage [87], and microfluidic devices. The highly-scattering or reflecting optical properties of many materials prohibit deep imaging penetration using OCT. Many material defects, however, originate at surface or interfacial boundaries, making the use of OCT a possibility for inspection and quality control. For transparent or translucent materials such as plastics or polymer composites, defect imaging at larger depths is feasible.

The increasing demand for compact, high-density, optical data storage has prompted the investigation of optical sources at shorter visible (blue) wavelengths. The optical ranging capabilities of OCT through scattering materials has the potential for increasing the data storage capacity by assembling multiple layers of optically-accessible data [87,88]. Figure 13.20 illustrates a schematic for the use of

OCT in optical data readout as well as images comparing OCT and direct optical detection from a compact disk [87]. The short-coherence length light sources used in OCT can be used for reading high volume density data from multi-layer optical disks. Because OCT can share much of the existing hardware and utilize similar scanning techniques present in existing optical data storage system, a high likelihood exists for rapid integration.

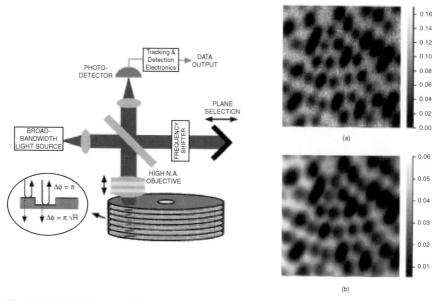

(a)

(b)

Fig. 13.20. Multi-layer optical data readout. Schematic illustrating the use of OCT for reading optical data from a multi-layer optical disk. Gray-scale images from a CD-ROM layer of optical data acquired with (**a**) OCT and (**b**) direct detection. Figure reprinted with permission [87]

13.5 Conclusions

The capabilities of OCT offer a unique and informative means of imaging biological specimens and non-biological samples. The non-contact nature of OCT and the use of low-power near-infrared radiation for imaging makes this technique non-invasive and safe. OCT imaging does not require the addition of fluorophores, dyes, or stains to improve contrast in images as in light, confocal, or multi-photon microscopy. Instead, OCT relies on the inherent optical contrast generated from variations in optical scattering and index of refraction. These factors permit the use of OCT for extended imaging over the course of hours, days, or weeks. OCT permits the cross-sectional imaging of tissue and samples and enables *in vivo* structure to be visualized in opaque specimens or in specimens too large for high-resolution confocal or light microscopy.

Imaging at cellular and subcellular resolutions with OCT is an important area of ongoing research. The *Xenopus* developmental animal model has been commonly used because its care and handling are relatively simple while allowing cells with a high mitotic index to be assessed. Cellular imaging in humans, particularly *in situ*, is a challenge because of the smaller cell sizes ($10-30\,\mu m$) compared to larger undifferentiated cells in developing organisms. To the advantage of scientists and physicians, poorly differentiated cells present in many neoplastic tissues tend to be larger, increasing the likelihood for detection using OCT at current imaging resolutions. With further advances in OCT technology, improved discrimination and imaging of more detailed structures should be possible. New laser sources at other wavelengths in the near-infrared can enhance tissue contrast as well as potentially provide functional information since tissue scattering and absorbance properties in specimens are wavelength dependent. Short coherence length short-pulse laser sources have been used to achieve higher axial resolutions on the order of 1–3 microns. Unfortunately, unlike superluminescent diode source, these high-speed and high-resolution systems utilize femtosecond lasers that are relatively large, complex, and costly. Developing compact, portable, turn-key optical sources at near-infrared wavelengths, with broad spectral bandwidths, and with high output powers is an area of active research.

Optical coherence tomography provides high-resolution morphological, functional, and cellular information of biological, medical, and material specimens and samples. OCT represents a multifunctional investigative tool which not only complements many of the existing imaging technologies available today, but over time, is also likely to become established as a major optical imaging modality.

Acknowledgments

I would like to acknowledge my colleagues for their contributions to this work. In particular, I would like to thank James Fujimoto from the Massachusetts Institute of Technology, Cambridge, MA, Brett Bouma and Gary Tearney from Massachusetts General Hospital, Boston, MA, Mark Brezinski from Brigham and Womens' Hospital, Boston, MA, and all of the investigators whose images and results are presented in this chapter.

References

1. D. Huang, E.A. Swanson, C.P. Lin, J.S. Schuman, W.G. Stinson, W. Chang, M.R. Hee, T. Flotte, K. Gregory, C.A. Puliafito, J.G. Fujimoto: Science **254**, 1178 (1991)
2. B.E. Bouma, G.J. Tearney, Eds.: *Handbook of Optical Coherence Tomography*. (Marcel Dekker, Inc. 2001)
3. M.R. Hee, J.A. Izatt, E.A. Swanson, D. Huang, C.P. Lin, J.S. Schuman, C.P. Lin, C.A. Puliafito, J.G. Fujimoto: Arch. Ophthalmology **113**, 325 (1995)

4. C.A. Puliafito, M.R. Hee, C.P. Lin, E. Reichel, J.S. Schuman, J.S. Duker, J.A. Izatt, E.A. Swanson, J.G. Fujimoto: Ophthalmology **102,** 217 (1995)
5. C.A. Puliafito, M.R. Hee, J.S. Schuman, J.G. Fujimoto: *Optical Coherence Tomography of Ocular Diseases.* (Slack, Inc, Thorofare, NJ 1995)
6. J.M. Schmitt, A. Knuttel, R.F. Bonner: Appl. Opt. **32,** 6032 (1993)
7. J.M. Schmitt, A. Knuttel, M. Yadlowsky, A.A. Eckhaus: Phys. Med. Biol. **39,** 1705 (1994)
8. J.G. Fujimoto, C. Pitris, S.A. Boppart, M.E. Brezinski: Neoplasia **2,** 9 (2000)
9. J.M. Schmitt, M.J. Yadlowsky, R.F. Bonner: Dermatology **191,** 93 (1995)
10. A. Sergeev, B. Gelikonov, G. Gelikonov, F. Feldchetin, K. Pravdenki, R. Kuranov, N. Gladkova, V. Pochinko, G. Petrova, N. Nikulin: High-spatial Resolution Optical Coherence Tomography of Human Skin and Mucus Membranes. In: *Conference on Lasers and Electro Optics* '95, *Washington, D.C.* OSA Technical Digest Series **15,** paper CThN4 (1995)
11. M.E. Brezinski, G.J. Tearney, B.E. Bouma, J.A. Izatt, M.R. Hee, E.A. Swanson, J.F. Southern, J.G. Fujimoto: Circulation **93,** 1206 (1996)
12. G.J. Tearney, M.E. Brezinski, S.A. Boppart, B.E. Bouma, N. Weissman, J.F. Southern, E.A. Swanson, J.G. Fujimoto: Circulation **94,** 3013 (1996)
13. M.E. Brezinski, G.J. Tearney, N.J. Weissman, S.A. Boppart, B.E. Bouma, M.R. Hee, A.E. Weyman, E.A. Swanson, J.F. Southern, J.G. Fujimoto: Heart **77,** 397 (1997)
14. I.K. Jang, B.E. Bouma, D.H. Kang, S.J. Park, S.W. Park, K.B. Seung, K.B. Choi, M. Shishkov, K. Schlendorf, E. Pomerantsev, S.L. Houser, H.T. Aretz, G.J. Tearney: J. Am. Coll. Cardiol. **39,** 604 (2002)
15. G.J. Tearney, M.E. Brezinski, J.F. Southern, B.E. Bouma, S.A. Boppart, and J.G. Fujimoto: Am. J. Gastroenterology **92,** 1800 (1997)
16. J.A. Izatt, M.D. Kulkarni, H-W. Wang, K. Kobayashi, M.W. Sivak: IEEE J. Selected Topics in Quant. Elect. **2,** 1017 (1996)
17. G.J. Tearney, M.E. Brezinski, J.F. Southern, B.E. Bouma, S.A. Boppart, J.G. Fujimoto: J. Urol. **157,** 1915 (1997)
18. E.V. Zagaynova, O.S. Streltsova, N.D. Gladkova, L.B. Snopova, G.V. Gelikonov, F.I. Feldchtein, A.N. Morozov: J. Urol. **167,** 1492 (2002)
19. S.A. Boppart, M.E. Brezinski, C. Pitris, J.G. Fujimoto: Neurosurg. **43,** 834 (1998)
20. B.W. Colston Jr., M.J. Everett, L.B. Da Silva, L.L. Otis, P. Stroeve, H. Nathal: Appl. Opt. **37,** 3582 (1998)
21. X. Clivaz, F. Marquis-Weible, R.P. Salathe: Elec. Lett. **28,** 1553 (1992)
22. B.E. Bouma, G.J. Tearney, S.A. Boppart, M.R. Hee, M.E. Brezinski, J.G. Fujimoto: Opt. Lett. **20,** 1486 (1995)
23. B.E. Bouma, G.J. Tearney, I.P. Biliinski, B. Golubovic, J.G. Fujimoto: Opt. Lett. **21,** 1839 (1996)
24. W. Drexler, U. Morgner, F.X. Kartner, C. Pitris, S.A. Boppart, X.D. Li, E.P. Ippen, J.G. Fujimoto: Opt. Lett. **24,** 1221 (1999)
25. G.J. Tearney, B.E. Bouma, S.A. Boppart, B. Golubovic, E.A. Swanson, J.G. Fujimoto: Opt. Lett. **21,** 1408 (1996)
26. G.J. Tearney, B.E. Bouma, J.G. Fujimoto: Opt. Lett. **22,** 1811 (1997)
27. A.M. Rollins, M.D. Kulkarni, S. Yazdanfar, R. Ung-arunyawee, J.A. Izatt: Opt. Express **3,** 219 (1998)
28. Z. Chen, T.E. Milner, S. Srinivas, X. Wang, A. Malekafzali, M.J.C. van Germert, J.S. Nelson: Opt. Lett. **22,** 1119 (1997)
29. J.A. Izatt, M.D. Kulkarni, S. Yazdanfar, J.K. Barton, A.J. Welch: Opt. Lett. **22,** 1439 (1997)

30. J.K. Barton, A.J. Welch, J.A. Izatt: Opt. Express **3**, 251 (1998)
31. J.F. de Boer, T.E. Milner, M.J.C. van Germert, J.S. Nelson: Opt. Lett. **22**, 934 (1997)
32. M.J. Everett, K. Schoenenberger, B.W. Colston Jr., L.B. Da Silva: Opt. Lett. **23**, 228 (1998)
33. K. Schoenenberger, B.W. Colston Jr., D.J. Maitland, L.B. Da Silva, M.J. Everett: Appl. Opt. **37**, 6026 (1998)
34. G.J. Tearney, S.A. Boppart, B.E. Bouma, M.E. Brezinski, N.J. Weissman, J.F. Southern, J.G. Fujimoto: Opt. Lett. **21**, 543 (1996)
35. S.A. Boppart, B.E. Bouma, C. Pitris, G.J. Tearney, J.G. Fujimoto, M.E. Brezinski: Opt. Lett. **22**, 1618 (1997)
36. X. Li, C. Chudoba, T. Ko, C. Pitris, J.G. Fujimoto: Opt. Lett. **25**, 1520 (2000)
37. G.J. Tearney, M.E. Brezinski, B.E. Bouma, S.A. Boppart, C. Pitris, J.F. Southern, J.G. Fujimoto: Science **276**, 2037 (1997)
38. B.E. Bouma, G.J. Tearney, C.C. Compton, N.S. Nishioka: Gastrointest. Endosc. **51**, 467 (2000)
39. A. Das, M.V. Sivak Jr., A. Chak, R.C. Wong, V. Westphal, A.M. Rollins, J. Willis, G. Isenberg, J.A. Izatt: Gastrointest. Endosc. **54**, 219 (2001)
40. X.D. Li, S.A. Boppart, J. Van Dam, H. Mashimo, M. Mutinga, W. Drexler, M. Klein, C. Pitris, M.L. Krinsky, M.E. Brezinski, J.G. Fujimoto: Endoscopy **32**, 921 (2001)
41. F.I. Feidchtein, G.V. Gelikonov, V.M. Gelikonov, R.V. Kuranov, A.M. Sergeev, N.D. Gladkova, A.V. Shakhov, N.M. Shakhova, L.B. Snopova, A.B. Terenteva, E.V. Zagainova, Y.P. Chumakov, I.A. Kuznetzova: Opt. Express **3**, 257 (1998)
42. K. Takada, I. Yokohama, K. Chida, J. Noda.: Appl. Opt. **26**, 1603 (1987)
43. A.F. Fercher, K. Mengedoht, W. Werner: Opt. Lett. **13**, 186 (1988)
44. C.K. Hitzenberger: Ophthalmol. Vis. Sci. **32**, 616 (1991)
45. A.M. Rollins, J.A. Izatt: Opt. Lett. **24**, 1484 (1999)
46. B.E. Bouma, G.J. Tearney: Opt. Lett. **24**, 531 (1999)
47. M.L. Osowski, T.M. Cockerill, R.M. Lammert, D.V. Forbes, D.E. Ackley, J.J Coleman: IEEE Photon. Tech. Lett. **6**, 1289 (1994)
48. H. Okamoto, M. Wada, Y. Sakai, T. Hirono, Y. Kawaguchi, Y. Kondo, Y. Kadota, K. Kishi, Y. Itaya: J. Lightwave Tech. **16**, 1881 (1998)
49. E.A. Swanson, S.R. Chinn, C.W. Hodgson, B.E. Bouma, G.J. Tearney, J.G. Fujimoto: Spectrally Shaped Rare-earth Doped Fiber ASE Sources for Use in Optical Coherence Tomography. In: *Conference on Lasers and Electro Optics CLEO '96, Washington, D.C.,* OSA Technical Digest Series **9**, paper CTuU5 (1996)
50. V. Chernikov, J.R. Taylor, V.P. Gapontsev, B.E. Bouma, J.G. Fujimoto: A 75 nm, 30 mW Superfluorescent Yttirbium Fiber Source Operation Around 1.06 µm. In: *Conference on Lasers and Electro Optics CLEO 97, Washington D.C.* OSA Technical Digest Series **11**, paper CTuG8 (1997)
51. B.E. Bouma, L.E. Nelson, G.J. Tearney, D.J. Jones, M.E. Brezinski, J.G. Fujimoto: J. Biomedical Opt. **3**, 76 (1998)
52. M. Bashkansky, M.D. Duncan, L. Goldberg, J.P. Koplow, J. Reintjes: Opt. Express **3**, 305 (1998)
53. W. Denk, J.H. Strickler, W.W. Webb: Science **248**, 73 (1990)
54. D.W. Piston, M.S. Kirby, H. Cheng, W.J. Lederer, W.W. Webb: Appl. Opt. **33**, 662 (1994)
55. E. Beaurepaire, L. Moreaux, F. Amblard, J. Mertz: Opt. Lett. **24**, 969 (1999)
56. U. Morgner, W. Drexler, F.X. Kartner, X.D. Li, C. Pitris, E.P. Ippen, J.G. Fujimoto: Opt. Lett. **25**, 111 (2000)
57. I. Daubechies: IEEE Trans. Inf. Theory **36**, 961 (1990)

58. R. Leitgeb, M. Wojtkowski, A. Kowalczyk, C.K. Hitzenberger, M. Sticker, A.F. Fercher: Opt. Lett. **25**, 820 (2000)
59. K.F. Kwong, D. Yankelevich, K.C. Chu, J.P. Heritage, A. Dienes: Opt. Lett. **18**, 558 (1993)
60. L. Giniunas, R. Danielius, R. Karkockas: Appl. Opt. **38**, 7076 (1999)
61. M. Lai: Engineering & Laboratory Notes, Supplement to Optics & Photonics News, **13**, 10 (2002)
62. S.A. Boppart, G.J. Tearney, B.E. Bouma, J.F. Southern, M.E. Brezinski, J.G. Fujimoto: Proceedings of the National Academy of Sciences USA **94**, 4256 (1997)
63. A.G. Podoleanu, J.A. Rogers, D.A. Jackson, S. Dunne: Opt. Express **7**, 292 (2000)
64. S. Bourquin, V. Monterosso, P. Seitz, R.P. Salathe: Opt. Lett. **25**, 102 (2000)
65. J. Rogers, A. Podoleanu, G. Dobre, D. Jackson, F. Fitzke: Opt. Express **9**, 533 (2001)
66. J.A. Izatt, M.R. Hee, G.M. Owen, E.A. Swanson, J.G. Fujimoto: Opt. Lett. **19**, 590 (1994)
67. B. Hoeling, A. Fernandez, R. Haskell, E. Huang, W. Myers, D. Petersen, S. Ungersma, R. Wang, M. Williams, S. Fraser: Opt. Express **6**, 136 (2000)
68. S.A. Boppart, B.E. Bouma, C. Pitris, G.J. Tearney, J.F. Southern, M.E. Brezinski, J.G. Fujimoto: Radiology **208**, 81 (1998)
69. Y. Pan, H. Xie, G.K. Fedder: Opt. Lett. **26**, 1966 (2001)
70. P.D. Nieuwkoop, J. Faber: *Normal Table of Xenopus Laevis* (Garland Publishing, Inc., New York 1994)
71. S.A. Boppart, M.E. Brezinski, B.E. Bouma, G.J. Tearney, J.G. Fujimoto: Developmental Biology **177**, 54 (1996)
72. S.A. Boppart, M.E. Brezinski, G.J. Tearney, B.E. Bouma, J.G. Fujimoto: J. Neuroscience Methods **70**, 65 (1996)
73. S.A. Boppart, B.E. Bouma, C. Pitris, J.F. Southern, M.E. Brezinski, J.G. Fujimoto: Nat. Med. **4**, 861 (1998)
74. Q. Li, A.M. Timmers, K. Hunter, C. Gonzalez-Pola, A.S. Lewin: Invest. Ophthalmol. Vis. Sci. **42**, 2981 (2001)
75. S.A. Boppart: Surgical Diagnostic, Guidance, and Intervention Using Optical Coherence Tomography. PhD Thesis, Massachusetts Institute of Technology, Cambridge, MA (1998)
76. S.A. Boppart, W. Drexler, U. Morgner, F.X. Kartner, J.G. Fujimoto: Ultrahigh Resolution and Spectroscopic Optical Coherence Tomography Imaging of Cellular Morphology and Function. In: *Proc. Inter-Institute Workshop on In Vivo Optical Imaging at the National Institutes of Health.* ed. by A.H. Gandjbakhche, 56 (1999)
77. A.M. Sergeev, V.M. Gelikonov, G.V. Gelikonov, F.I. Feldchtein, R.V. Kuranov, N.D. Gladkova, N.M. Shakhova, L.B. Snopova, A.V. Shakov, I.A. Kuznetzova, A.N. Denisenko, V.V. Pochinko, Y.P. Chumakov, O.S. Streltzova: Opt. Express **1**, 432 (1997)
78. S.A. Boppart, A.K. Goodman, C. Pitris, C. Jesser, J.J. Libis, M.E. Brezinski, J.G. Fujimoto: *Brit. J. Obstet. Gyn.* **106**, 1071 (1999)
79. C. Pitris, A.K. Goodman, S.A. Boppart, J.J. Libus, J.G. Fujimoto, M.E. Brezinski: *Obstet. Gynecol.* **93**, 135 (1999)
80. C. Zhong-wei, Y. Dong-yue, C. Di-Sheng: *Microsurgery.* (Springer Verlag, New York, NY 1982)
81. M.D. Rooks, J. Slappey, K. Zusmanis: Microsurgery **14**, 547 (1993)
82. M.E. Brezinski, G.J. Tearney, S.A. Boppart, E.A. Swanson, J.F. Southern, J.G. Fujimoto: J. Surg. Res. **71**, 32 (1997)

83. S.A. Boppart, J.M. Herrmann, C. Pitris, D.L. Stamper, M.E. Brezinski, J.G. Fujimoto: J. Surg. Res. **82,** 275 (1999)
84. S.A. Boppart, J.M. Herrmann, C. Pitris, D.L. Stamper, M.E. Brezinski, J.G. Fujimoto: Comput. Aided Surg. **6,** 94 (2001)
85. M.D. Duncan, M. Bashkansky, J. Reintjes: Opt. Express **2,** 540 (1998)
86. F. Xu, H.E. Pudavar, P.N. Prasad, D. Dickensheets: Opt. Lett. **24,** 1808 (1999)
87. S.R. Chinn, E.A. Swanson: Opt. Lett. **21,** 899 (1996)
88. S.R. Chinn, E.A. Swanson: Optical Memory & Neural Networks **5,** 197 (1996)

14 Near-Field Optical Microscopy and Application to Nanophotonics

Motoichi Ohtsu

14.1 Introduction

Near-field optics deals with the local electromagnetic interaction between optical field and matters in a nanometric scale. Due to the size-dependent localization and size-dependent resonance, optical near-field is completely free from diffraction of light. The initial idea of near field optics as a super-resolution imaging tool, in a somewhat primitive form, has been proposed in 1928 [1]. After about half a century of silence, basic research has been started early 1980's by several groups in Europe, US and Japan almost independently. A short comment has been written in the author's laboratory notebook on February 26, 1982, pointing out that *the chemical etching of fiber for fabricating a probe is not straightforward, and thus, requires a continuous effort to establish a nano-fabrication technology* [2]. However, after continuous efforts, study of near field optics showed a lot of progress. For example, the single string of DNA molecule has been successfully imaged. The linewidth of the image has been as narrow as 4 nm or even less [3]. As a result of this progress it can be claimed that a new field, the so called "nano-photonics" has been started as a basic technology to support the society of the 21st century. Further, a novel field of "atom-photonics" has also been started, which controls the thermal and quantum motions of atoms precisely by the optical near-field [4,5].

Three examples of technical trends and technical problems for the present and future optical industry are presented here. (1) It has been estimated that the society of the year of 2010 requires the high-density optical storage and readout technology, with the storage density of 1 Tb/in^2 and data transmission rate of 1 Gbps. Since the mark-length for 1 Tb/in^2 is as short as 25 nm, writing such a small pit is far beyond the diffraction limit of light. (2) The progress of DRAM technology requires the drastic improvement of photo-lithography. It is estimated that the size of the fabricated pattern should be as narrow as 50 nm or even narrower. Though novel methods using excimer lasers, EUV, and SR light sources have been developed, this required linewidth is far beyond the diffraction limit of the methods using a conventional visible light source. (3) The progress of optical fiber transmission system requires the improvement of the photonic integration technology. It is estimated that the photonic matrix switching device should be sub-wavelength in its size in order to realize more than 1000 × 1000 element integration of the devices for the 10 Tbps switching capability in the year of 2015. Since the conventional photonic devices, e.g., diode lasers and optical waveguides have to confine the light in them, their minimum sizes

Török/Kao (Eds.): Optical Imaging and Microscopy, Springer Series in Optical Sciences Vol. 87 – © Springer-Verlag, Berlin Heidelberg 2003

are limited by the diffraction of light. Thus, the size required above is beyond the diffraction limit.

By the three examples shown above, it can be easily understood that a novel optical nanotechnology is required to go beyond the diffraction limit to support the science and technology of the 21st century. Near field optics can be used as such a novel one to meet this requirement. This paper reviews the recent progress of author's works on near-field optical microscopy and its application to chemical vapor deposition, nano-photonic switching, and optical storage/readout by optical near-field for nano-photonics.

14.2   Nano-Scale Fabrication

There has been an interest in application to nano-structure fabrication because of the possibility of realizing nano-photonic integration [6]. For the realization of a device which uses the optical near-field as a carrier for signal transmission, various materials with nanometric size must be integrated laterally on a substrate. For this integration, we need an advanced nano-structure fabrication technique, able to realize spatially high resolution, precise control of size and position, and be applicable for various materials. Photo-enhanced chemical vapor deposition (PE-CVD) method has been attracting attention. Based on photochemical reaction, PE-CVD offers not only the possibility for the lateral integration of different in a truly single growth run [7,8] but also the option of chemical selective growth by varying the wavelength of the light source used. Since the optical near-field energy is concentrated within nanometric dimensions smaller than the wavelength of light [9,10], we are able to deposit various materials of nanometric dimensions by utilizing the photo-decomposition of chemical gases. The combination of optical near-field technology with the PE-CVD, i.e., a near-field optical CVD (NFO-CVD) thus appears to be the most suitable technology for the integration of nanometer scale elements, because it not only allows us to fabricate nanostructures but also gives dual advantage of *in-situ* measurement of the optical properties of the fabricated nanostructures.

14.2.1   Depositing Zinc and Aluminum

Let us discuss the deposition of Zn as an example of NFO-CVD by using a diethlzinc (DEZ) as a parent gas. As NFO-CVD is based on a photodissociation reaction, it is necessary for a reactant molecule to absorb photons with a higher energy than its dissociation energy. This molecule absorbs light with a photon energy higher than $4.59\,eV$ ($\lambda < 270\,nm$), and the photodissociation reaction occurs as: $Zn(C_2H_5)_2 + 2.256\,eV \rightarrow ZnC_2H_5 + C_2H_5$, and $ZnC_2H_5 + 0.9545\,eV \rightarrow Zn + C_2H_5$. Thus the second harmonic light (SH light) of an Ar^+ laser ($\lambda = 244\,nm$) and an ArF excimer laser ($\lambda = 193\,nm$) were used as the light source for photodissociation of DEZ gas. For the purpose of generating the UV optical near-field with sufficient power density to decompose DEZ gas, an UV fiber with transmission loss as low as $1.1\,dB/m$ at $244\,nm$ was used to fabricate a probe [11]. An UV probe was coated with 200 nm thick Al film after being tapered by chemical etching. The throughput of the probe was 1×10^{-4} and the power density at the probe tip with a sub-100 nm aperture for 1 mW of incident UV light was as high as $1 \, kW/cm^2$. Figure 14.1a, b shows two methods of NFO-CVD, i.e., direct gas phase photodissociation method and prenucleation method, respectively. The advantage of the direct gas phase photodissociation method Fig.14.1a lies in the possibility of selective deposition of various materials by changing parent gases, which is useful for lateral integration [12]. One disadvantage of this method is that the probe tip is also gradually covered with the depositing materials while it is fabricating a nano-structure on the substrate. In our experience, however, this only becomes a problem after a few hours of operation, while only a few seconds are necessary to fabricate a nano-structure. It is therefore not a serious problem. In order to examine the effect of optical near-field energy on the growth, an experiment was performed by varying the illumination time at a constant gas pressure of 1 mTorr and an input SH light power of 15 mW. The value of the optical near-field energy was evaluated by measured SH light power × illumination time. It was confirmed that the dot size depends on the spatial distribution of the optical near-field in a direction lateral to the substrate surface, while depending on the optical near-field energy in the normal direction. One of the most attractive features of this technique is its high spatial resolution. The lateral size of the fabricated pattern depends on the spatial distribution of the optical near-field energy and its reproducibility also depends on the reproducibility of fabricating probes. Figure 14.2 shows the shear-force image of dots. Two dots with a diameter of 52 nm and 37 nm (FWHM of the cross-sectional profile) were

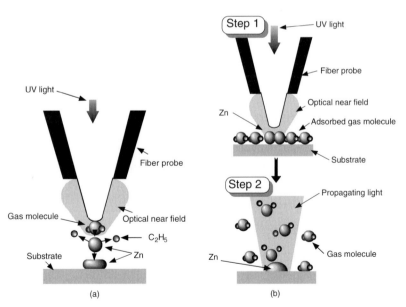

Fig. 14.1. Principles of NFO-CVD. (**a**) Gas phase direct photodissociation method. (**b**) Prenucleation method

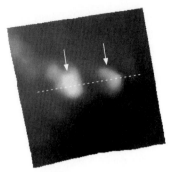

300nm x 300nm   **Fig. 14.2.** Shear-force image of Zn dots

fabricated at a very close distance of 45 nm. Since the measured diameter of the dot image includes the resolution of a vacuum shear-force microscope(VSFM) [13], the intrinsic diameter may be smaller than the value estimated from these figures.

Figure 14.3 shows the shear-force image of the loop-shaped Zn pattern on a glass substrate produced by the prenucleation method Fig. 14.1b [14]. The vacuum chamber was evacuated to below 10^{-5} Torr prior to the prenuclei fabrication stage, then filled with about 10 Torr of DEZ gas, maintaining the pressure for 20 min. Next, the chamber was re-evacuated to the pressure of 10^{-5} Torr, leaving a few adsorbed monolayer on the substrate surface. Prenucleation was performed by delivering the SH light on the substrate covered with adsorbed molecules using a probe. Nuclei of Zn were formed by the decomposition of DEZ gas adsorbed on the substrate with the optical-near field at the probe tip. In the growth stage after nuclei fabrication, DEZ gas was refilled in the chamber with a few Torr and the unfocused ArF excimer laser with the maximum energy of 10 mJ was directly irradiated on the prenucleated

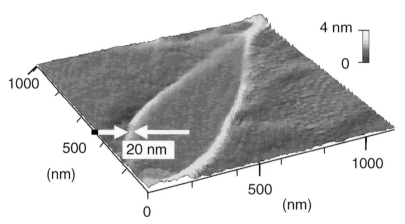

Fig. 14.3. Shear-force image of a loop-shaped Zn deposited on a glass substrate by the prenucleation method

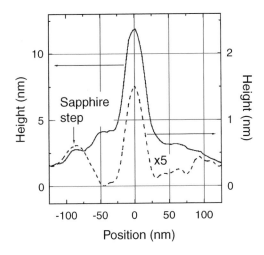

Fig. 14.4. Cross sectional profiles of Zn dots deposited on a sapphire step substrate deposited by the light of $\lambda = 244$ nm (*solid curve*) and $\lambda = 488$ nm (*broken curve*)

substrate. Then growth proceeded only on the pre-existing nuclei. As this figure shows, the minimum width of the pattern is as small as 20 nm. The width achieved here is two orders smaller than the minimum width reported so far by conventional PE-CVD using a far-field light [15]. Since the measured width includes the resolution of VSFM, the intrinsic width can be smaller than the value estimated from this figure. This method has the advantage of being free from the deposition at the probe tip, but otherwise has the drawback that the lateral integration of various materials is not easy due to using the propagating light in the second step. The optical near-field has unexpected unique properties, which was examined for the first time by NFO-CVD of nanometer-sized Zn dots with nonresonant light [16]. In order to investigate the deposition effect by nonresonant light, the fiber probe without metal coating on it, i.e., a bare fiber probe, was used for deposition. Therefore, the far-field light leaked to circumference of the fiber probe. The solid curve in Fig. 14.4 shows the cross-sectional profile of a Zn dot on the sapphire step substrate deposited by using the 244 nm-wavelength light, i.e., the light resonant to the absorption spectrum. It has tails of 4 nm height on both side of the dot. The tails correspond to the deposition by the leaked far-field light. The broken curve in this figure shows the profile of the dot deposited by using the light of $\lambda = 488$ nm, i.e., the nonresonant light. This curve has no tails since, in the case of the far-field PE-CVD, Zn deposition using the nonresonant light is not possible [17]. We discuss the possible mechanism of DEZ dissociation and deposition by the nonresonant optical near-field. In the case of using the far-field light, the dissociation of DEZ is induced by photo-absorption and predissociation. In contrast, in the case of using the optical near-field, the dissociation can take place even under the nonresonant condition. Its first reason is two-photon absorption process due to the high energy density of the optical near-field at the apex of the high throughput fiber probe. The second is the induced transition to the dissociation channel by the apex of the fiber. The third involved the direct coupling between the optical near-field and the dissociation molecular vibration mode.

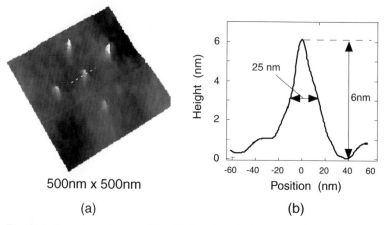

500nm x 500nm

(a)

Position (nm)

(b)

Fig. 14.5. Shear-force image of five Al dots deposited on a sapphire substrate

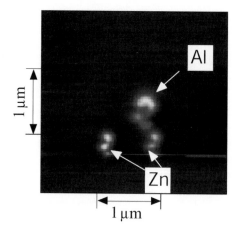

Fig. 14.6. Zn and Al dots deposited closely on a sapphire substrate

This nonresonant deposition method has high potential for the application to the NFO-CVD of gas sources for which far-field PE-CVD is unrealizable. NFO-CVD also allows us to fabricate nanostructures of several metals (Al, Cr, W, etc.) as well as Zn. For example, Fig. 14.5 shows shear-force image of five Al dots deposited on a sapphire substrate by dissociating the trimethylalminium (TMA) using the light of $\lambda = 244$ nm [18]. The TMA partial pressure and the light power incident into the fiber probe were 10 mTorr and 1 mW, respectively. The FWHM of the dots are typically 25 nm, which is comparable to the apex diameter of the used probe. As demonstrated by the deposition of Zn and Al, various materials can be deposited selectively by only changing the reactant gases. Furthermore, there is no limitation in regard to substrate and deposited materials. Zn and Al were deposited on an insulator substrate like sapphire. These are the advantage of this technique. Another of the most attractive points of this technique is the possibility for in-situ lateral integration of nano-scale structures of different materials with ease, which is difficult in conventional technique. In order to demonstrate the possibility, Zn dots were deposited at the first step, under the DEZ pressure of 10 mTorr. As the second step, Al dots were deposited under TMA pressure of 10 mTorr after evacuating the deposition chamber to 5×10^{-6} Torr. Figure 14.6 shows the result, i.e., shear-force image of deposited Zn and Al dots at closely position on a sapphire substrate. The distance between the Zn and Al dots was as close as 100 nm.

14.2.2   Depositing Zinc Oxide

One of the advantages of this NFO-CVD is the fact that there is no limitation in regard to substrate and deposited materials. That is, nanostructure of oxides, insulators, and semiconductors are deposited as well as metals. As an example, let us demonstrate the nanofabrication by NFO-CVD of ZnO on a sapphire substrate as insulator [19]. As a preliminary experiment, ZnO films were deposited by the PE-CVD method by using a propagating far-field light. Here, we used the reaction between oxygen and DEZ conveyed by a carrier gas (Ar) into the chamber during the irradiation by the propagating SH light ($\lambda = 244$ nm) of an Ar^+ laser. A (0001) sapphire was used as a substrate for the epitaxial growth of ZnO. The reaction chamber was filled with the reactant gases at the ratio of DEZ : O of 1 : 10 at a working pressure of 10 mTorr. The chamber pressure was maintained at 10 mTorr during the growth. In order to find the optimal growth conditions, the crystallinity, stoichiometry, optical transmission, and photoluminescence were evaluated. The PE-CVD of ZnO was carried out for 10 min within a range of the substrate temperature from room temperature to 300 °C. The energy density of the laser source and the spot size were 10 mW and 600 μm, respectively. Crystalline films were deposited at substrate temperatures over 100 °C and the films with c-axis oriented crystalline. For films deposited at a substrate temperature above 150 °C, the atomic ratio of Zn : O

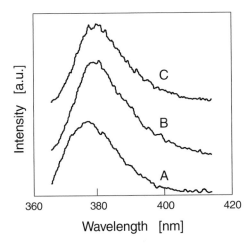

Fig. 14.7. Photoluminescence spectra of ZnO films deposited at the substrate temperature of 150 °C (A), 200 °C (B), and 300 °C (C). They were measured at room temperature

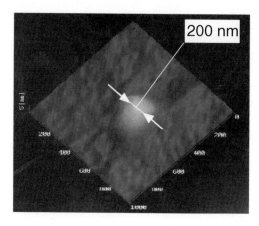

Fig. 14.8. Shear-force image of a ZnO dot deposited on the (0001) sapphire substrate by NFO-CVD

was 1.00 : 1.00 within an accuracy of a few percent. The optical properties were also investigated. Transmission fell off steeply at around 380 nm, a characteristic of high-quality ZnO film. From the plot of the wavelength vs. the absorption coefficient, optical band gap energies ranging from 3.26 to 3.31 eV were estimated, which is identical to the value recorded for high-quality ZnO films [20]. The photoluminescence spectra were measured using the 325 nm line of a CW He-Cd laser. Figure 14.7 shows the emission spectra measured at room temperature from films deposited at the substrate temperature from 150 °C to 300 °C. The emission peak position is coincident with the expected energy of the free exciton, and a strong free exciton emission at 380 nm can be clearly observed even at room temperature [21,22]. This confirms that if we use PE-CVD method at a low temperature, a ZnO film emitting UV light at room temperature can be fabricated. This was the first observation of room temperature UV emission from ZnO films deposited by PE-CVD. Under the growth conditions studied by the preliminary experiments mentioned above, the ZnO nanostructure fabrication was carried out by NFO-CVD. The fabrication was performed by introducing the SH light of an Ar^+ laser onto the sapphire substrate surface through the fiber probe. Figure 14.8 shows the VSFM image of fabricated nanometric scale ZnO. The dot was of a 200 nm diameter and a 5 nm height. The diameter is smaller than the wavelength of the irradiating light source. However, because this value includes broadening due to the resolution of VSFM, the real diameter should be much smaller than that observed.

14.3   Nanophotonic Devices and Integration

Future optical transmission systems require an advanced photonic integrated circuit (IC) for increasing speed and capacity. To meet this requirement, its size should become much smaller than that of a conventional diffraction-limited photonic IC. The concept of such a nano-photonic IC is shown in Fig. 14.9, where metallic wires, light emitters, optical switches, input/output terminals, and photo-detectors are all

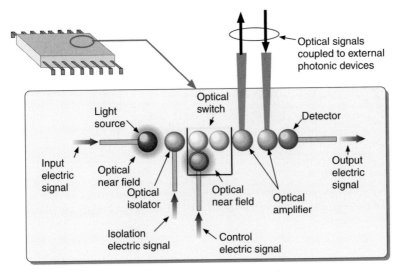

Fig. 14.9. Concept of a planar nano-photonic integrated circuit

controlled by nano-scale single dots and lines [6]. These devices use the optical near-field as a carrier for signal transmission.

As has been demonstrated in the previous sections, the NFO-CVD constitutes a very promising tool for in-situ patterning of nanostructures for this integration because this technique exhibits extraordinarily high controllability and reproducibility in fabricating nanostructures at desired position. This has never been demonstrated by any other conventional self-organized growth technique for semiconductor nanostructures. What is excellent is that as it is based on a photodissociation reaction, selective growth of various materials, i.e., metals, insulators, and semiconductors, can be accomplished by the choice of light source. It allows us to realize a nano-photonic IC composed of nanostructures. This section reviews our recent works on nano-switching devices, which is a key device for nano-photonic integration.

14.3.1 Switching by Nonlinear Absorption in a Single Quantum Dot

In order to demonstrate, e.g., a nonlinear optical switching capability of a single quantum dot (QD), we measured the nonlinear absorption change in a self-assembled single InGaAs QD grown on a (100) GaAs substrate by gas-source molecular beam epitaxy [23]. The average QD diameter was 30 nm and the height was 15 nm. The QD density was about 2×10^{10} dots/cm^2. These QDs were covered with cap layers with a total thickness of 180 nm. As schematically explained in Fig. 14.10a, a probe light (λ = 900–980 nm) was introduced into the back of the sample, and the transmitted light was collected and detected by a high throughput fiber probe with a double tapered structure [24] placed in the vicinity of the sample surface. When the sample was illuminated by a pump light(λ = 635 nm) passing

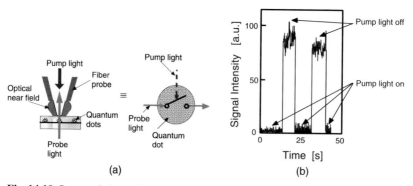

Fig. 14.10. Saturated absorption spectroscopy of a semiconductor quantum dot. (**a**) Schematic explanation of the principle. (**b**) An experimental result of optical switching properties of a single InGaAs quantum dot

through the fiber probe, carriers were generated in the barrier layer, and flew into the QDs. The ground states of the QDs were occupied by the carriers. The resultant reduction of the absorption of the QDs was measured by detecting the transmission change in the probe light. Experiments were carried out in a liquid helium cryostat, where a double modulation/demodulation technique was used for detecting very weak signals. Figure 14.10b shows an experimental result demonstrating a deep modulation of the transmitted probe light power due to irradiation of the pump light. This result confirms that a single QD works like an optical switch, and moreover, the switching operation can be detected by a conventional optical signal detection technique, which is advantageous for application to nano-photonic IC.

14.3.2 Switching by Optical Near-Field Interaction Between Quantum Dots

We have proposed another novel approach towards a nano-photonic switch for both elementary and functional photonic devices [25]. The building blocks of the proposed device consist of three nanometric QDs as illustrated in Fig. 14.11. The discrete energy levels of each dot are described as

$$E_n = E_B + (h^2/8Ma^2)(n_x^2 + n_y^2 + n_z^2), \quad (n_x, n_y, n_z = 1, 2, 3, \cdots), \tag{14.1}$$

where the mass and size are denoted as M and a, respectively. When the sizes of dots 1, 2, 3 are respectively chosen as $a/2$, $a/\sqrt{2}$, and a, two neighboring dots with the distance r_{ij} ($i, j = 1, 2, 3$) have the same excited energies denoted as E_1, E_2, and E_3 in Fig. 14.11. The energy level E_3 of dot 1, e.g., specified by $(n_x, n_y, n_z) = (1, 1, 1)$, is resonant with states specified by both $(2, 1, 1)$ for dot 2 and $(2, 2, 2)$ for dot 3. Dot 1 is coupled to the input light, which is transmitted by the inter-dot Yukawa interaction [26] to dot 2 that is connected to the output light. Dot 3 is coupled to the control light that governs the switching mechanism. When the control light is on, optical near-field interaction of dot 3 with the other dots are forbidden, since the level E_1

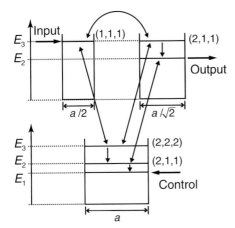

Fig. 14.11. Three-dot configuration as fundamental block of the proposed device. Optical near-field is abbreviated by ONF

of dot 3 is occupied and the intra-dot relaxation time is much faster than the inter-dot transfer time. In contrast, when the control light is off, those interactions are allowed, which results in the population difference in dot 2 producing a difference in the transmission signal. It should be noted that the fast intra-dot relaxation time guarantees a single direction for the signal transmission, and that the frequency conversion from input to output avoids irrelevant cross-talk.

In order to analyze dynamic properties of the device, we solved the following master equations for the population $P_n^j(t)$ of dot j to remain the excited states E_n at time t. We adopted CuCl dots for the following case study though any arbitrary material system is, in principle, suitable. These QDs have discrete energy levels due to quantum size effects, as a result of exciton confinement [27]. The energy levels are described as in (14.1), which means the energy structure reflecting each size of the QDs. Figure 14.12 shows an example of the time evolution of $P_{n=2}^{j=2}(t)$, where the largest dot size is assumed to be 10 nm and the distance between neighboring dots is 10 nm. The intra-dot relaxation time is set as 1 ps. The curve with circles represents the ON condition while the curve with triangles shows the OFF condition.

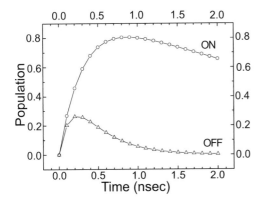

Fig. 14.12. Population of dot 2 as a function of time when control light is ON and OFF

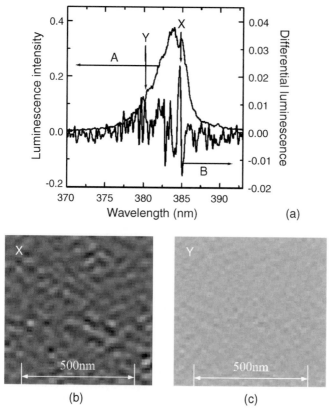

(a)

(b)   (c)

Fig. 14.13. Spectral and spatial information of a sample. (**a**) Curves A and B show the far-field and near-field photoluminescence spectra, respectively, of CuCl quantum dots embedded in NaCl matrix at 15 K. (**b**) and (**c**) Spatial distributions of the photoluminescence peaks X and Y, respectively, marked in a. The spatial distributions for peaks X and Y correspond to the spatial distributions of the quantum dots with sizes of 4.3 nm and 3.4 nm, respectively

We estimate from this figure that an optical near-field nano-switch can be operated within a few hundred picoseconds, and expect that it would, in principle, be applicable to other material systems like ZnO and GaAs. In order to experimentally verify the switching mechanism and operation proposed, we need to overcome several kind of difficulties. The first step is to identify a specific QD with a desirable size or energy, after observing the spatial and spectral distribution of QDs. Then it is necessary to verify the excitation energy transfer between two QDs, i.e. to detect the desirable frequency conversion from ω_3 to ω_2. Finally we have to show the switching operation of the proposed device. Here we review experimental results on the first step, using CuCl QDs embedded in NaCl matrix. The spatial and spectral characteristics of the sample were investigated by using a near-field optical spectrometer [28]. The curves A and B of Figure 14.13a show the far-field and near-field photoluminescence spectra, respectively, of a sample at 15 K that was excited by a He-Cd laser($\lambda = 325$ nm). The far-field spectrum is inhomogeneously broadened by the size distribution of QDs while the near-field spectrum has the very fine structure that ideally corresponds to a vertical scan at a specific spatial position of the sample.

Spatial distributions of photoluminescence with energies X and Y are shown in Figs. 14.13b and c, respectively. These values of X and Y are respectively equivalent to the dot sizes of 4.8 nm and 3.4 nm, which correspond to a pair of dots 2 and 1, or dots 3 and 2. The spatial resolution of the images is close to the aperture diameter of the fiber probe used. These results indicate that we have established how to identify QDs selected both spatially and spectrally.

14.4   Optical Storage and Readout by Optical Near-Field

The use of an optical near-field for realization of a high-density optical storage system has attracted a great deal of attention. This section propose and demonstrate a new contact slider with a high throughput ratio of near-field intensity for realization high recording density and fast readout to the phase-change medium [29]. Schematics of the slider structure and the data storage system are illustrated in Fig. 14.14. A pyramidal silicon probe array is arranged on the rear pad of the slider. The advantages of such a slider are as follows: (1) The high refractive index of the silicon ($n = 3.67$ at $\lambda = 830$ nm) leads to a short effective wavelength inside the probe, which results in higher throughput and smaller spot size than those of conventional fiber probes made of silica glass [30]. (2) The height of the pyramidal silicon probe array is fabricated to be less than 10 μm so that sufficiently low propagation loss in the silicon is maintained. Furthermore, the probe array has high durability because it

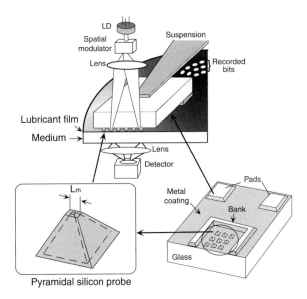

Fig. 14.14. Schematic of the data storage/readout system with a contact slider and a pyramidal silicon probe array (L_{m}: mesa length)

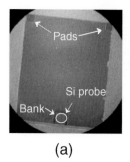
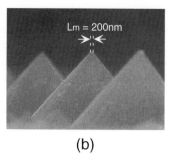

(a)   (b)

Fig. 14.15. A contact slider. (a) An optical image. (b) A SEM image of the pyramidal silicon probe array

is bonded to a thick glass substrate. (3) Compared with those of previously reported pyramidal probes fabricated by use of a focused ion beam [31] or by the transfer mold technique in a pyramidal silicon groove [32], ultrahigh homogeneity in the heights of the probes and pads can be obtained, since the flatness of the probe tip and the upper surface of the pads are determined by the uniformity of the thickness of silicon wafer. (4) Use of a probe array with many elements increases the total data transmission rate by parallel readout [32,33]. In this system the incident light is spatially modulated by an electro-optics method, and the scattered light from a different probe can be read out as a time-sequential signal. Since the key issue in realizing a pyramidal silicon probe array is high homogeneity in the heights of the probes, the probe array is fabricated from a (100)-oriented silicon-on-insulator (SOI) wafer. Figure 14.15a, b show an optical image of the contact slider and a scanning electron microscopic image of the pyramidal silicon probe array fabricated on the slider, respectively. The height dispersions of the probes and pads should be less than 10 nm, because these dispersions are determined by the uniformity of thickness of the SOI wafer. Here the slider is designed by use of the design criteria for a contact-type hard-disk head so that its jumping height over the phase-change medium is maintained at less than 10 nm. Furthermore, since the phase-change medium is fragile, we designed the bank so that the contact stress becomes 100 times weaker than the yield stress of the magnetic disk at a constant linear velocity (CLV) of 0.3 m/s, corresponding to a data transmission rate of 10 MHz for a data density of 1 Tbit/in². To increase the readout speed 100 times, i.e., to realize a 1 Gbit/s data transmission rate for data density of 1 Tbit/in², we fabricated 100 probe elements on the inner part of the bank for parallel readout.

In recording and readout experiments with the fabricated contact slider, we compared signals transmitted through phase-change marks recorded with a single element of the probe array and focused propagating light. The experimental setup is shown in Fig. 14.16. The contact slider was glued to a suspension. The slider was in contact with a phase-change medium coated with a thin lubricant film (Fomblin Z-DOL). A laser beam ($\lambda = 830$ nm) was focused on one element of the probe array on the slider, where the frequency of the rectangularly modulated signal with 50% duty was changed from 0.16 to 2.0 MHz at a CLV of 0.43 m/s. Then the light transmitted through the recording medium was detected with an APD. We used an

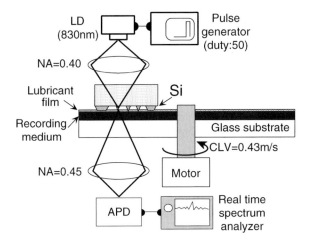

Fig. 14.16. Experimental setup for phase-change storage/readout by the contact slider

as-deposited AgInSbTe film as a recording medium. The optical recording powers for a pyramidal silicon probe with a mesa length L_m of 150 nm (see Figs. 14.1 and 14.17a) and a focused propagating light with an object lens (NA= 0.4) were 200 mW and 15 mW, respectively, which in both cases is the lowest recording power of which we are aware. The optical throughput of the pyramidal silicon probe with a 30 nm thick aluminum coating is 7.5×10^{-2}, which is estimated from the ratio of the optical powers for near- and far-field recordings. Readout was carried out at a CLV of 0.43 m/s, and the constant reading optical powers for the pyramidal silicon

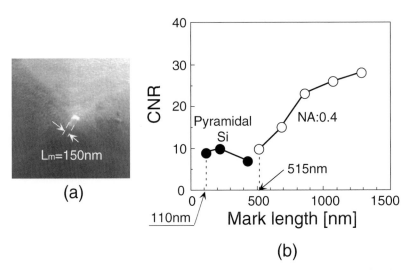

Fig. 14.17. Experimental results of the optical storage and read out. (**a**) Magnified SEM image of the pyramidal silicon probe tip used for the experiment. (**b**) Dependence of the CNR on mark length for the optical near-field (*closed circles*) and propagating light focused by an objective lens (*open circles*)

probe and the focused propagating light were 20 mW and 3.6 mW, respectively. The resolution bandwidth was fixed at 30 kHz.

The dependence of the carrier-to-noise ratio (CNR) on mark length is shown in Fig. 14.17b. In this figure one can see that shorter crystalline phase-changed marks beyond the diffraction limit were recorded and read out by an optical near field generated on the pyramidal silicon probe. The shortest mark length was 110 nm at a CLV of 0.43 m/s, corresponding to a data transmission rate of 2.0 MHz. This is, to our knowledge, the first phase-change recording-reading with a contact slider. Since this slider has 100 elements in the probe array, we expect a 100-fold increase in the data transmission rate by parallel readout. Furthermore, a higher CLV can be expected, since we did not observe any damage on the probe tip or the recording medium after a series of experiments. The constant CNR of the pyramidal silicon probe seen in Fig. 14.17b might be due to the small spot size for recording-reading and the narrow recorded mark width, which are as small as L_m of the pyramidal silicon probe. These results indicate that an increased CNR and decreases mark length will be achieved by means of tracking during readout. Furthermore, it is expected that the recording density can be increased to as high as 1 Tbit/in^2 by optimization of the interference characteristics of the guided modes in the pyramidal silicon probe [34].

14.5 Conclusion

This paper reviewed the recent progress of application of interactions between optical near-fields and nanoscale materials. Photochemical vapor deposition of nanometeric Zn and Al were realized by using an UV optical near-field. Deposition of nanoscale ZnO was also shown. Optical near-field technology offers the opportunity of modifying surfaces and developing new nanostructures that may exhibit a quantum effect due to their extremely small size. Utilizing the very advanced potential of this technology, the concept of nano-photonic IC was proposed. The optical switching operation of a single InGaAs quantum dot was shown to be able to be used for nano-photonic devices. Nano-photonic switching operation utilizing optical near-field interaction was also proposed and related spectroscopy of CuCl quantum dots were demonstrated. High density storage and read-out by optical near-field was also demonstrated. By combining the technique reviewed here with atom manipulation by the optical near-field [5,35], further progress in depositing novel materials and operating more advanced photonic IC can be expected.

Acknowledgements

The author would like to thank Prof. H. Ito, Drs. M. Kourogi, Y. Yamamoto, H. Fukuda (Tokyo Inst. Tech.), Drs. K. Kobayashi, S. Sangu, T. Yatsui, T. Kawazoe, H. Aiyer, V. Polonski, G.H. Lee (ERATO), Drs. T. Saiki and S. Mononobe (KAST) for their collaborations and valuable discussions.

References

1. E.A. Synge: Philos. Mag. **6**, 356 (1928)
2. M. Ohtsu: 'Overview', In: *Near-Field Optics: Principles and Applications*, ed. by X. Zhu, M. Ohtsu (World Scientific, Singapore 2000) pp. 1–8
3. Uma Maheswari, S. Mononobe, K. Yoshida, M. Yoshimoto, M. Ohtsu: Jap. J. Appl. Phys. **38**, 6713 (1999)
4. H. Ito, T. Nakata, K. Sakaki, M. Ohtsu, K.I. Lee, W. Jhe: Phys. Rev. Lett. **76**, 4500 (1996)
5. M. Ohtsu, K. Kobayashi, H. Ito, G.H. Lee: Proc. IEEE **88**, 1499 (2000)
6. M. Ohtsu: P. Soc. Photo–Opt. Inst. **3749**, 478 (1999)
7. E. Maayan, O. Kreinin, G. Bahir, J. Salzman, A. Eyal, R. Beserman: J. Cryst. Growth **135**, 23 (1994)
8. D. Bauerle: Appl. Surf. Sci. **106**, 1 (1996)
9. M. Ohtsu: *Near-Field Nano/Atom Optics and Technology* (Springer-Verlag, Berlin 1998)
10. M. Ohtsu, H. Hori: *Near-Field Nano-Optics* (Kluwer Academic/Plenum Publishers, New York 1999)
11. S. Mononobe, T. Saiki, T. Suzuki, S. Koshihara, M. Ohtsu: Opt. Commun. **146**, 45 (1998)
12. Y. Yamamoto, M. Kourogi, M. Ohtsu, V. Polonski, G.H. Lee: Appl. Phys. Lett. **76**, 2173 (2000)
13. V.V. Polonski, Y. Yamamoto, J.D. White, M. Kourogi, M. Ohtsu: Jap. J. Appl. Phys. **38**, L826 (1999)
14. V.V. Polonski, Y. Yamamoto, M. Kourogi, H. Fukuda, M. Ohtsu: J. Microsc. – Oxford **194**, 545 (1999)
15. D. Ehrlich, R.M. Osgood Jr., T.F. Deutch: J. Vac. Sci. Technol. **21**, 23 (1982)
16. T. Kawazoe, Y. Yamamoto, M. Ohtsu: Appl. Phys. Lett. **79**, 1184 (2001)
17. M. Shimizu, H. Kamei, M. Tanizawa, T. Shiosaki, A. Kawabata: J. Cryst. Growth **89**, 365 (1988)
18. Y. Yamamoto, T. Kawazoe, G.H. Lee, T. Shimizu, M. Kourogi, M. Ohtsu: 'In-situ Lateral Fabrication of Zinc and Aluminum Nanodots by Near Field Optical Chemical Vapor Deposition', In: *Tech. Digest of the Photonic Switching*, (IEEE LEOS, Piscataway 2001) pp. I520–I521
19. G.H. Lee, Y. Yamamoto, M. Kourogi, M. Ohtsu: Thin Solid Films **386**, 117 (2001)
20. T.Y. Ma, G.C. Park, K.W. Kim: Jap. J. Appl. Phys. **35**, 6208 (1996)
21. Z.T. Tang, G.K.L. Wong, P. Yu: Appl. Phys. Lett. **72**, 3270 (1998)
22. D.B. Bagnall, Y.F. Chen, Z. Zhu, T. Yao: Appl. Phys. Lett. **70**, 2230 (1997)
23. T. Matsumoto, M. Ohtsu, K. Matsuda, T. Saiki, H. Saito, K. Nishi: Appl. Phys. Lett. **75**, 3246 (1999)
24. T. Saiki, S. Mononobe, M. Ohtsu, N. Saito, J. Kusano: Appl. Phys. Lett. **68**, 2612 (1996)
25. K. Kobayashi, T. Kawazoe, S. Sangu, J. Lim, M. Ohtsu: 'Theoretical and Experimental Study on a Near-field Optical Nano-switch', In: *Photonics in Switching, OSA Tech. Digest*, (OSA LEOS, Washington, DC 2001) p. 27
26. K. Kobayashi, S. Sangu, H. Ito, M. Ohtsu: Phys. Rev. A **63**, 013806 (2001)
27. N. Sakakura, Y. Masumoto: Phys. Rev. B **56**, 4051 (1997)
28. T. Kawazoe, K. Kobayashi, J. Lim, Y. Narita, M. Ohtsu: 'Verification of Principle for Nano-meter Size Optical Near-field Switch by Using CuCl Quantum Cubes', In: *Tech. Digest of the Photonic Switching*, (IEEE LEOS, Piscataway 2001) pp. I194–I195
29. T. Yastui, M. Kourogi, K. Tsutsui, M. Ohtsu, J. Takahashi: Opt. Lett. **25**, 1279 (2000)
30. H.U. Dangebrink, T. Dziomba, T. Sulzbach, O. Ohlsson, C. Lehrer, L. Frey: J. Microsc. – Oxford **194**, 335 (1999)

31. F. Isshiki, K. Ito, S. Hosaka: Appl. Phys. Lett. **76**, 804 (2000)
32. Y.J. Kim, K. Kurihara, K. Suzuki, M. Nomura, S. Mitsugi, M. Chiba, K. Goto: Jap. J. Appl. Phys. **39**, 1538 (2000)
33. T. Yatsui, M. Kourogi, K. Tsutsui, J. Takahashi, M. Ohtsu: P. Soc. Photo–Opt. Inst. **3791**, 76 (1999)
34. M. Kourogi, T. Yatsui, M. Ohtsu: P. Soc. Photo–Opt. Inst. **3791**, 68 (1999)
35. H. Ito, K. Sakaki, M. Ohtsu, W. Jhe: Appl. Phys. Lett. **70**, 2496 (1997)

15 Optical Trapping of Small Particles

Alexander Rohrbach, Jan Huisken, Ernst H.K. Stelzer

15.1 Introduction

That light carries momentum and that it can exert forces when being reflected or refracted had already been postulated by Kepler in the 17th century. He tried to explain the phenomenon that comet tails always point away from the sun. Later, Sir William Crookes suggested that light pressure was responsible for the rotation of a light mill. The light mill consists of four vanes, which are blackened on one side and silvered on the other. The vanes are attached to the arms of a rotor, which is balanced on a vertical support and turns with very little friction. The light mill can be encased in a glass bulb with a high, but not perfect, vacuum. When light shines on the apparatus, the vanes rotate. Crookes was wrong, however, to attribute the rotation to the transfer of the photons' momentum: in fact it is a heating effect [1]. Although both Kepler and Crookes were mistaken, they were nevertheless among the first to deal with the phenomenon of light pressure.

In 1873 James Clerk Maxwell provided a quantitative description of optical forces using his electromagnetic theory. Later Lebedev was able to verify these predictions experimentally [2]. However, the calculated forces were so small, that there seemed to be no application for optical pressure. John H. Poynting said in 1905: "A very short experience in attempting to measure these light forces is sufficient to make one realize their extreme minuteness – a minuteness which appears to put them beyond consideration in terrestrial affairs." Only in astronomy, in extraterrestrial space where light intensities and distances are huge and the effects of gravity are small, could one imagine any significant influence of optical forces on the movement of small particles.

The invention of the laser changed the scientific world in many ways, also in the field of optical manipulation. The high intensities and high gradients in the electromagnetic field of a laser beam suddenly made it possible to generate forces large enough to accelerate and to trap particles. With lasers, particles as small as atoms and as large as whole organisms have been trapped and manipulated in various media.

Arthur Ashkin, considered by many to be the pioneer of optical trapping and manipulation, performed his first experiments in the beginning of the 1970s, studying the optical forces on small glass beads. Ashkin's first publication on optical forces described the acceleration of beads in water by a horizontal laser beam [3]. He also constructed a simple trap using two focused, counter-propagating beams: particles were trapped in between two foci. When the beam was directed upwards, he caused the particles to float in air [4]. This feat, termed optical levitation, was later also performed in vacuum [5] and with various particles. An overview of the historical development of optical traps can be found in reviews by A. Ashkin (see [6,7]).

The developments of Ashkin culminated in the invention of the single-beam trap [8]. Whereas in optical levitation gravity is necessary to confine particles, this is not the case for a single-beam trap. In the single-beam trap, strong gradients in a highly focused beam exert a gradient force that counters the scattering force, creating a three-dimensional trap with a single beam. Because of the need for a high numerical aperture lens to strongly focus the laser beam, it was straightforward to realize an optical trap in a microscope: a laser beam coupled into a conventional microscope generates an optical trap in the focus of the objective. Later called optical tweezers, the single-beam trap is now a highly developed and heavily applied tool in biology, physics and chemistry. Optical tweezers have become the tool of choice for many applications in which a gentle and remotely controllable manipulation is needed.

The following sections will cover optical levitation and optical trapping, including photon momentum transfer and the arising trapping forces. The way optical tweezers function is outlined as well as multiple applications in biophysics and atom physics. An advancement of optical tweezers, the Photonic Force Microscope (PFM) will be introduced as an example of a high-resolution scanning probe microscope. The theory section sketches highly focused fields, particle scattering, the position detection technique, optical forces, and thermal noise analysis. This is followed by a description of the experimental setup of a Photonic Force Microscope. The last section describes two typical applications of the PFM in biophysics pointing out future directions in optical trapping of small particles.

15.2   Optical Levitation

As mentioned, the history of optical manipulation of microscopic particles has mainly been dominated by two techniques: optical levitation and optical tweezers. Both techniques use optical forces to confine particles spatially. However, optical levitation is generally used for the lateral confinement and axial transport of particles. Either gravity is counteracting the axial forces, the particle is pushed against a surface or it is simply pushed from one location to another. Particles are often so large that in addition to the optical forces (describable with ray optics) radiometric forces must be taken into account to explain the observed levitation phenomena [9]. In this section, we start by studying momentum transfer to understand the processes involved in optical levitation. We then describe an experimental setup for realizing levitation and its applications.

15.2.1   Momentum Transfer

The phenomenon of optical trapping is based on the transfer of momentum. Photons carry a momentum, which is changed when the photons are deflected. Any reflection or refraction or, more generally speaking, any deflection of a beam of light leads to a change in the photons' momentum. Since the total momentum in a closed system is conserved, the difference in momentum before and after the scattering event is transferred to the particle. The same holds true for the photon's energy and the total energy, which is also conserved. The particle experiences a force, which pushes it along the direction of the difference vector of the total photons' momentum before and after the scattering event. This can be expressed in the pressure of light p, which in the case of a fully absorbing surface is $p = I/c$, with I being the intensity of the incoming light and c the speed of light in the medium. If the ray is not fully absorbed, but rather partially reflected and refracted, one can regard this as an absorption followed by reemission in the directions of the reflected and refracted rays.

To levitate and to confine a particle in a defined volume, certain conditions concerning the particle's and the beam's characteristics must be satisfied. Gradients in the beam, as well as the shape and optical properties of the particle determine the particle's behavior in the beam. We assume a collimated beam with a lateral Gaussian intensity profile incident on a spherical particle with a higher refractive index than the surrounding medium. These assumptions are often made, primarily because most lasers operate in a TEM_{00} mode and the most common commercially available microparticles are spherical. Secondly, the formalism is particularly simple but still a valid approximation for the basic principles of optical levitation. This is also due to the fact that the beam can be assumed to be collimated over a distance comparable to the particle's diameter. In more thorough calculations, one has to take the Gaussian nature of the beam and its divergence into account.

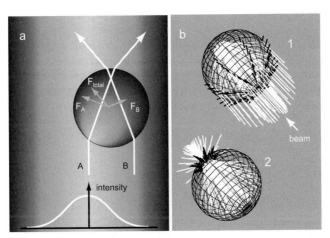

Fig. 15.1. (a) Bead levitated in a Gaussian laser beam. Two rays (*A* and *B*) are shown as they traverse through the bead. The resulting force pushes the particle onto the axis and along the beam. (b) To calculate the total force one has to integrate over all the rays that are incident on the bead. Shown is a bundle of rays as it (*1*) enters and (*2*) exits the bead. The resulting forces on each of the points of intersection are shown in black

In Fig. 15.1 (a) a transparent bead is shown in a laser beam with a lateral intensity gradient. The beam propagates along the vertical axis from the bottom to the top in the picture. The bead sits slightly off-axis. The two rays (A and B), which are shown as examples for all the rays hitting the sphere, are refracted as they traverse the bead. The resulting forces F_A and F_B are illustrated in the figure with F_A being larger than the counterpart for ray B because the intensity is higher in the center of the beam than off-axis. The resulting total force F_{total} has a component pushing the bead forward (also termed the scattering force) and a component pushing it onto the axis (the gradient force). The latter vanishes as soon as the bead reaches the on-axis position. In a detailed calculation one integrates over the bead's surface to obtain the total force [8]. This is illustrated in Fig. 15.1 (b). A bundle of rays is shown as the rays traverse the particle. At each point of intersection (i.e., when the rays enter and when they exit the particle) a force and a torque are exerted on the particle. The calculation can be stopped after the second intersection because the subsequent forces do not contribute significantly. This ray optics approach is justified as long as the particle's diameter is much larger than the laser's wavelength. A theory for smaller particles is presented later in the study of optical tweezers.

If we now assume that the laser beam is pointing upwards (i.e., anti-parallel to the gravitational force) a bead placed somewhere in the beam will be drawn onto the axis, and will be either pushed upwards or slowly fall down depending on the laser power. Only if the scattering force matches the gravitational force exactly will the bead float or levitate. Since the laser power is hard to control so accurately that this condition is met, in practice the beam is often slightly focused. If the bead now enters an oversized beam somewhere above the focus, it will fall until the power incident on the bead is high enough to compensate gravity and will levitate the bead. It must be emphasized that focusing is not essential for the levitation: it only facilitates the experiment. This is contrary to the situation in single-beam traps, as explained in Sect. 15.3. In optical levitation the particle is only trapped laterally by the beam's gradient; axially, the gravitational force or the negative buoyancy enable an equilibrium with the optical forces.

15.2.2   Experimental Setup

The basic components of a levitation apparatus are a laser, a lens, and a chamber. A typical setup is shown in Fig. 15.2 (a). The beam of a laser (e.g., $\lambda = 488$ nm, power $P = 25$ mW) is expanded and then directed upwards with a mirror. A low NA lens ($f = 50$ mm) focuses the beam into a chamber, which is filled with air or a liquid. The particles to be levitated are now either dropped into the beam from the top or they are lifted from the bottom of the chamber. The latter solution requires piezos to shake the chamber and to overcome any adhesion forces between bead and surface. Once a bead is trapped it can be moved either by moving the lens or, more elegantly, by moving the beam and changing the laser power. To observe the particle and its behavior in the trap, a microscope is needed that has a long working distance to look into the chamber from the outside.

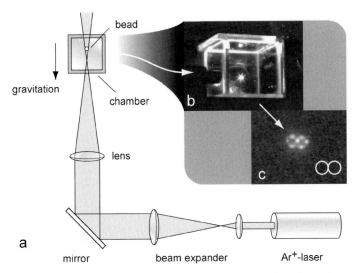

Fig. 15.2. (**a**) Optical setup of a levitation apparatus. A laser beam is expanded and directed upwards. It is focused in a chamber in which particles can be levitated. (**b**) Picture of a particle levitated in an air-filled glass chamber (2 cm cube). (**c**) Levitated cluster of two beads with diameters of 4 µm observed with a microscope objective. The beads can be identified by four glare spots surrounding each bead. For a better understanding of the picture, a schematic representation of the two beads is drawn in the lower-right corner

Once a particle is trapped it can be detected by its strong scattering. As seen in Fig. 15.2 (b) a cluster of two beads (diameter each 4 µm) is trapped inside an air-filled glass chamber (2 cm cube). The glass chamber protects the particle from convection that can easily move the bead out of the trap. Figure 15.2 (c) shows a picture of the cluster as seen with a microscope from the side. It was taken with a long working distance objective lens. The magnification is high enough to resolve the individual particles. Since the beads are transparent, only two glare spots of each illuminating lamp (on top and on the side of the setup) are seen for each bead; the laser light is blocked by a filter. In the experiment the cluster rotates around the vertical axis, which is the beam axis.

15.2.3   Applications

Optical levitation is ideal for holding a probe without mechanical contact in front of a microscope. Not only can the particle's dimensions and its behavior in the trap be studied, also any microscopy or spectroscopy technique can be easily applied to investigate the particle under observation.

The main advantage of levitation over optical tweezers is the long working distance, i.e. the distance by which particles can be moved. This is due to the fact that the trapping beam is weakly focused. Even if the term "optical levitation" is only justified as long as the beam is counteracting gravity or more general negative buoyancy, the theory still holds true if the beam is tilted or even horizontally aligned. This method of micromanipulation is sometimes termed "optical transportation technology" or "all-optical guiding". In principle, particles can be moved any distance along the axis as long as the lateral forces confine the particle on the beam axis and the laser power is high enough to push the particle along. Laterally, particles can be moved small distances by moving the beam and over long distances by moving the lens (and the mirror). This is exploited in several recent experiments.

Gauthier and co-workers have applied optical levitation as a delivery system to selectively transfer particles into a dual beam fiber trap, in which a particle is trapped in between the counter-propagating beams [10]. The spheres in a sample chamber can be individually selected and then delivered into the trap volume by centering the bead on the upward propagating levitation beam. This is an elegant way to deposit particles in the small trapping region in between two fibers (100 μm apart). The beams from an optical fiber are weakly focused by a microlens and can be used to push particles. By using an arrangement of fibers, particles can be trapped and moved by simply moving the fibers. Taguchi et al. exploited this in their experiments [11]. Two fibers were pointed down obliquely into a sample chamber onto a microsphere laid at the bottom. The bead could be levitated against gravity and could be transferred in 3D by exploiting the lateral gradient forces that counteracted axial and gravitational forces. Koyanaka and co-workers investigated the behavior of particles of various materials in a laser beam and evaluated its application for the transport and the separation of particles of different refractive index, shape or size [12]. The long-range movement of micron-sized particles was simulated and experimentally verified.

15.3   Optical Trapping

15.3.1   Principles

Photons are able to encode information by their wavelength, their density, their polarization, their propagation direction, or their relative phase to other photons (coherence). A never-ending list of applications in optics in general and especially in microscopy results from these properties of photons. However, the momentum p connected with the wavelength λ of a photon has been rarely exploited in microscopy, because it was considered as negligibly weak. The momentum, i.e. the force F that the photon exerts over a time Δt, is

$$p = \hbar k = h/\lambda = F\Delta t \,, \tag{15.1}$$

with k-number $k = 2\pi/\lambda$ and with Planck's constant $h = \hbar 2\pi = 6.626 \times 10^{-34}$ Nms. The momentum of a single photon is, for example, $p \approx 6.6 \times 10^{-28}$ Ns at a wavelength of $\lambda = 1$ μm in air. In other words, 1.5×10^{18} photons, corresponding to an optical power of $P = 0.3$ W, would exert a force of $F = 1$ nN (nanoNewton) provided the particle absorbs all photons. This force is significantly more than those forces that drive cell organelles in biological cells. For a dielectric particle, which does not absorb light, but scatters light in different directions, a momentum due to radiation pressure is transferred with pressure efficiency $Q_{pr} \ll 1$. However this results in an optical force of $F = Q_{pr}Pn/c$ (c is the speed of light), which is still in the range of several pN (picoNewton). For example, a small glass sphere with a diameter of $\lambda/2$ embedded in water (refractive index $n = 1.33$) has a pressure efficiency $Q_{pr} = 0.035$, and thus feels an optical force of $F = 35$ pN when all 1.5×10^{18} photons hit the sphere.

However, to impinge a maximum amount of photons onto a small particle, light must be focused to a small spot where the smallest diameter that can be achieved is $\Delta x \approx \lambda$. Especially for coherent laser light, strong intensity gradients arise around the point of maximum intensity. Assuming that a particle consists of a cluster of dipoles, these dipoles oscillate as a response to the incident electromagnetic field. As a consequence, the dipoles feel a Lorentz force towards the point of the highest intensity if the light frequency ν is below the dipole's resonance frequency ν_0, if $\nu > \nu_0$ the dipole is repulsed in the opposite direction. This optical force is called the gradient force F_{grad}, while the force that is due to momentum transfer is called the scattering force F_{scat}. The gradient force pulls a particle, i.e. the dipoles, into the center of the focus, while the scattering force pushes the particle away from the focus in the direction of light propagation. For this reason, lateral optical forces are much stronger than axial forces due to the strong lateral intensity gradients and the small momentum transfer in the lateral direction. These characteristics are outlined in Fig. 15.9 for a small latex sphere. The trapping position, an equilibrium point for all forces, occurs behind the geometric focus, with the property that any displacement of a particle from this point results in a restoring force.

At this point we must distinguish between different particle sizes located in highly focused beams. A scatterer with a diameter ($2a$) much smaller than the incident wavelength, i.e. $2a \ll \lambda$, is called a Rayleigh scatterer and experiences a gradient force that is much stronger than the scattering force. Rayleigh scatterers, such as atoms or small molecules are stably trapped in the center of the focus. Larger particles, which are smaller or equal to the incident wavelength, in addition "feel" a scattering force pushing the particle behind the geometric focus until an equilibrium of both forces is reached. With increasing particle diameter, the gradient force averages out because different dipoles inside the scatterer experience different forces in different directions. The $\lambda/2$ diameter glass sphere embedded in water experiences a maximum backward force of $F_{grad} + F_{scat} \approx 27$ pN at a transmitted laser power of $P = 0.3$ W, although a maximum force of $F_{scat} = 10$ pN pushes the sphere in the opposite direction out of the focus. For particles larger than both the incident wavelength and the dimensions of the focus, the gradient force averages out and becomes negligible, while momentum is transferred to the front and the back surface of the scatterer. These particles, often referred to as Mie-scatterers, can be treated with ray-optics. They experience approximately the same optical force independently of their diameter. Figure 15.3 illustrates the momentum transferred to a particle in the axial direction as the difference between the mean momentum vector before and

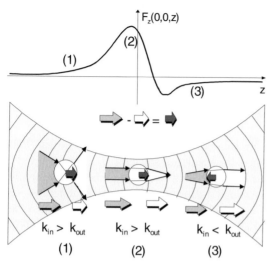

Fig. 15.3. Momentum transfer and optical forces along the optical axis. Three positions (*1*), (*2*), (*3*), of a spherical particle characterize the force profile $F_z(0,0,z)$. The momentum transferred to the particle is the difference between the incident mean momentum k_{in} and the outgoing mean momentum k_{out} of all photons. This is indicated by light gray and white k-vectors, respectively. The difference, indicated by a dark gray vector within the particle, is positive both before and in the geometrical focus, resulting in an optical force that pushes the particle in the direction of light propagation. If $k_{in} < k_{out}$, momentum is transferred in backward direction and the particle is pulled back towards the geometrical focus (negative force at position (*3*)). Stable trapping occurs behind the focus where the optical force is zero ($F_z(0,0,z) = 0$)

after scattering. It is shown, that in front of as well as in the geometric focus, the incident momentum is larger than the outgoing. Only at a certain distance behind the geometric focus, is the outgoing mean axial momentum larger than the incident.

15.3.2 Optical Tweezers

Optical tweezers exploit the principle that optical forces arising in highly focused beams can generate a stable three-dimensional optical trap. A particle captured in an optical trap can be moved to different positions in the three-dimensional space by moving the trap, or the particle can be held at a fixed position by maintaining the position of the trap. Particles undergo small position fluctuations inside the optical trap due to thermal noise. These thermal fluctuations become stronger as the size of the trapped particle or the laser power decrease. However, because of the relatively small position changes of the particle's Brownian motion, the optical trap can be switched on and off or can be switched to another position to manipulate another particle simultaneously. By time-multiplexing in the 100 Hz range, the untrapped particle is not able to leave the trapping area. The idea of moving particles

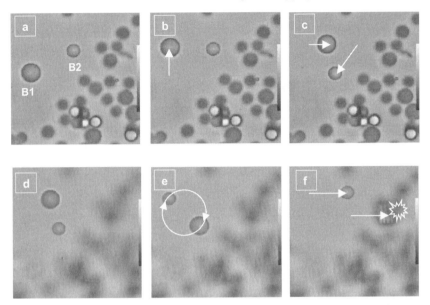

Fig. 15.4. Simultaneous three-dimensional trapping of two beads indicated by *B1* and *B2*. Beads adhering to the coverslip are in a different focal plane and are visible as dark spots. *White arrows* indicate the movements of the beads *B1* and *B2*. (**a**) Beads *B1* and *B2* are trapped by time-multiplexed optical tweezers. (**b, c**) Beads *B1* and *B2* are moved laterally one after the other. (**d**) Beads *B1* and *B2* are simultaneously moved further away from the coverslip. The beads adhering to the coverslip are out of focus. (**e**) Beads *B1* and *B2* are simultaneously rotated. (**f**) Beads *B1* and *B2* are simultaneously moved to the right until bead *B1* is pushed out of the trap by a bead attached to the coverslip

in three dimensions with optical tweezers is visualized in Fig. 15.4 by single frames extracted from a movie. Here two beads are simultaneously trapped and moved laterally and upwards.

The main advantages of optical tweezers are that they are able to exert small forces in the pN range and that manipulation of particles is realized without mechanical contact. For comparison, atomic force microscopes [13] use a fine tip that touches the sample and exerts or measures much higher forces in the μN range. The first demonstration of optical tweezers in biology was realized by Ashkin et al. in 1987 by trapping viruses and E. coli bacteria [14,15]. A large number of experiments with this manipulative technique providing new insights especially in cell biology followed. For example, Block et al. [16] measured the torsional compliance of bacterial flagella. The technique of tweezing was extended to laser cutting ("Laser scissors") of cells and organelles: Greulich and co-workers cut and manipulated pieces of chromosomes for gene isolation [17]; and Berns and co-workers brought cells into contact in order to effect cell fusion by cutting the common wall [18]. In fertility studies, three-dimensional manipulation of live sperm cells [19] and even insertion of selected sperms into eggs [20] were performed with optical tweezers.

Holding a particle in a trap means to prevent it from moving. A particle that tries to leave the optical trap exerts a force that can be measured. To do this, the trap must be calibrated, i.e. the optical force profile or the slopes of the force profiles (the force constants f_x, f_y, or f_z, see Fig. 15.9) must be determined. This idea was used to study the forces of molecular motors that move along microtubules or actin filaments of cells. Here a small bead is used as a handle that is linked to a motor protein. The bead is trapped, and the motor, activated by ATP or GTP, tries to pull the bead out of the trap by moving forward [21–23]. In addition, it became possible to observe the exact position of a moving motor protein: Svoboda et al. measured the motion of a kinesin molecule along a microtubule using an interferometric technique with subnanometer resolution [24]. By using feedback techniques, the precision of force and position measurements could be further increased, which was applied for studying the interaction of myosin-actin systems [25,26]. Stretching macro-molecules such as DNA by exerting optical forces enabled studies of the micro-mechanical properties, but also of the dynamical behavior of DNA polymers [27–29].

Numerous applications grew up during the last years using optical traps in chemistry, polymer, and colloidal physics to measure surface and colloidal forces or the dynamics and the visco-elastic properties of polymer chains.

15.3.3   Photonic Force Microscopy

It was believed for a long time that the thermal motion of a trapped particle limited the resolution of force measurements and the precision of position measurements. This became especially apparent when an optically trapped particle was used as a probe to scan and image soft surfaces. The first scanning probe microscope was reported by Hertz and co-workers, who captured a 290 nm diameter silica particle, which was scanned around the object at a distance of more than 10λ [30]. The light scattered by the trapped particle was altered by the object, which consisted of thin, freestanding wires. By the early 90s, it became apparent that the key to a high-resolution scanning instrument was the position detection of the trapped particle, when it is also used as a probe.

Several detectors have since been developed to record and to analyze the fluctuations in the position of a trapped bead. Denk and Webb measured the lateral thermal motion of microscopic objects with a two-dimensional detector [31]. Their instrument, based on a modified differential interference contrast microscope, was used to investigate the motion of hair bundles. Ghislain and Webb used a trapped stylus (a small glass shard) to scan a soft surface [32]. The probe's axial position changed according to the scanned surface profile and was measured with a photo detector. The detector signal, determined by the intensity of the forward scattered light, was found to be linearly dependent on the particle's displacement within a certain range. At the same time Kawata et al. presented a laser trapped probe scanning microscope [33], where the illumination of the sample with evanescent waves was exploited. However, determining the axial position of the probe turned out to be difficult, even by using a spot illumination [34]. Friese et al. [35,36] evaluated the intensity of light reflected from a trapped sphere, which was used to probe three-dimensional surfaces. Florin et al. [37] scanned a trapped fluorescent latex bead across a soft surface. Here two-photon fluorescence of a 1064 nm Nd:YAG laser was exploited to determine the axial position of the probe, which was displaced by contacting the surface. The scanning could be performed either in a 'rolling mode' or a 'tapping mode'. This device, which represented the first version of the Photonic Force Microscope (PFM) [38], did not yet exploit thermal motion at this time. All the techniques described so far required high laser powers to achieve a good spatial resolution by minimizing thermal noise fluctuations, and, therefore, presented a high risk of photo damage to a living specimen.

A major step towards a modest illumination strength of the biological sample and to a precise and flexible three-dimensional position detection of the probe was to take advantage of the coherent nature of the trapping light. The interference of unscattered and forward scattered light produces an interference pattern, which can be recorded with a (quadrant) photodiode in the back-focal plane of the detection lens. This technique provides not only a lateral position signal similar to that obtained by Denk and Webb [31] and Ghislain and Webb [32], but also encodes the axial position of the particle [39] due to the phase anomaly of focused beams [40]. With this detection technique, position displacements of a particle in x,y and z can be tracked at a rate of more than 0.5 MHz with nm spatial precision. Hence, only recently, it became possible to perform three-dimensional topography scans around three-dimensional objects by exploiting the Brownian motion of the probe inside the trap [41].

The three basic principles driving a PFM, optical forces, thermal fluctuations, and position detection of the probe, are outlined in Fig. 15.5. An objective lens generates a focus with Airy-disc diameter $\Delta x \approx \lambda$, which is necessary for the generation of strong intensity gradients in all three dimensions. A particle is drawn into the focal region and is trapped stably provided the applied laser power is in the milliWatt range. Due to stochastic collisions with the surrounding fluid, the particle undergoes a random walk (Brownian motion) and thus scans a specific volume within a certain time period. The volume in which these thermal fluctuations are detected is the volume of the optical trap (see random track and trap border outlined in Fig. 15.5). When the laser power and thus the optical forces decrease, the trapping volume increases. If the optical forces become too weak, the collision forces can kick the particle out of the trap. The position fluctuation, i.e. the rate with which the particle changes its direction, is influenced by the viscous drag γ and the force profile of the optical trap. The optical trap can be calibrated by determining the force profile at known viscous drag γ [42].

The local environment (as outlined in Fig. 15.5) influences the position fluctuations of the trapped probe in a specific manner. By analyzing the altered fluctuations, it is possible to determine the coupling of the probe with external forces.

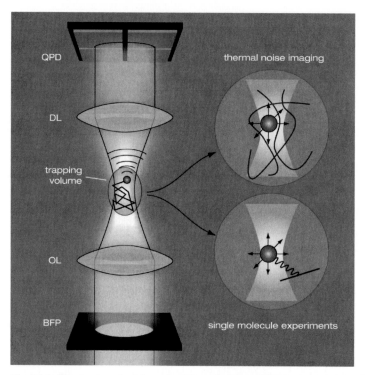

Fig. 15.5. The basic principles of a Photonic Force Microscope. A laser beam passes the aperture at the back focal plane (*BFP*) of the objective lens (*OL*). A particle in the focal region interacts with the light. The arising optical forces generate a three-dimensional optical trap. The particle scans the trapping volume due to Brownian motion. Light scattered by the particle and unscattered light generate an interference pattern, which is projected by a detection lens (*DL*) on e.g. a quadrant photodiode (*QPD*), which is placed in the *BFP* of the detection lens. The particle's position is determined with nanometer precision in all three dimensions at a rate of more than 0.5 MHz. The detector signals caused by the fluctuating particle are analyzed for thermal noise imaging or single molecule experiments. For imaging experiments the trap is moved stepwise along a grid and fluctuation data is acquired for every position of the trap

These forces might be due to mechanical linkage or contact with macromolecules or motor proteins [43], due to viscosity changes or local flow gradients [44–46], due to the presence of electrostatic or magnetostatic potentials, or due to reflective surfaces [47]. By moving the trap volume along a grid and repeating the procedure, it is possible to establish complete multi-dimensional maps of the parameters described above [38,41].

How is it possible to measure these specific position fluctuations of the probe? A condensor used for detection, which can be another objective lens, collects the unscattered light and the light scattered at the particle in the forward direction (see Fig. 15.5). A quadrant photodiode or a fast camera placed in the back focal plane of

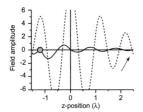 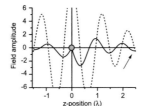 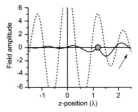

Fig. 15.6. Axial line scans of the electric field around the geometric focus ($z = 0$) illustrate the effect of the Gouy-phase shift on different particle positions. The incident field is indicated by the *dotted curve* and identical in all three figures. The scattered wave, indicated by the *solid curve*, is emitted at three positions of the particle (*gray spot*) where the amplitude of the incident field is maximal. The interference between the scattered and the unscattered field (see *arrow marker* on the right) depends on the position of the particle

the detection lens, records the interference pattern as a function of time and particle position [39,48]. Analyzing the position signals determines the particle's fluctuations and positions in three dimensions at a rate of 500 kHz with an accuracy of better than 5 nm [49]. Alternative methods, such as measuring the power of the back scattered light [35], the forward scattered light [50] or exploiting evanescent waves [51], provide the same information, but with a lower spatial precision and less flexibility. Another method is to exploit the strong axial intensity gradient of a two-photon focus that generates a signal that is also approximately linear with the intensity [37]. This method is rather insensitive to phase distortions induced by the specimen, but requires higher laser powers.

The interference intensity of the scattered and the unscattered beam at the detector is usually not a ring pattern, but the overlap of smoothly decaying spots, a bright spot from the unscattered wave and a dim spot from the scattered wave. The spot from the scattered wave changes in shape and magnitude depending on the scatterer. For a scatterer located off-axis, the center of the smaller spot moves in the opposite direction. If a position detector consists of at least 4 diodes, the sum of two adjacent detector element intensities is different than the sum of the other two pixel intensities. Lateral displacements of the scatterer from the optical axis can be measured with nanometer-precision applying this technique. Surprisingly, axial displacements of the scatterer can be measured when the sum of all 4 pixel intensities is recorded [39]. This only works for coherent scattering processes, where the Gouy phase shift (also phase anomaly) is exploited, which is inherent in every non-plane wave. The Gouy phase shift is a phase shift of π along the region of the maximum intensity of a focused beam and results in a longer wavelength and a shorter momentum vector in the focus. This effect is illustrated in Fig. 15.6, where axial line scans of the electric fields of an incident focused beam and a scattered wave is shown for three different axial positions of a scatterer. The abscissas of the plots are scaled in λ, but the distances between field maxima are larger than λ. The interference intensity at the detector plane depends on the phase difference of the scattered and the unscattered wave. This phase difference depends on the z-position of the scatterer and is non-constant due to the Gouy phase shift.

15.3.4 Atom Traps

Since the presentation of the first optical trap [52], the field of atom trapping increased in significance in an impressive manner and resulted in several Nobel prizes. Atoms absorb a net momentum from photons with a rate depending on the frequency of the incident field. The deceleration of atoms flying in opposite direction to the photons was used for optical cooling. With a red detuned laser, i.e. a laser with frequency below the atomic resonance frequency, atoms with reduced absorption frequency due to the Doppler-shift absorbed a net momentum. Using totally six lasers in three orthogonal directions it became possible to generate an 'optical molasses', consisting of atoms cooled down to a few microKelvin [53]. By using focused laser light, the arising gradient forces acting on dipoles were used to realize three-dimensional atom traps [7]. The polarizability α of the atoms could be adjusted by the used laser frequency, resulting in an optical force acting on the atoms towards ($\alpha > 0$) or away from ($\alpha < 0$) the center of the focus. This simple trap laid the foundations for the field of Bose-Einstein-Condensates (BEC) [54], i.e. an atom vapor that is coherent in a single ground state and reveals the fascinating properties inherent to bosons.

15.4   Theory

In order to understand the interaction of photons with a particle and the interaction of the trapped probe with its local environment, one has to appreciate the following points theoretically: the fields in a highly focused beam, the scattering of various particles therein, the interference of the scattered and the unscattered fields for the particle's position detection, the arising trapping forces and the interplay with the Brownian motion. We want to mention that neither ray optics calculations for trapping forces of particles smaller than or equal to the wavelength provide correct results, nor do Gaussian optics describe highly focused beam correctly.

15.4.1   Arbitrary Focused Fields

A flexible and elegant method to calculate arbitrary focused fields is the following approach: we start with a certain field distribution $\tilde{E}_0(k_x, k_y)$ in the back focal plane of an objective lens. Since we use lasers, this typically will be an expanded Gaussian beam in the TEM$_{00}$ mode. This distribution will be altered by passing the aperture stop (usually of a circular shape) and in the focus will deviate from a Gaussian shape. We define an arbitrary focused field as an electromagnetic field $\tilde{E}_i(k_x, k_y)$ with an arbitrary magnitude and phase across the aperture of the back focal plane of the lens. The field distribution in the focal region of an ideal lens can be obtained by taking the three-dimensional Fourier transform of the field in the back focal plane

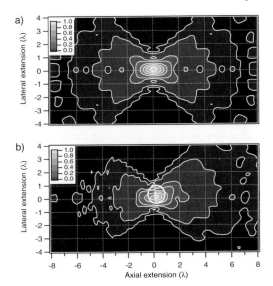

Fig. 15.7. Intensity distributions in the xz-plane. The scaling is in wavelengths ($\lambda = \lambda_0/n_m$) of the laser in the medium. (**a**) Intensity in the focus resulting from an x-polarized Gaussian beam with a two-fold over-illumination of the back focal plane of an ideal lens with $NA = n_m 0.9$. (**b**) Same focusing but with a dielectric scatterer of 1.2λ diameter at position $b = (\lambda/2, 0, \lambda/8)$ from the geometric focus (the scatterer is outlined by a *white circle*). The intensity distribution is altered due to scattering; momentum is transferred to the scatterer. Nearfields around the scatterer have been neglected

$\tilde{E}_i(k_x, k_y)$ [55,56]. $E(x, y, z)$ can be written as a decomposition of plane waves with directions (k_x, k_y):

$$E_i(x, y, z) = (2\pi)^{-3} \iiint_{k \leq k_0 NA} \tilde{E}_i(k_x, k_y, k_z) \exp[-i\mathbf{k}\mathbf{r}] dk_x \, dk_y \, dk_z \tag{15.2}$$

$$= (2\pi)^{-2} \iint_{k \leq k_0 NA} \tilde{E}_i(k_x, k_y) \exp\left[-i\left(k_x x + k_y x + \sqrt{k_n^2 - k_x^2 - k_y^2}\, z\right)\right] dk_x \, dk_y \ .$$

$\tilde{E}_i(k_x, k_y, k_z)$ is called the pupil function and lies on a spherical surface. $NA = n_0 \sin \alpha$ is the numerical aperture of the objective lens and $k_\perp = (k_x^2 + k_y^2)^{1/2}$ is the lateral component of the k-vector with length $k_n = k_0 n_0 = 2\pi n_0/\lambda_0$. Furthermore $\tilde{E}_i(k_x, k_y, k_z)$ is fully described by its two-dimensional projection, the angular momentum representation $\tilde{E}_i(k_x, k_y)$. This spectrum can be further split into a product that describes the field strength $E_0(\mathbf{k})$, the transmission and phase $A(\mathbf{k})$, the polarization $P(\mathbf{k})$ and the apodization $B(\mathbf{k})$ of the incident electric field at the back focal plane (BFP) [57].

$$\tilde{E}_i(k_x, k_y) = E_0(k_x, k_y) A(k_x, k_y) P(k_x, k_y) B(k_x, k_y) \ . \tag{15.3}$$

The apodization function obeys the sine-condition and is written as $B(\mathbf{k}) = \sqrt{(k_n/k_z)} = 1/\sqrt{\cos(\theta)}$. The polarization function $P(\mathbf{k}) = (P_x(\mathbf{k}), P_y(\mathbf{k}), P_z(\mathbf{k}))$ describes the components of the electric field vector of a polarized beam that changes its direction from $k = (0, 0, 1)$ to $k = (k_x, k_y, k_z)$. A calculation result of the intensity of (15.2) in the yz-plane is shown in Fig. 15.7 (a).

There are other calculation methods that deliver the same results by solving integrals corresponding to that of (15.2) in space domain [58], but become more complicated when accounting for aberrations in amplitude and phase. A method frequently used to calculate trapping forces starts with a small-angle approximation equation and then corrects with higher orders of a Gaussian beam [59]. This method delivers correct widths of a highly focused beam in the lateral direction, but produces differences of up to 60% for the slope of the axial focus profile in comparison to a realistic, aperture limited focus calculated with our method. The correct description of the electromagnetic fields in the focus is required in order to calculate the correct trapping forces.

15.4.2   Scattering by Focused Fields

It is about a century ago that Rayleigh investigated the scattering of the sunlight by point-like particles. His theory was extended by Gans and by Debye for particles of arbitrary shape with sizes of up to the wavelength, but with relatively small refractive indices [60]. At about the same time, Mie derived exact solutions for spherical particles of arbitrary size and refractive index. Their calculations took account of the scattering of a single plane wave by a particle.

However, using the superposition principle it is possible to extend their theories to arbitrary incident fields, which can be decomposed into a spectrum of plane waves with directions (k_x, k_y). The concept is the following: the scatter function $T^\pm(k_x, k_y)$ is determined for a specific particle and for an incident plane wave (the superscript $\pm$ indicates forward and backward scattering). Multiplication with $D_x = \pm(k_n^2 - k_x^2)^{1/2}/k_n$ takes into consideration the characteristic dipole emission, which is zero in the x-direction when the incident wave is polarized in x. Multiplication with $P(k_x, k_y)$ generates the three vector components $\tilde{E}_{sx}, \tilde{E}_{sy}$, and $\tilde{E}_{sz}$ of the scattered field. A rotation matrix $M^{-1}(k_{ix}, k_{iy})$ is used to rotate the resulting distribution about the Euler angles [61] defined by (k_{ix}, k_{iy}) for an incident plane wave with magnitude $\tilde{E}_i(k_{ix}, k_{iy})$. The resulting superposition of all scattered waves can thus be written as [49]:

$$\tilde{E}_s^\pm(k_x, k_y) = \frac{\pm ik_n^2}{4\pi k_z} \iint_{k_{i\perp} \le NAk_0} \tilde{E}_i(k_{ix}, k_{iy}) T^\pm(k_x', k_y') D_x(k_x', k_y') P(k_x', k_y') dk_{ix}\, dk_{iy} \; .$$

(15.4)

The primed coordinates

$$\left(k_x', k_y', (k_n^2 - k_x'^2 - k_y'^2)^{1/2}\right) = M^{-1}(k_{ix}, k_{iy})\left(k_x, k_y, (k_n^2 - k_x^2 - k_y^2)^{1/2}\right)$$

depend on the incident components (k_{ix}, k_{iy}) and $k_{i\perp} = (k_{ix}^2 + k_{iy}^2)^{1/2} \le NA\, k_0$ defines the integration limit for the angular spectrum of the incident field.

If the particle is shifted to a position $\boldsymbol{b} = (b_x, b_y, b_z)$ relative to the center of the beam $(0, 0, 0)$, it simplifies the calculations to leave the particle at its position $(0, 0, 0)$ and to shift the incident beam by $-\boldsymbol{b}$, (i.e. to modulate the spectrum of the incident field). The resulting intensities scattered in the forward and backward direction $I_s^+(k_x, k_y)$ and $I_s^-(k_x, k_y)$, respectively, are:

$$I_s^\pm(k_x, k_y, \boldsymbol{b}) = \varepsilon_0 c \left(\left| \tilde{E}_{sx}^\pm(k_x, k_y, \boldsymbol{b}) \right|^2 + \left| \tilde{E}_{sy}^\pm(k_x, k_y, \boldsymbol{b}) \right|^2 + \left| \tilde{E}_{sz}^\pm(k_x, k_y, \boldsymbol{b}) \right|^2 \right) \quad (15.5)$$

$$= I_{sx}^\pm(k_x, k_y, \boldsymbol{b}) + I_{sy}^\pm(k_x, k_y, \boldsymbol{b}) + I_{sz}^\pm(k_x, k_y, \boldsymbol{b}) \ .$$

The interference intensity of the scattered and the unscattered fields in the yz-plane is displayed in Fig. 15.7 (b). The calculation, performed by applying (15.2) and (15.4), does not consider near fields.

15.4.3   Position Detection

The angular intensity distribution of scattered and unscattered light is determined up to an angle α_D in the back focal plane (BFP) of a detection lens with $NA_D = n_m \sin \alpha_D$. The interference pattern is projected directly onto a quadrant photodiode where the quadrants are numbered 1 to 4. An aperture stop regulates the NA_D and hence the contribution of high-angle rays to the position signal. The interference pattern is described as follows:

$$I_D(k_x, k_y, \boldsymbol{b}) = \varepsilon_0 c \left| \tilde{E}_i(k_x, k_y) + \tilde{E}_s^+(k_x, k_y, \boldsymbol{b}) \right|^2 \text{step} \left(k_0 NA_D - \sqrt{k_x^2 + k_x^2} \right). \quad (15.6)$$

The signal $I_D(k_x, k_y, \boldsymbol{b})$ is delimited in its spatial extent by the diameter R_D of the BFP. This diameter R_D corresponds to the NA_D of the detection lens, and can be reduced by using an aperture stop. In a next step the intensity is integrated over the area of each quadrant with index $m \in [1, 4]$, where the integration limits A, C are either 0 or k_n and B, D are either $-k_n$ or 0, respectively:

$$S_m(\boldsymbol{b}) = \int_B^A \int_D^C I_D(k_x, k_y, \boldsymbol{b}) dk_x \, dk_y \ . \quad (15.7)$$

The three components of the signal $S(\boldsymbol{b}) = (S_x, S_y, S_z)$ for a particle located at $\boldsymbol{b}$ are then obtained as follows:

$$S_x(\boldsymbol{b}) = [(S_1(\boldsymbol{b}) + S_2(\boldsymbol{b})) - (S_3(\boldsymbol{b}) + S_4(\boldsymbol{b}))]/S_0 \quad (15.8)$$
$$S_y(\boldsymbol{b}) = [(S_1(\boldsymbol{b}) + S_3(\boldsymbol{b})) - (S_2(\boldsymbol{b}) + S_4(\boldsymbol{b}))]/S_0$$
$$S_z(\boldsymbol{b}) = [(S_1(\boldsymbol{b}) + S_2(\boldsymbol{b})) + (S_3(\boldsymbol{b}) + S_4(\boldsymbol{b}))]/S_0 \ .$$

S_0 is a constant value, for example the total intensity incident on the diode when no scatterer is present.

For particle displacement $\boldsymbol{b}$ along only one axis through the center of the trap, a linear relationship between $S(\boldsymbol{b})$ and $\boldsymbol{b}$ can be expected. It can be shown that scattering into high angles $(\sin(\theta) > 0.6)$ leads to destructive interference of the scattered

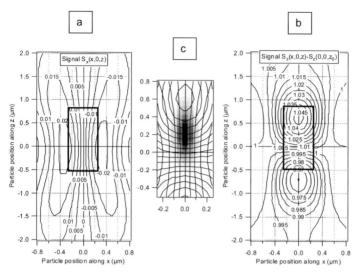

Fig. 15.8. Detector signals for particle positions in the xz-plane. The particle is a 300 nm latex sphere. The central regions of (**a**) and (**b**) are magnified and overlaid in an additional contour plot (**c**). The probability density $p_B(x, 0, z)$ of finding the trapped particle at a certain position in the xz-plane is shown by the gray level image in the center plot. A darker gray level indicates a higher probability to find the particle within this volume element. The diode signals S_x and S_z are displayed by labeled contour lines. Each line represents a constant signal value at the detector. Vertical contour lines in the S_x-plot indicate a signal that is independent of the axial particle position; horizontal contour lines in the S_z-plot represent a signal that is independent of the lateral particle position (a, b, c)

and unscattered fields, whereas for small angles the interference is constructive [49]. Combining both the small-angle and the high-angle signals has no positive effect on the determination of the axial particle position. As a compromise we chose a capture angle $\sin(\alpha) = 0.5$ of the detection lens in the following calculations. For lateral displacements, however, a large aperture radius provides the best sensitivity.

An important point is that decreasing the NA_D of the condenser lens increases the extent of the detection point-spread function, which improves its overlap with the trapping volume. For the above reasons, we recommend to adjust the aperture stop of the detection lens to values of $\sin(\alpha) = 0.3$ to 0.5.

The signals $S_x(\mathbf{r})$, $S_y(\mathbf{r})$ and $S_z(\mathbf{r})$ can be approximated by the first order of a Taylor-series around the position $\mathbf{r}_0 = (x_0, y_0, z_0)$, which is typically the center of the focus $= (0, 0, 0)$ or a stable trapping position $\mathbf{r}_0 = (0, 0, z_0)$. In the latter case the linear dependency for the i-th component of the signal is ($i = x, y, z$):

$$S_i(\mathbf{r}) \approx \left.\frac{\partial S_i(\mathbf{r})}{\partial x_i}\right|_{r=r_0} x_i + S_i(\mathbf{r}_0) = \partial_i S_i x_i + S_i(\mathbf{r}_0) . \tag{15.9}$$

How strong is the coupling between the lateral and the axial position signals when particles are displayed off-axis? In Fig. 15.8 the signal distributions $S_x(x, 0, z)$ and

$S_z(x, 0, z)$ are shown for a whole range of particle positions $b = (x, 0, z)$. Figure 15.8 displays the detector signals by contour lines, while the probability density is shown in a gray level image (center) for different particle displacements. Here darker gray levels indicate a higher probability to find the particle at this position. A particle moving along a contour line gives a constant detector signal. Vertical contour lines in the S_x-plot indicate a signal that is independent of the axial particle position; horizontal contour lines in the S_z-plot represent a signal that is independent of the lateral particle position. Magnifications of the central regions of $S_x(x, 0, z)$ and $S_z(x, 0, z)$ reveal that nearly all lines are curved. The curvature increases with the displacement from the point $(0, 0, z_0)$. The signals become non-linear for smaller off-axis displacements but are still unique. This is not the case for larger displacements. Since nearly all curves are closed, a straight line scan in one direction intersects a contour line twice and the two particle positions deliver the same diode signal.

With help of the simulations it is possible to estimate the positioning error in nanometer between the actual particle position r and the position r_{rec} returned by the detector signal $S(r)$. This estimation can be done for displacements of a particle from the center of the focus or for trapping experiments, where the trap and the detector were calibrated by thermal noise analysis. For a correction of the position error it is required to record all three position signals and relate them to their correct location. Such a correction is not possible when only two position signals are known.

15.4.4 Trapping Forces

The electromagnetic force density $f(r, t)$ acting on a small polarizable particle or volume element with dipole moment density $P(r, t)$ is given by [62]

$$f(r, t) = (P(r, t)\nabla) E_m(r, t) + \frac{\partial P(r, t)}{\partial t} \times B_m(r, t), \qquad (15.10)$$

where E_m and B_m are the electric and the magnetic fields in a homogeneous medium with refractive index n_m.

The dipole moment density $P(r, t) = \varepsilon_m \alpha E_m(r, t)$ is the polarization of the particle and $\alpha = 3(m^2 - 1)/(m^2 + 2)$ is the polarizability of the volume element according to Clausius-Mossotti [63]. The relative refractive index is $m = n_s/n_m$. After rearranging and integrating over all volume elements dV, the total optical force on a particle can be expressed by two components, the gradient force F_{grad} and the scattering force F_{sca} [57]:

$$F(r) = F_{grad}(r) + F_{sca}(r) = \iiint_V \frac{\alpha n_m}{2c} \nabla I_i(r') dV' + \frac{n_m}{c} I_0 C_{sca}(g/k_n). \qquad (15.11)$$

$\nabla I_i(r)$ is the gradient of the intensity inside the scatterer. For a particle with a diameter smaller than the wavelength and a refractive index not much larger than 1, the Born approximation is valid and the intensity distribution inside the particle is proportional to the distribution without scatterer $\nabla I_i(r) \propto \nabla I_{internal}(r)$. The transfer vector g points in the direction of momentum transfer, which is the z-direction for symmetric incident fields and for on-axis spheres. The magnitude $|g/k_n| = \langle\cos(\theta_i)\rangle - \langle\cos(\theta_s)\rangle$ is obtained by the difference of the mean k-direction cosines from the incident and the scattered field. The power of the scattered light $I_0 C_{sca}$ is defined by the intensity in the focus of the incident beam $I_0 = \varepsilon_0 c|E_i(0,0,0)|^2$. The scattering cross-section C_{sca} is computed by the scattered intensity $\tilde{I}_s^\pm(k_x, k_y, b)$ (see (15.5)).

Equation (15.11) can be reduced for optical forces on a Rayleigh scatterer [64] to

$$F(r) = \frac{n_m}{2c}\nabla I_i(r)\alpha V + \frac{n_m}{6\pi c}I_i(r)k_n^3\langle k_i\rangle(\alpha V)^2 , \tag{15.12}$$

with volume $V = 4/3a^3\pi$ (a is the radius of the scatterer), $C_{sca} = (2/3)^3\pi a^6 k_n^4\alpha^2$ and with momentum transfer $g = \langle k_i\rangle$. In the center of the focus, $\langle k_i\rangle$ is the mean k-vector, which is shortened due to the Gouy-phase shift.

Another way of modifying (15.10) leads to the following integral for the optical force:

$$F(r) = \mathrm{Re}\iint_A \varepsilon\left(E^*(\hat{n}\cdot E) - \frac{1}{2}\hat{n}|E|^2\right) + \left(B^*(\hat{n}B) - \frac{1}{2}\hat{n}|B|^2\right)dA \tag{15.13}$$

The integrand is the Maxwell stress tensor $\hat{T}$ acting on the vector $\hat{n}$, which is normal on the surface element dA. The surface integral surrounds the scatterer in an arbitrary manner. The surface should have a large distance to the scatterer so that near fields can be neglected. The electric field $E = E_i + E_s$ as well as the magnetic field $B = B_i + B_s$, are the sums of the respective scattered and the incident fields. All fields must be evaluated on the enclosing surface. The asterix * indicates the complex conjugate field. Although (15.13) delivers the correct optical force for every particle, the integral is less intuitive and rather complicated to solve [65,66].

This approach is different from the two components approach, which describes the scattering force and the gradient force separately. The scattering force results from a momentum transfer from the incident photons onto the scatterer; the gradient force is the dipole force acting on each volume element by drawing it towards the point of the highest intensity. The gradient force therefore enables stable trapping, whereas the scattering force pushes the scatterer out of the trap in the direction of light propagation.

The trapping force $F(r)$ can be described independently of the total incident optical power P by the trapping efficiency $Q = (Q_x, Q_y, Q_z)$. It is defined as follows (c/n_m is the speed of light in the immersion medium):

$$Q(r) = F(r)\frac{c}{Pn_m} . \tag{15.14}$$

The effect of the two components is illustrated by the axial force profile in Fig. 15.9 left, where the point $F_z(r_0) = 0$ is behind the center of the focus. The profiles are plotted with the dimensionless trapping efficiency $Q(r)$. For particles smaller than the wavelength, the scattering force is always positive and the gradient force is

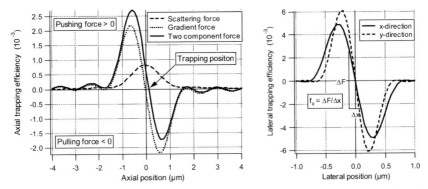

Fig. 15.9. Force profiles calculated with the two component approach for a 216 nm latex bead ($n = 1.57$) in axial and lateral directions. The trapping forces are for a water immersion lens with $NA = 1.2$, illuminated with a Gaussian beam with waist $w_0 = 2R$ ($R =$ aperture radius), i.e. the aperture is over-illuminated, the wavelength is $\lambda_0 = 1.064\,\mu m$, the field is linearly polarized in x. The axial profile (*left*) represents the sum of the scattering and the gradient force. The lateral profiles (*right*) differ due to the field polarization, the difference in the slopes f_x and f_y is nearly 60%

always bipolar. In Fig. 15.9, a small bead is trapped in water with a water immersion lens, the incident wave was polarized in x. The lateral force profiles on the right hand side clearly show that in the direction of polarization both the maximum force and the force constant are weaker. The force constant is the linear slope of the force profile at the point $F(r_0) = 0$ and defines the stiffness of the trap. In the center of the optical trap r_0, the trapping force can thus be approximated by

$$F(\mathbf{r} - \mathbf{r_0}) \approx \left(f_x(x - x_0), f_y(y - y_0), f_z(z - z_0) \right) ,\tag{15.15}$$

where f_i is the force constant in direction i ($i = x, y, z$). The force constants f_i are connected in a nonlinear relation to the maximum (backward) force and the depth of the trapping potential. The latter is decisive for the stability of an optical trap. The three force constants f_i are strongly affected by spherical aberrations, which often occur in trapping experiments due to refractive index mismatch [67].

15.4.5   Thermal Noise and Trap Calibration

Assume a small particle of mass m in a viscous medium where the friction γ damps the particle thus that $\gamma\,dx/dt \gg m\,d^2x/dt^2$. In addition to that, the driving force for the particle's motion is assumed to be a random function of time. This situation is described by the Langevin equation [68]:

$$\gamma \dot{\mathbf{r}}(t) + \mathbf{F}_{opt}(\mathbf{r}) + \mathbf{F}_{ext}(\mathbf{r}) = \mathbf{F}_{rand}(t) .\tag{15.16}$$

The viscous drag of the immersion fluid (also friction constant) is $\gamma = 6\pi a \eta$ (a = sphere radius, η = viscosity) for a spherical particle, which is Stokes law. The diffusion constant, defined by the Einstein-relation:

$$D = k_B T/\gamma , \tag{15.17}$$

determines the strength of the collision forces, i.e. the maximum magnitude of the random force is $|F_{rand}| = (6D/\tau)^{1/2}\gamma$. Here k_B is the Boltzmann constant and T is the temperature in Kelvin. When the optical force F_{opt} and the external force F_{ext} are zero, the position fluctuations $r(t)$ can be described as a random walk and the correlation between $r(t)$ and $r(t + \tau)$ is zero for $\tau \gg m/\gamma$. Then, for larger times $\tau \gg m/\gamma$, the mean square displacement $\langle r^2(\tau) \rangle = 6D\tau$ is determined by the diffusion coefficient. This situation changes when the optical force F_{opt} is non-zero. In most cases it can be approximated by (15.15). The autocorrelation function for the i-th component of the position $r = (x_1, x_2, x_3)$ then becomes [42,69]

$$\langle x_i(t)x_i(t + \tau) \rangle = x_i^2 \exp(-\tau/\tau_i) . \tag{15.18}$$

The decay constant is $\tau_i = \gamma/f_i$ ($i = x, y, z$). In other words, the mean force constant f_i of the optical trap increases with a shorter auto-correlation time between two subsequent positions $x_i(t)$ and $x_i(t + \tau)$. The mean square fluctuation of the particle is $x_i^2 = k_B T/f_i$. The trap stiffness f_i can also be obtained by evaluating the power spectrum of (15.15), where the frequency $1/\tau_i = f_i/\gamma$ at half amplitude corresponds to the auto-correlation time τ_i.

The detector signal $S(r)$ is proportional to r (see (15.9)) for small displacements $r - r_0$. Then the trap stiffness $f_i = \gamma/\tau_i$ in direction $i = x, y, z$ can be determined by $\langle S_i(t)S_i(t + \tau) \rangle$ as defined in (15.18). This relationship is illustrated in Fig. 15.10 for the movement of a small latex bead in an ideal optical trap. The three position signals are shown in Fig. 15.10 (a), where the two lateral signals $S_x(t)$ and $S_y(t)$ are bipolar and the axial signal $S_z(t)$ has an offset that is determined by the intensity of the incident light. The corresponding auto-correlation functions are displayed in Fig. 15.10 (b) where the auto-correlation times are in the millisecond range, i.e. $dt \ll \tau_i$ guarantees a sufficiently fine temporal sampling. It is remarkable that, due to the polarization, the two lateral force constants f_x and f_y differ by about 60%. At the same time, the histograms of the position signals $S_i(t)$ in Fig. 15.10 (c) have a Gaussian shape, reflecting the harmonic trapping potential and the linear detector response. However, the difference in the widths σ'_x and σ'_y of the histograms are less pronounced, which means that the detector compensates the polarization effect on f_x and f_y.

The assumption of a linear relationship between the trapping force and displacement as expressed with (15.15), entails the assumption of harmonic trapping potentials $W(x_i) = 1/2f_i \cdot (x_i - x_{i0})^2 + \text{const}$. If we further assume that the position distribution of a trapped particle obeys Boltzmann statistics, then the probability density to find a particle at position r depends on the trapping potential $W(x, y, z)$ as follows:

$$p_B(r) = p_0 \exp\left(-\frac{W(r)}{k_B T}\right) \approx p_0 \exp\left(-\frac{f_x x^2}{2k_B T} - \frac{f_y y^2}{2k_B T} - \frac{f_z z^2}{2k_B T}\right) . \tag{15.19}$$

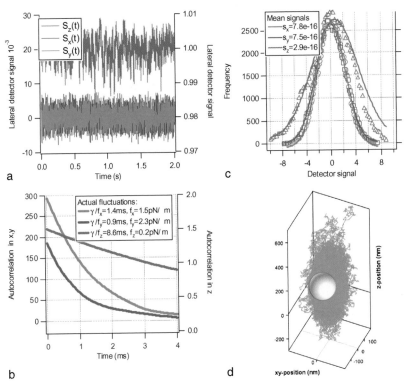

Fig. 15.10. (a) The three detector signals S_x, S_y, and S_z as a function of time for a 216 nm latex bead diffusing in an ideal optical trap at a laser power of $P = 10$ mW, at temperature $T = 300$ K, with viscosity $\eta_{\text{water}} = 0.001025$ N/(m²s) and with a time sampling of $dt = 20$ µs. The z-signal is centered around 1, the two lateral signals are centered around zero. (b) The corresponding auto-correlation functions of the fluctuations have a weak decay in z-direction (top) and the strongest decay in y-direction (bottom). (c) The signal histograms derived from the data shown in (a) reveal the same amplitudes for the signals S_x and S_y, although the trap stiffnesses are different. (d) A three-dimensional position track plot (100000 points in 2 seconds) of the same bead at $P = 10$ mW. All graphs are based on a Brownian dynamics simulation

The parameter p_0 normalizes the distribution p_B, which is a three-dimensional distribution with widths σ_x, σ_y, and σ_z. Figure 15.10 (d) shows the position track $r(t)$ of the bead. The three-dimensional distribution of positions corresponds to the three-dimensional probability density. Tracks at the very edge of the distribution (see upper right corner of Fig. 15.10 (d)) reveal high thermal energies of the diffusing particle, which can achieve about $8k_B T$ within a time period of 1–2 seconds.

The optical trap can be calibrated for the force by exploiting the relationship of (15.18) for the detector signals S_x, S_y, and S_z. Because the force constant changes its magnitude for different positions inside the trap, only mean force constants f_x, f_y, and f_z are obtained with this method. To calibrate the trap for the particle positions, the signal histograms have to be evaluated to obtain the calibration factor g_i ($i = x, y, z$) where $S_i(\mathbf{r}) = g_i x_i$. The widths σ_x, σ_y, σ_z ($\sigma_i = (2k_BT/f_i)^{1/2}$) obtained from (15.19) are assumed to be proportional to the widths σ'_x, σ'_y, σ'_z from the signal histograms. Hence, the calibration factor connects the two histograms by their widths as $g_i = \sigma'_i/\sigma_i$. Therefore, we obtain for the position in direction ($i = x, y, z$):

$$S_i(\mathbf{r}) = g_i x_{i.} = \sigma'_i/(2k_BT/f_i)^{1/2}x_{i.} . \tag{15.20}$$

It can be seen that for an arbitrary position $\mathbf{r}$ inside the trap a constant slope g_i is approximated and that the position signal $S_i(\mathbf{r})$ only changes in the direction of x_i. The resulting position errors, are nevertheless smaller than 5 nm in average for most trapping configurations [49].

If the local probe interacts with its environment, the position fluctuations are influenced by a third force F_{ext} as expressed in (15.16). The effect of the external forces can be rather complicated and kinematic models have to be established. A simple model for F_{ext} is described in [70] for the problem of a tethered motor protein as explained in Sect. 15.6.2.

15.5   Experimental Setup

15.5.1   Mechanics and Optics

The more recent Photonic Force Microscopes at EMBL make use of an easily accessible microscope stand, i.e. in the third generation PFM the frame and most mounts are made of steel and cast iron to achieve maximum mechanical stability. The distance between objective lens and detection lens is kept constant, and the object's fine positioning is achieved with a xyz-piezo scanner.

Figure 15.11 summarizes the main mechanical and optical components of a PFM. A low noise Nd:YAG-laser operating in the near infrared ($\lambda_0 = 1064$ nm) is used for optical trapping and position detection. The IR-beam passes an acousto-optic modulator (AOM), a beam steering device (BSD), the beam expander (Exp) and a dichroic beam splitter (BS). The aperture of the BSD is imaged onto the back focal plane (BFP) of the objective lens (OL, infrared water immersion, $NA = 1.2$). The object plane (OP), defined by a standard glass cover slip, can be moved with an xyz-scanner (stage), consisting of a coarse manual xy-translation stage and a fine xyz-piezo scanner. Micromanipulation is performed in the open chamber with a patch-pipette (PP). A detection lens (DL, water immersion, $NA = 0.75$) collects scattered and unscattered light of the trapping laser, which is projected with a lens system (L) and by a dichroic beam splitter onto the InGaAs quadrant photodiode (QPD). The visible light of an Argon-Ion laser passes an acousto-optic tunable filter (AOTF), a beam expander, and dichroic beam splitters to illuminate the object plane. Fluorescence light excited in the object plane is imaged with the objective lens and the tube lens (TL) onto a CCD-camera. The visible light of a halogen lamp is linearly polarized (P) and passes both a mirror (M) and the dichroic beam splitter. A Wollaston prism (WP) near the BFP of the detection lens splits the beam for DIC

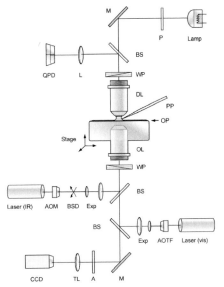

Fig. 15.11. Photo and schematic of a 3rd generation Photonic Force Microscope. Light paths from three light sources (infrared laser, visible laser and halogene lamp) pass several optical and mechanical elements: acousto-optical modulator (*AOM*), acousto-optical tunable filter (*AOTF*), beam steering device (*BSD*), beam expander (*Exp*), dichroic beam splitters (*BS*), objective lens (*OL*), *xyz*-scan stage (*stage*), object plane (*OP*), patch-pipette (*PP*), detection lens (*DL*), mirrors (*M*), lens (*L*), quadrant photodiode (*QPD*), polarizer (*P*), Wollaston prisms (*WP*), analyzer (*A*), tube lens (*TL*), CCD-camera (*CCD*)

microscopy. After transmitting the object, a second Wollaston prism behind the objective lens recombines the two beams. After passing a mirror and the DIC analyzer (A), the light is focused by the tube lens onto the CCD-camera.

15.5.2   Traps and Probes

Trapping wavelengths of $\lambda_0 = 0.8\,\mu m$ and $\lambda_0 = 1.06\,\mu m$ were found to be most appropriate because cell damage is relatively low in comparison to visible light [71]. Although a wavelength $\lambda_0 = 800\,nm$ results in stronger optical forces, there is a lack of suitable laser sources, whereas Nd:YAG lasers at $\lambda_0 = 1064\,nm$ offer a good performance at reasonable prices. Typically, the particle is spherical, with a diameter between 0.05 and 2 times the incident wavelength and a refractive index sufficiently large in comparison to that of its environment. The immersion medium is typically water, aqueous solution or cytoplasm with indices $n_{med} = 1.33–1.39$. Refractive indices n_s of dielectric scatterers are 1.39–1.48 for silica, about 1.57–1.60 for polystyrene (latex) and 3.5 for silicon. For metallic gold particles, such as gold beads, the complex refractive index is $n_s = 0.4 + 7.4i$ at $\lambda_0 = 1.05\,\mu m$ [72]. The large imaginary part corresponds to a strong absorption of light and leads to additional radiometric forces. Only small metallic scatterers with a diameter of less than $\lambda/10$ are trapped in the lateral center of the focus, larger metal spheres are trapped off-axis [73,74].

15.5.3   Electronics

Besides all necessary controllers that drive scanners, acousto-optical devices, cameras and lasers, the most interesting electronic component is required for amplifying and acquiring the position data from the quadrant photodiode. An InGaAs pin photodiode with a photosensitivity of 0.67 A/W is used for our PFMs. Pre-amplification of the signals by 20 V/mA (up to 850 kHz) leads to a voltage of 13.4 V per 1 mW laser power at the detector. The output noise of the pre-amplifier is 1 mV. Intensity noise of the diode-pumped infrared laser is < 0.03% at 10 Hz–1 MHz. If necessary, this noise can be subtracted from the signal by simultaneously measuring the laser's intensity noise.

The performance limit of the detector due to electronic noise can be estimated by taking account of the scattering efficiency of a weak scatterer (radius $a = 150$ nm, index $n = 1.39$), which is 0.3%. A mean axial sensitivity of 2.5% /μm for this scatterer can be estimated (see Table I in Rohrbach and Stelzer [49]). Assuming that 2% of the light in the focus is projected onto the diode, a laser power of 40 mW in the focus will be reduced to 0.8 mW at the diode. Pre-amplification provides a voltage of 0.8 mW $\times$ 13.4 V/mW $= 10.7$ V. A change of 2.5% /μm results in 268 mV/μm. The minimum voltage due to noise is 1 mV, hence the axial resolution limit due to noise is 1 μm/268 = 3.7 nm. For a scatterer with $a = 108$ nm, $n = 1.57$, $P = 30$ mW this value is 1 μm/313 = 3.1 nm. In the lateral direction, the sensitivity is better by a factor 2.5 to 4 for all investigated scatterers.

The three pre-amplified signals for the x-, y-, and z-positions are further amplified by a factor of about 100, depending on the IR-light transmission of the detection lens. A fast data acquisition PC-card converts simultaneously up to 4 voltages at a rate of up to 5 MHz to 12-bit signals, which are then analyzed as described in the last part of the Sect. 15.4.

15.6   Applications of Photonic Force Microscopy

15.6.1   3D Thermal Noise Imaging

Conventional scanning microscopes such as the scanning tunneling microscope (STM) or the atomic force microscope (AFM) [13] achieve excellent spatial resolution due to their precise positioning of the probe. However, their applicability is in general restricted to flat surfaces, because of the mechanical connection of probe and actuator. First attempts to hold the probe with optical tweezers were only applied to two-dimensional imaging. Recently, scanning with an optically trapped probe was extended to volumes. A three-dimensional picture of an object was obtained by moving a local probe volume-wise through a three-dimensional structure. The technique was called thermal noise imaging [41], because the scanning of the

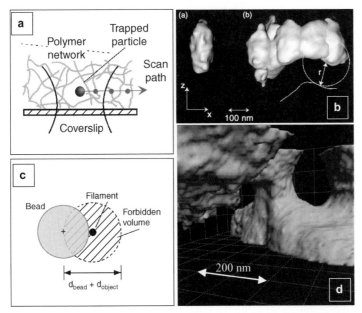

Fig. 15.12. (a) 3D thermal noise scanning of a polymer network with a trapped particle. The scan path describes a 3D volume. (b) Frequency isosurfaces of 5 counts per voxel of a 216 nm bead trapped by optical tweezers in solution (left) and in the network (right). The position of the trap was laterally scanned inside the network with a stepwidth of 80 nm. The object's surface (*white line*) was calculated taking the radius of the bead into account (*white circle*). (c) Mechanical amplification effect. A mechanical stiff filamentous object with a diameter d_{object} will appear in a thermal noise scan as a cylinder with a minimal diameter of $d_{min} = d_{bead} + d_{object}$. (d) 3D thermal noise image of an agar network. The image shows the volume's surface, that is not accessible to the bead

probe within each volume element (the trapping volume) was performed by the thermal fluctuations of the probe (see Fig. 15.12 (a)).

A particle trapped by optical tweezers explores the trapping potential due to thermal fluctuations. The probability of finding the particle at a given volume element in an optical trap is described by the Boltzmann distribution. The distribution can be measured by recording the position of the particle over a time interval and calculating a position histogram. The maximal energy level explored depends on the total recording time and the position autocorrelation time. Typical levels are on the order of several $k_B T$. Figure 15.12 (b) (left) shows a frequency isosurface of a position histogram at a level of five counts per voxel for a 216 nm bead. The center position of the bead was recorded for 0.8 s with a data acquisition rate of 100 kHz and the histogram was calculated with a binwidth of 8 nm in each direction. The bead explored a volume of about ±60 nm in the lateral and about ±150 nm in the axial (z) direction. The roughness of the isosurface reflects the counting statistics in the histogram at low frequencies.

Objects brought into the probe volume will be randomly probed from all directions. Large objects can be explored by moving the central trap position around the outside of the object. The presence of an object restricts the volume accessible to the bead, which is shown in Fig. 15.12 (b) (right). In this experiment, the bead was captured in the sample, an agar network, and the optical trap was then moved in 80 nm steps along the x-axis. The thermal position fluctuations were recorded for 0.8 s at each trap position. The corresponding frequency isosurfaces show a clear constraint at their lower end, which results from the interaction of the bead with an agar filament. This surface is the result of a convolution of the bead with the object. The actual position of the object's surface has to be calculated by taking the radius of the bead into account. The white line and the circle in Fig. 15.12 (b) (right) illustrate this.

A 3D thermal noise image is displayed in Fig. 15.12 (d). A larger volume around an agar network was scanned. The isosurfaces border that region to which the center of the bead had no access. The minimal diameter of the objects in Fig. 15.12 (d) is about the diameter of the bead (216 nm) and demonstrates a mechanical amplification effect that results from the probe-sample convolution (Fig. 15.12 (c)). When a bead probes a mechanically stiff filamentous object with a diameter d_{object}, then the radius of the cylinder in the three-dimensional scanning probe image is $r_{bead} + r_{object}$. Consequently, even a molecular thin filament will appear to have at least the diameter of the bead. Nevertheless, since the bead radius is known, the objects extension $r_{object} = r_{image} - r_{bead}$ can be measured with high accuracy.

15.6.2   Micro-mechanical Properties of Single Molecules

Understanding the elasto-mechanical properties of the motor protein kinesin can help to understand how structural changes of this protein influences its molecular function. Depending on the type of kinesin, it can consist of 1 or 2 heads, which are connected via hinges to a stalk-like part, which is the tail of the molecule. Also this part may consist of subunits, which may have specific mechanical properties and functions.

A coated latex or glass bead is trapped close to a cover slip. Once the trapped bead is connected to the stalk-like molecule kinesin and the motor protein is bound to a fixed microtubule, the bead's thermal fluctuations will be restricted in a specific manner [75]. This situation is displayed in Fig. 15.13 (a). The microtubule (MT) is fixed to the coverslip, the anchoring position of the kinesin on the MT and the position of the laser focus define an angle θ. By recording the beads position fluctuations with the PFM and by applying Boltzmann statistics, a three dimensional energy landscape is obtained. The 3D energy iso-surface (see Fig. 15.13 (b)), looks like the segment of a spherical shell. The maximum size of the segment is determined by the sum of the length of the kinesin tail (about 70 nm) and the bead radius (100–500 nm). The thickness of the shell (in the normal direction) is stipulated by the spring constant of the elastic kinesin tail and the optical forces of the trap. The trapping forces also influence the tangential change of the 3D energy landscape, together with the restoring forces of the kinesin's hinge. The normal z'-direction and

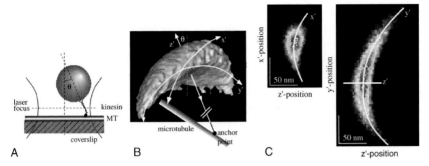

Fig. 15.13. Investigating elasto-mechanical properties of a motorprotein. (**a**) A bead, trapped in a laser focus, is linked to the stalk-like kinesin. The motorprotein is bound to a microtubule (*MT*), which is fixed to the cover slip. The anchor position of the kinesin on the MT and the position of the laser focus define an angle θ. (**b**) A surface of constant thermal energy, evaluated from the beads position fluctuations, has a specific shape and orientation relative to the anchor point and the MT. The directions x', y' and z' are indicated by double headed arrows. (**c**) Two slices of the 3D energy landscape cut in the $x'z'$-plane and in the $y'z'$-plane. Different colors/shades indicate a different energy level. Analyzing the energy along normal and tangential lines (indicated by *curved lines*) reveals specific elasto-mechanical properties of the kinesin

the tangential x'- and y'-directions are defined by the anchor point and the orientation of the microtubule as indicated in Fig. 15.13 (b). Specific information can be extracted from the 3D energy distribution with help of 2D slices or 1D line scans as shown in Fig. 15.13 (c). The force constants $f_K(r')$ along lines r' can be derived from the energy profiles $W_K(r')$ by a twofold differentiation of $W_K(r')$. $W_K(r')$ was corrected for the optical trapping forces and surface adhesion effects. The normal direction (z') describes the kinesin's axial stiffness $f_K(z')$, whereas the tangential directions $(x'$ and $y')$ provide information about the strength of the molecule's hinge.

The distance between the center of the trap and the surface influences the diffusion of the bead inside the trap, because the viscous drag increases close to an interface. By calibrating the trap first far away and then close to the coverslip, it is possible to consider the influence of the non-constant viscous drag.

15.7   Future Aims in Photonic Force Microscopy

According to the principal mechanisms driving a Photonic Force Microscope, the future directions are defined by further research on local probes, on special trapping techniques, and on the position detection of the probes, on new scanning strategies, and on an analysis that takes all possible aspects of the probe's motion into account. Although latex and silica spheres have been successfully used for a large variety of trapping applications, further efforts must be undertaken to develop new kinds of local probes. These probes should have different properties concerning the interaction with light. Different optical forces, torques, scattering characteristics, or adhesion forces should be determinable by the probes' material, their shape or their labeling. New kinds of optical traps must be engineered to control the behavior of the local probes by exploiting amplitude and phase filters in the back focal plane of the objective lens. At the same time, the detection system must be extended so that orientations of asymmetric particles can also be measured. The detection system can also be improved by increasing the detection volume, i.e. the volume in which the position of the probe can be measured with high accuracy. It is also a challenge to extend this detection technique for two or more particles that are trapped simultaneously. The precision of the detection system is also determined by the procedure of analyzing the diode's signals. It is desirable to extend the linear approximation between position r and signal $S(r)$ and force constant $f(r)$ used so far to a dependency considering all nonlinearities.

Measuring multi-dimensional energy landscapes due to interaction potentials between different binding partners enlarges the spectrum of applications for Photonic Force Microscopy in biology, chemistry, and physics. Boltzmann statistics encode these energy landscapes under certain conditions by determining the thermal fluctuations of a probe. The fluctuations of a probe connected to single molecules may be confined to specific paths by special optical traps such that single molecules are elongated or twisted in specific directions to reveal further micro-mechanical properties.

The phase of the incident wave is disturbed while propagating inside scattering media. This results in a weaker trapping of the probe. It is still unclear to what extent this effect can be compensated by increasing the laser power without inducing photo-damage in the sample. There are still few applications where tweezers have been used to manipulate processes inside living cells or even tissues. It is a major challenge to disturb a biological system not only by holding or moving a particle microscopically (optical tweezing), but to use the confined thermal motion of a particle to perform manipulations in (thermo-)dynamic systems at a sub-microscopic level. The PFM meets the requirements for entering new levels of detail.

Acknowledgements

The authors thank James Jonkman for a thorough reading of the manuscript and Ernst-Ludwig Florin, Christian Tischer, Silvia Jeney and Heinrich Hörber for providing data.

References

1. J. C. Maxwell: Philos. T. R. Soc. London **170**, 231 (1879)
2. P. Lebedev: Ann. Physik **4**, 433 (1901)
3. A. Ashkin: Phys. Rev. Lett. **24**, 156 (1970)
4. A. Ashkin and J. M. Dziedzic: Appl. Phys. Lett. **19**, 283 (1971)
5. A. Ashkin: Science **210**, 1081 (1980)
6. A. Ashkin: IEEE Journal of Selected Topics in Quantum Electronics **6**, 841 (2000)

7. A. Ashkin, J. M. Dziedzic, J. E. Bjorkholm and S. Chu: Opt. Lett. **11**, 288 (1986)
8. T. C. Bakker Schut, G. Hesselink, B. G. de Grooth and J. Greve: Cytometry **12**, 479 (1991)
9. J. Huisken and E. H. K. Stelzer: Opt. Lett. **27**, 1223 (2002)
10. R. C. Gauthier and A. Frangioudakis: Appl. Optics **39**, 26 (2000)
11. K. Taguchi, K. Atsuta, T. Nakata and R. Ikeda: Opt. Commun. **176**, 43 (2000)
12. S. Koyanaka and S. Endoh: Powder Technology **116**, 13 (2001)
13. G. Binnig and H. Rohrer: Reviews of Modern Physics **71**, S324 (1999)
14. A. Ashkin and J. M. Dziedzic: Science **235**, 1517 (1987)
15. A. Ashkin, J. M. Dziedzic and T. Yamane: Nature **330**, 769 (1987)
16. S. M. Block, D. F. Blair and H. C. Berg: Nature **338**, 514 (1989)
17. S. Seeger, S. Monajembashi, K. J. Hutter, G. Futterman, J. Wolfrum and K. O. Greulich: Cytometry **12**, 497 (1991)
18. R. W. Steubing, S. Cheng, W. H. Wright, Y. Numajiri and M. W. Berns: Cytometry **12**, 505 (1991)
19. Y. Tadir, W. H. Wright, O. Vafa, T. Ord, R. H. Asch and M. W. Berns: Fertil Steril **52**, 870 (1989)
20. K. Schutze, A. Clement Sengewald and A. Ashkin: Fertil Steril **61**, 783 (1994)
21. S. M. Block, L. S. Goldstein and B. J. Schnapp: Nature **348**, 348 (1990)
22. A. Ashkin, K. Schutze, J. M. Dziedzic, U. Euteneuer and M. Schliwa: Nature **348**, 346 (1990)
23. S. C. Kuo and M. P. Sheetz: Science **260**, 232 (1993)
24. K. Svoboda, C. F. Schmidt, B. J. Schnapp and S. M. Block: Nature **365**, 721 (1993)
25. J. T. Finer, R. M. Simmons and J. A. Spudich: Biophys. J. **66**, A353 (1994)
26. R. M. Simmons, J. T. Finer, S. Chu and J. A. Spudich: Biophys. J. **70**, 1813 (1996)
27. T. T. Perkins, D. E. Smith and S. Chu: Science **264**, 819 (1994)
28. S. B. Smith, Y. Cui and C. Bustamante: Science **271**, 795 (1996)
29. M. D. Wang, H. Yin, R. Landick, J. Gelles and S. M. Block: Biophys. J. **72**, 1335 (1997)
30. L. Malmqvist and H. M. Hertz: Opt. Commun. **94**, 19 (1992)
31. W. Denk and W. W. Webb: Appl. Optics **29**, 2382 (1990)
32. L. P. Ghislain and W. W. Webb: Opt. Lett. **18**, 1678 (1993)
33. S. Kawata, Y. Inouye and T. Sugiura: Japanese Journal of Applied Physics Part 2-Letters **33**, L1725 (1994)
34. T. Sugiura, T. Okada, Y. Inouye, O. Nakamura and S. Kawata: Opt. Lett. **22**, 1663 (1997)
35. M. E. Friese, H. Rubinsztein-Dunlop, N. R. Heckenberg and E. W. Dearden: Appl. Optics **35**, 7112 (1996)
36. M. E. J. Friese, A. G. Truscott, H. Rubinsztein-Dunlop and N. R. Heckenberg: Appl. Optics **38**, 6597 (1999)
37. E.-L. Florin, J. K. H. Hörber and E. H. K. Stelzer: Appl. Phys. Lett. **69**, 446 (1996)
38. E.-L. Florin, A. Pralle, J. K. H. Hörber and E. H. K. Stelzer: Journal of Structural Biology **119**, 202 (1997)
39. A. Pralle, M. Prummer, E.-L. Florin, E. H. K. Stelzer and J. K. H. Hörber: Microscopy Research and Techniques **44**, 378 (1999)
40. M. Born and E. Wolf: *Principles of optics, 7.ed.* (Cambridge University Press, New York 1999) pp. 497-499
41. C. Tischer, S. Altmann, S. Fisinger, J. K. H. Hörber, E. H. K. Stelzer and E.-L. Florin: Appl. Phys. Lett. **79**, 3878 (2001)
42. E.-L. Florin, A. Pralle, E. H. K. Stelzer and J. K. H. Hörber: Appl. Phys. A **66**, S75 (1998)

43. C. Veigel, M. L. Bartoo, D. C. S. White, J. C. Sparrow and J. E. Molloy: Biophys. J. **75**, 1424 (1998)
44. F. Gittes, B. Schnurr, P. D. Olmsted, F. C. MacKintosh and C. F. Schmidt: Phys. Rev. Lett. **79**, 3286 (1997)
45. A. Pralle, E.-L. Florin, E. H. K. Stelzer and J. K. H. Hörber: Appl. Phys. A **66**, S71 (1998)
46. A. Pralle, P. Keller, E.-L. Florin, K. Simons and J. K. H. H. Hörber: The Journal of Cell Biology **148**, 997 (2000)
47. A. Jonas, P. Zemanek and E. L. Florin: Opt. Lett. **26**, 1466 (2001)
48. F. Gittes and C. F. Schmidt: Opt. Lett. **23**, 7 (1998)
49. A. Rohrbach and E. H. K. Stelzer: J. Appl. Phys. **91**, 5474 (2002)
50. L. Ghislain, N. Switz and W. Webb: Review Of Scientific Instruments **65, N9 (Sep)**, 2762 (1994)
51. K. Sasaki, M. Tsukima and H. Masuhara: Appl. Phys. Lett. **71**, 37 (1997)
52. A. Ashkin: Opt. Lett. **9**, 454 (1984)
53. A. Ashkin and J.-M. Dziedzic: Phys. Rev. Lett. **54**, 1245 (1985)
54. J. R. Anglin and W. Ketterle: Nature **416**, 211 (2002)
55. C. W. McCutchen: J. Opt. Soc. Am. **54**, 240 (1964)
56. J. W. Goodman: *Introduction to Fourier optics, 1st ed.* (McGraw-Hill, San Francisco 1968) pp. 48-56
57. A. Rohrbach and E. H. K. Stelzer: J. Opt. Soc. Am. A **18**, 839 (2001)
58. P. Török, P. Varga, Z. Laczik and G. R. Booker: J. Opt. Soc. Am. **12**, 325 (1995)
59. J. P. Barton and D. R. Alexander: J. Appl. Phys. **66**, 2800 (1989)
60. C. Bohren and D. R. Huffman: *Absorption and scattering of light by small particles* (Wiley Science Paperback, New York 1998) p. 158
61. P. Török, P. D. Higdon, R. Juskaitis and T. Wilson: Opt. Commun. **155**, 335 (1998)
62. J. P. Gordon: Phys. Rev. **8**, 14 (1973)
63. C. Bohren and D. R. Huffman: *Absorption and scattering of light by small particles* (Wiley Science Paperback, New York 1998) p. 137
64. Y. Harada and T. Asakura: Opt. Commun. **124**, 529 (1996)
65. J. S. Kim and S. S. Lee: J. Opt. Soc. Am. **73**, 303 (1983)
66. J. P. Barton, D. R. Alexander and S. A. Schaub: J. Appl. Phys. **66**, 4594 (1989)
67. A. Rohrbach and E. H. K. Stelzer: Appl. Optics **41**, 2494 (2002)
68. F. Reif: *Fundamentals of statistical and thermal physics* (McGraw-Hill, Auckland 1985) pp. 560–565
69. R. Bar-Ziv, A. Meller, T. Tlusty, E. Moses, J. Stavans and S. A. Safran: Phys. Rev. Lett. **78**, 154 (1997)
70. A. Rohrbach, E. L. Florin and E. H. K. Stelzer: Proc. SPIE, in Photon Migration, Optical Coherence,Tomography, and Microscopy **4431**, 75 (2001)
71. K. C. Neuman, E. H. Chadd, G. F. Liou, K. Bergman and S. M. Block: Biophys. J. **77**, 2856 (1999)
72. K. Svoboda and S. M. Block: Opt. Lett. **19**, 930 (1994)
73. S. Sato, Y. Harada and Y. Waseda: Opt. Lett. **19**, 1807 (1994)
74. P. C. Ke and M. Gu: J. Mod. Optics **45**, 2159 (1998)
75. S. Jeney, E.-L. Florin and J. K. H. Hörber: 'Use of Photonic Force Microscopy to study single-motor-molecule mechanics', In: *Kinesin Protocols*, ed. by I. Vernos (Humana Press, Totowa, NJ 2000) pp. 91–107

Index

Printing (Computer to Plate): Saladruck Berlin
Binding: Stürtz AG, Würzburg

Springer Series in
OPTICAL SCIENCES

New editions of volumes prior to volume 60

Published titles since volume 60

Springer Series in
OPTICAL SCIENCES